STRUCTURAL HEALTH MONITORING TECHNOLOGIES and NEXT-GENERATION SMART COMPOSITE STRUCTURES

Composite Materials: Analysis and Design

Series Editor
Ever J. Barbero

PUBLISHED

STRUCTURAL HEALTH MONITORING TECHNOLOGIES and NEXT-GENERATION SMART COMPOSITE STRUCTURES

Jayantha Ananda Epaarachchi
Gayan Chanaka Kahandawa

CRC Press
Taylor & Francis Group
Boca Raton London New York

CRC Press is an imprint of the
Taylor & Francis Group, an **informa** business

CRC Press
Taylor & Francis Group
6000 Broken Sound Parkway NW, Suite 300
Boca Raton, FL 33487-2742

First issued in paperback 2019

© 2016 by Taylor & Francis Group, LLC
CRC Press is an imprint of Taylor & Francis Group, an Informa business

No claim to original U.S. Government works

ISBN-13: 978-1-4822-2691-1 (hbk)
ISBN-13: 978-0-367-86938-0 (pbk)

Contents

Series Preface

Half a century after their commercial introduction, composite materials are of widespread use in many industries. Applications such as aerospace, windmill blades, and highway bridge retrofit require designs that ensure safe and reliable operation for 20 years or more. Using composite materials, virtually any property, such as stiffness, strength, thermal conductivity, and fire resistance, can be tailored to the user's needs by selecting the constituent material, their proportion and geometrical arrangement, and so on. In other words, the engineer is able to design the material concurrently with the structure. Also, modes of failure are much more complex in composites than in classical materials. Such demands for performance, safety, and reliability require that engineers consider a variety of phenomena during the design. Therefore, the aim of this *Composite Materials: Analysis and Design* book series is to bring to the design engineer a collection of works written by experts on every aspect of composite materials that is relevant to their design.

Variety and sophistication of material systems and processing techniques have grown exponentially in response to an ever-increasing number and type of applications. Given the variety of composite materials available as well as their continuous change and improvement, the understanding of composite materials is by no means complete. Therefore, this book series serves the practicing engineer as well as the researcher and student who are looking to advance the state-of-the-art knowledge in understanding material and structural response and developing new engineering tools for modeling and predicting such responses.

Thus, the series is focused on bringing to the public existing and developing knowledge about material–property relationships, processing–property relationships, and the structural responses of composite materials and structures. The series scope includes analytical, experimental, and numerical methods that have a clear impact on the design of composite structures.

Preface

Over the past few decades there have been many breakthroughs in the development of smart materials and miniaturization of sensors. The innovations and advancement of micro- and nanoscale sensor technologies have brought the development of smart structures closer to reality, attracting enormous research attention. The focus of this technology is to bring to fruition diagnosis/self-prognosis and self-healing capabilities in composite structures.

Increased usages of fiber-reinforced composites in advanced engineering fields have created an overwhelming demand for smart/intelligent composite materials that can be utilized in designing next generation products. Moreover, the rising concern about possible failures in composite materials is leading to another unavoidable expectation for intelligent and live diagnostics systems in so-called smart composite structures. This is especially important for aerospace vehicles that need all of their constituent parts to be at optimal performance. Globally, the development of structural health monitoring (SHM) systems and smart structures are being pushed forward by the physical phenomena/properties inherited by new materials as well as new adaptive sensing and control measures. The stress/strain response, acoustic emission, wave propagation, vibration, and damping of materials are some of the physical phenomena/properties that are widely used in the SHM field. The use of electrically activated polymers and shape memory alloys are good examples of advanced smart materials. Although there has been a significant involvement, smart structures are not limited to the use of smart materials and integrated SHM systems. The smart structures are being developed with innovative design concepts using the smart materials' inherent physical and mechanical properties.

High-end space, aerospace, energy, automobile, and civil infrastructure industries demand advanced design and manufacturing methods as well as maintenance management systems because of the global awareness of recent advances in SHM systems, smart structures and smart/intelligent materials, and nanotechnologies. For this reason, many engineering disciplines, such as aerospace, materials, mechanical, electrical/electronic, and civil engineering, have engaged in extremely difficult research work for finding SHM systems and smart materials for new generation products and structures. Each field of engineering has distinct limitations and challenges with varying levels of difficulty in replacing traditional materials with smart materials and adopting principles of SHM in its place. In this context, exchanges of specific information related to SHM systems and smart structures are bound by intellectual properties issues and commercially confidential environments. Thus, further advancement of SHM systems and smart structures is facing severe drawbacks.

This book provides insights into recent advancements of unbound applied research in SHM, smart materials fields, and futuristic smart structures. Chapters 1 through 3 provide recent developments in general embedded sensor technologies for SHM. Chapters 4 through 6 provide general sensor technologies and the developments in sensor response interrogation and data communication.

Chapters 7 through 13 contain the developments in damage matrix formulation, damage mechanics and analysis, smart materials/structures, and specific SHM examples in aerospace applications.

We express our gratitude to the contributing authors for their tireless work in preparing the chapters and to Professor John Canning, the director of the Interdisciplinary Photonic Laboratory of the University of Sydney, Australia, for his continuous support, encouragements, and advice for our SHM research that has spanned more than a decade. We take this opportunity to acknowledge our research team members, research collaborators, and colleagues who directly and indirectly helped to complete this book. Finally, we appreciate the patience and support of our families and friends.

Jayantha Ananda Epaarachchi
University of Southern Queensland

Gayan Chanaka Kahandawa
University of Southern Queensland

MATLAB® is a registered trademark of The MathWorks, Inc. For product information, please contact:

The MathWorks, Inc.
3 Apple Hill Drive
Natick, MA 01760-2098 USA
Tel: 508 647 7000
Fax: 508 647 7001
E-mail: info@mathworks.com
Web: www.mathworks.com

Editors

Jayantha Ananda Epaarachchi earned a PhD at the University of Newcastle, Australia, in 2003. He earned a BSc (Eng) specializing in mechanical engineering at the University of Peradeniya, Sri Lanka. In 1990, he migrated to Australia and earned an MEngSc at the University of New South Wales, Australia.

He then worked in various Australian public and private sector industrial organizations for more than 10 years. In 2003, he joined the James Goldstein Faculty of Engineering and Physical Sciences at the Central Queensland University, Australia, as a lecturer. In 2006, he joined the University of Southern Queensland (USQ) and is teaching engineering mechanics, finite element analysis, and mechanical engineering design as a senior lecturer.

Dr. Epaarachchi is leading the Smart Structures and Structural Health Monitoring group at the Centre of Excellence in Engineered Fibre Composite at USQ. He has been continuously involved in a collaborative project with Boeing Research and Technology, Queensland, Australia, and Boeing Aircraft Company, Chicago, Illinois, and shares a US patent. Presently, he is supervising five PhD candidates. Four PhD projects have been completed under his supervision.

He has been a visiting fellow at the Department of Aerospace, University of Bristol, UK, and a research associate in the Inter-Disciplinary Photonic Laboratory at the University of Sydney, Australia.

Dr. Epaarachchi has been invited and delivered many talks on structural health monitoring (SHM) at international conferences, and he was a cochair of the 4th International Conference on Smart Materials and Nano-Technology in Engineering, Gold Coast, Australia, in 2013. He has organized and chaired many special sessions on SHM at international conferences and has served as a member of international technical committees as well as organizing committees.

Dr. Epaarachchi has authored an invited book chapter and edited four special issues of SHM and material journals. He has published over 50 research articles in journals and conferences. His research interests include SHM, fatigue and fracture, and smart materials.

Gayan Chanaka Kahandawa earned a PhD in mechanical engineering at the University of Southern Queensland (USQ), Australia, in 2013. He earned a BSc at the Department of Production Engineering, University of Peradeniya and worked as a lecturer in the same department for 3 years. He earned an MSc in industrial automation at the University of Moratuwa, Sri Lanka, in 2009. He worked as a lecturer at Uva Wellassa University of Sri Lanka for 1 year, where he pioneered the establishment of a mechatronics engineering degree course in Sri Lanka. Upon completion of his PhD, he worked as a postdoctoral research fellow for one and a half years at the Centre of Excellence in Engineered Fiber Composites at USQ. He joined the Federation University Australia as a lecturer of mechatronics in 2014.

His PhD project "Monitoring Damage in Advanced Composite Structures Using Fibre Optic Sensors" was funded by the Boeing Company, Chicago, Illinois, and the project addressed some of the unsolved questions in structural health monitoring (SHM) technologies for composite aerospace structures. The groundbreaking outcomes of the project earned him a US patent. His research work received the Research Excellence Award by USQ in 2012.

He currently holds two patents and has published more than 30 research articles as journal and conference papers. His research expertise includes mechatronics systems, structural health monitoring, fiber-optic sensors, and artificial intelligence.

His current research interests include use of fiber-optic sensors for monitoring concrete structures and soil movement. In addition, he is working on mechatronics systems and robotics.

Contributors

W.S. Al Azzawi
Centre of Excellence in Engineered
  Fibre Composites
Faculty of Health, Engineering, and
  Sciences
University of Southern Queensland
Toowoomba, Queensland, Australia

and

College of Engineering
University of Diyala
Baqubah, Iraq

Sudhirkumar V. Barai
Department of Civil Engineering
Indian Institute of Technology
Kharagpur, India

John Canning
Interdisciplinary Photonics Laboratories
School of Chemistry
Faculty of Science
University of Sydney
Sydney, New South Wales, Australia

Ramadas Chennamsetti
Composites Research Centre
Research and Development
  Establishment
Defence Research and Development
  Organisation
Pune, India

Kiyoshi Enomoto
Aircraft Materials Division
SOKEIZAI Center
Tokyo, Japan

Jayantha Ananda Epaarachchi
Centre of Excellence in Engineered
  Fibre Composites
School of Mechanical and Electrical
  Engineering
Faculty of Health, Engineering, and
  Sciences
University of Southern Queensland
Toowoomba, Queensland, Australia

Ranjan Ganguli
Department of Aerospace Engineering
Indian Institute of Science
Bangalore, India

Dongyue Gao
School of Materials Science and
  Engineering
Dalian University of Technology
Dalian City, Liaoning Province,
  People's Republic of China

Rahim Gorgin
State Key Laboratory of Structural
  Analysis for Industrial Equipment
School of Aeronautics and Astronautics
Dalian University of Technology
Dalian City, Liaoning Province,
  People's Republic of China

Jayavardhana Gubbi
Department of Electrical and Electronic
  Engineering
University of Melbourne
Parkville, Victoria, Australia

Kazuo Hotate
Department of Electrical Engineering
and Information Systems
University of Tokyo
Tokyo, Japan

M. Mainul Islam
Centre of Excellence in Engineered
Fibre Composites
Faculty of Health, Engineering, and
Sciences
University of Southern Queensland
Toowoomba, Queensland, Australia

Ratneshwar (Ratan) Jha
Advanced Composites Institute
and
Raspet Flight Research Laboratory
Mississippi State University
Starkville, Mississippi

Hongli Ji
State Key Laboratory of Mechanics and
Control of Mechanical Structures
College of Aerospace Engineering
Nanjing University of Aeronautics and
Astronautics
Nanjing, China

Gayan Chanaka Kahandawa
School of Engineering and Information
Technology
Faculty of Science and Technology
Federation University
Churchill, Victoria, Australia

Ayad Arab Ghaidan Kakei
Centre of Excellence in Engineered
Fibre Composites
Faculty of Health, Engineering, and
Sciences
University of Southern Queensland
Toowoomba, Queensland, Australia

and

College of Engineering
University of Kirkuk
Kirkuk, Iraq

Masato Kishi
Department of Electrical Engineering
and Information Systems
University of Tokyo
Tokyo, Japan

Yoshihiro Kumagai
Yokogawa Electric Co.
Research and Development Division
Tokyo, Japan

Simon Laflamme
Department of Civil, Construction,
and Environmental Engineering
Iowa State University
Ames, Iowa

Alan Lau
Department of Mechanical Engineering
Hong Kong Polytechnic University
Hong Kong, China

Jinsong Leng
Center of Composite Materials and
Structures
Harbin Institute of Technology
Harbin, China

Satoshi Matsuura
Yokogawa Electric Co.
Research and Development Division
Tokyo, Japan

Priyan Mendis
Department of Infrastructure
Engineering
University of Melbourne
Parkville, Victoria, Australia

Viviana N. Meruane
Department of Mechanical Engineering
University of Chile
Santiago, Chile

Tuan Ngo
Department of Infrastructure
    Engineering
University of Melbourne
Parkville, Victoria, Australia

Marimuthu Palaniswami
Department of Electrical and Electronic
    Engineering
University of Melbourne
Parkville, Victoria, Australia

Gang-Ding Peng
Joint National Fibre Facility
Photonic and Optical Communication
School of Electrical Engineering and
    Telecommunications
University of New South Wales
Sydney, New South Wales, Australia

Jinhao Qiu
State Key Laboratory of Mechanics and
    Control of Mechanical Structures
College of Aerospace Engineering
Nanjing University of Aeronautics and
    Astronautics
Nanjing, China

Aravinda S. Rao
Department of Electrical and Electronic
    Engineering
University of Melbourne
Parkville, Victoria, Australia

Nozomi Saito
Structural Mechanics Research
    Laboratory
Nagoya Research and Development
    Center
Technology and Innovation
    Headquarters
Mitsubishi Heavy Industries, Ltd.
Tokyo, Japan

Yinan Shan
State Key Laboratory of Structural
    Analysis for Industrial Equipment
School of Aeronautics and Astronautics
Dalian University of Technology
Dalian City, Liaoning Province,
    People's Republic of China

Rani Warsi Sullivan
Department of Aerospace Engineering
Mississippi State University
Starkville, Mississippi

Filippo Ubertini
Department of Civil and Environmental
    Engineering
University of Perugia
Perugia, Italy

Hongyuan Wang
State Key Laboratory of Mechanics and
    Control of Mechanical Structures
College of Aerospace Engineering
Nanjing University of Aeronautics and
    Astronautics
Nanjing, China

Zhanjun Wu
State Key Laboratory of Structural
    Analysis for Industrial Equipment
School of Aeronautics and Astronautics
Dalian University of Technology
Dalian City, Liaoning Province,
    People's Republic of China

Takashi Yari
Structural Mechanics Research
    Laboratory
Nagoya Research and Development
    Center
Technology and Innovation
    Headquarters
Mitsubishi Heavy Industries, Ltd.
Tokyo, Japan

Chao Zhang
State Key Laboratory of Mechanics and
    Control of Mechanical Structures
College of Aerospace Engineering
Nanjing University of Aeronautics and
    Astronautics
Nanjing, China

Jinling Zhao
State Key Laboratory of Mechanics and
    Control of Mechanical Structures
College of Aerospace Engineering
Nanjing University of Aeronautics and
    Astronautics
Nanjing, China

1 Scalable Sensing Membrane for Structural Health Monitoring of Mesosystems

Simon Laflamme and Filippo Ubertini

CONTENTS

INTRODUCTION

Structural health monitoring (SHM) is the process of automating condition assessment of structures by observing and interpreting their in-service responses [4]. Typically conducted on structures under normal operational conditions, SHM has the potential to enable early damage detection and condition-based maintenance, thus limiting the needs for periodic inspections and permitting an optimal allocation of public funds dedicated to infrastructure maintenance [16,20,33,40].

Despite the wide variety of sensors and signal processing tools that have been proposed in recent years for SHM of civil infrastructures, most SHM applications are either at a numerical or laboratory proof-of-concept stage. This is mainly attributed to the lack of scalability of the proposed solutions, either due to economical and/or technical constraints. For example, fiber optics [6,12,27,30] typically require expensive embedment and data acquisition systems, and are brittle. Conversely, vibration sensors, such as accelerometers, are often used for global vibration-based monitoring [1,7,31,33], but result in a complex link signal-to-damage (e.g., diagnosis and localization). A promising alternative is found in large-area electronics (LAEs) [5,17,18,21,28,38,44]. These technologies are designed to detect local damages over global surfaces, analogous to biological skin.

The technological progresses in LAEs have been stimulated by recent advances in conductive polymers [10]. Conductivity is typically achieved by dispersing highly conductive nanoparticles into the polymeric matrix. A popular choice in SHM application is to fabricate resistive strain sensors using carbon nanotubes (CNT) [11,29] due to their high mechanical strength and excellent electrical properties [14,32]. However, CNTs are associated with high fabrication costs and difficult scalability, because they require special procedures (e.g., sonication) to be properly dispersed, and these procedures are difficult to apply over large quantities of material. Alternatives fillers include carbon nanofibers and carbon black, which are orders of magnitude less expensive than CNTs, but result in lower strength and conductivity. Strain sensors exploiting the variation of electrical capacitance with deformation have also been proposed, including applications for measuring local strain [2,37], pressure [28], tri-axial force [9], and humidity [13,15].

This chapter gives a global overview of a scalable sensing membrane recently proposed by the authors to measure mesosurface strain. The basic principle of this sensing solution consists of monitoring local strain over global surfaces by deploying a network of soft elastomeric capacitors (SECs), a type of LAEs. Potential applications include, but are not limited to, transportation infrastructures (e.g., roads and bridges), civil structures (e.g., buildings and dams), energy structures (e.g., wind turbines and power houses), and aerospace systems (e.g., aircrafts). The proposed SEC has very large elasticity, is mechanically robust, is inexpensive to fabricate, can be easily scaled up, and can be deployed over complex surfaces.

The chapter is organized as follows. The section "Sensing Material" presents the sensing material of the SEC. This includes the fabrication process, robustness, temperature behavior, and recent advances in developing an environmental friendly version using castor oil. The section "Electromechanical Model" describes the sensing principle of the SEC. The electromechanical model of the sensor and a dedicated signal processing algorithm for strain decomposition are derived. The section "Experimental Results and Simulations" reports the results of various proof-of-concept experiments conducted using SECs. These include static and dynamic characterizations, crack detection, vibration signature identification, and strain map and deflection shape reconstruction.

SENSING MATERIAL

The SEC is a synthetic metal, obtained by mixing organic and inorganic particles. The challenge in its fabrication lies in the selection of inexpensive particles of appropriate mechanical and electrical properties to obtain a scalable soft capacitor. The dielectric of the SEC is composed of an Styrene–ethylene/butylene–styrene (SEBS) matrix filled with titania (TiO_2). SEBS is a block copolymer widely used for medical applications, because of its purity, softness, elasticity, and strength [43]. Titania is an inorganic particle

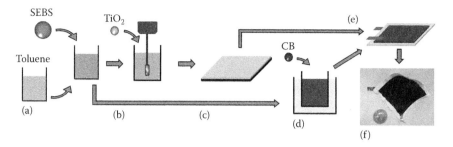

FIGURE 1.1 Solution cast fabrication process: (a) dissolution of SEBS, (b) dispersion of TiO_2 (sonication), (c) drop-casting of dielectric, (d) dispersion of CB (sonic bath), (e) painting of electrodes, and (f) drying. (Reprinted from Laflamme, S. et al., *J. Eng. Mech.*, 139, 879–885, 2012, with permission from ASCE.)

characterized by a high dielectric permittivity that increases the permittivity and durability of the SEBS matrix [36]. The dielectric is sandwiched between two electrodes. They are constituted from the same organic matrix, but filled with CB particles to create a conductive polymer. These CB particles are selected due to their high conductivity and low cost. The utilization of the same polymer matrix (SEBS) for both the electrodes and dielectric results in a strong mechanical bond between the layers that constitute the SEC.

The SEC used by the authors is fabricated using a solution cast process, shown in Figure 1.1, which is as follows:

1. SEBS (Mediprene Dryflex) particles are dissolved in toluene (Figure 1.1a).
2. TiO_2 rutile particles (Sachtleben R 320 D) are dispersed in part of the SEBS-toluene solution at a 15 vol% concentration using an ultrasonic tip (Fisher Scientific D100 Sonic Dismembrator) (Figure 1.1b).
3. SEBS-TiO_2 solution is drop casted on a 75×75 mm^2 glass slide and let drying for 48 h to allow evaporation of toluene (Figure 1.1c).
4. CB particles (Orion Printex XE 2-B) are dispersed in the remaining SEBS-toluene solution at a 10 vol% concentration and dispersed in a sonic bath over 24 h (Figure 1.1d).
5. SEBS-CB solution is painted onto the top and bottom surfaces of the dried dielectric. During the process, two conductive copper tapes are embedded into the liquid electrode layers to create mechanical connections for the wires linking the sensor to the data acquisition system (Figure 1.1e).
6. Resulting multilayer nanocomposite is let drying for 48 h to allow evaporation of toluene (Figure 1.1f).

Figure 1.2 shows the scanning electron microscope (SEM) image of the dielectric layer of the SEC. A good dispersion of the titania particles is observed.

Elasticity and Robustness

The utilization of the SEBS is motivated primarily by the high elasticity and robustness of the material, which makes it an excellent candidate for application to harsh environments (e.g., wind turbine blades, bridges, etc.). In particular, it enables a

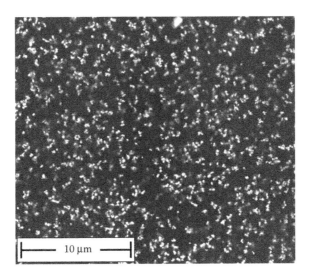

FIGURE 1.2   SEM image of the SEBS-TiO$_2$ nanocomposite.

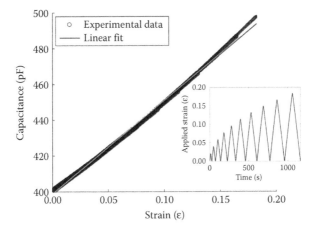

FIGURE 1.3   Elastic behavior under large strain.

linear electromechanical model of the sensor over a wide range of strain, and provides electrical robustness against mechanical tampering.

Figure 1.3 is a plot of the sensors' signal (capacitance) as a function of strain, for an applied strain time history over the range 0%–20% ε (bottom right corner plot in Figure 1.3). Results show an approximately linear behavior over the entire range.

The electrical robustness of the sensor is demonstrated in Figure 1.4. A portion of the sensing material is removed from the original sensor (Figure 1.4a) using a hole punch, as shown in Figure 1.4b. The measured capacitance of both sensing materials ($C + C_{cut} = 691.6$ pF) is approximately equal to the capacitance of the original sensor ($C_0 = 692.6$ pF). A stain step load was applied to both the original sensor (C_0) and the modified sensor (C). The time series of the load input is shown in Figure 1.4e.

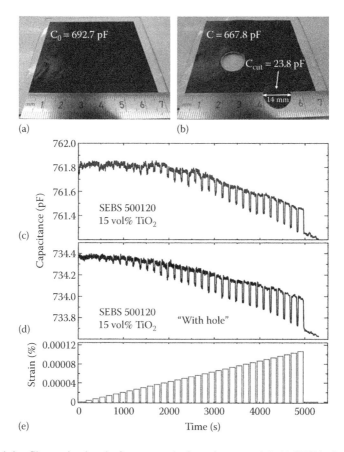

FIGURE 1.4 Change in signal after removal of sensing material: (a) SEC before removal of material, (b) SEC after removal of materials, (c) measured signal before removal of the material, (d) measured signal after removal of the materials, and (e) applied strain. (Reprinted from Laflamme, S. et al., *J. Eng. Mech.*, 139, 879–885, 2012, with permission from ASCE.)

The signal before and after material removal is shown in Figure 1.4c and d, respectively. The high value in capacitance is attributed to parasitic capacitance from the cable and prestretch of the sensor before loading. Note that the signal of the sensor (under a highly controlled environment) can be discerned at approximately 0.00001% strain (0.1 ppm), and is above noise beyond 0.00002% strain (0.2 ppm). The drift in the signal is attributed to the data acquisition system used in the experiment (ACAM PSA21-CAP).

The sensor robustness against mechanical tampering is shown in Figure 1.5. Five cuts have been induced in the sensor epoxied onto a wood substrate, using a knife. Each cut was made over approximately half the length of the sensor (35 mm), and capacitance data acquired over the length of the experiment (Figure 1.5b). Results show a step increase after each cut due to a change in the material's geometry, and a stabilization of the electrical signal after each cut.

The robustness of the sensing material with respect to temperature has also been assessed. An experiment was conducted in accordance with test method standard

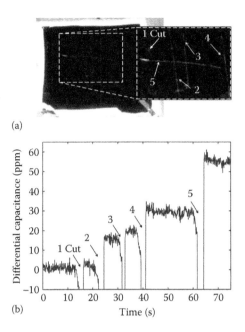

FIGURE 1.5 Stability of the signal after cuts: (a) picture of the SEC with five cuts and (b) measured signal after each cut. (Reprinted from Laflamme, S. et al., *J. Eng. Mech., 139,* 879–885, 2012, with permission from ASCE.)

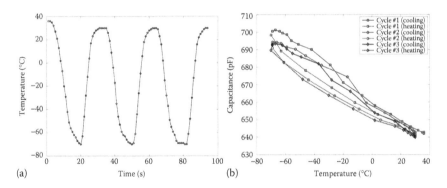

FIGURE 1.6 Temperature shock test: (a) applied temperature load and (b) measured signal. (Reprinted with permission from Saleem, H. et al., Static characterization of a soft elastomeric capacitor for nondestructive evaluation applications. In *Proceedings of the 40th Annual Review of Progress in Quantitative Nondestructive Evaluation: Incorporating the 10th International Conference on Barkhausen Noise and Micromagnetic Testing,* vol. 1581, AIP Publishing, Melville, NY, pp. 1729–1736, 2014. Copyright 2014 American Institute of Physics.)

MIL-STD-810G for temperature shocks. A single SEC was installed in a temperature chamber, and submitted to three temperature cycles ranging from −70°C to 30°C applied at a rate of 0.2°C/s (Figure 1.6a). Figure 1.6b is a plot of the capacitance versus temperature over the three cycles. The capacitance has an approximately linear dependence of −0.6 pF/°C, and decreases due to the increase in volume of the dielectric altering the distance between the electrodes, and also exhibits hysteresis due to the viscoelastic behavior of the material (high-temperature load rate). The downward shift in the hysteresis loops is attributed to the Mullins effect [8], and stabilizes after three cycles.

TOWARD ENVIRONMENTALLY FRIENDLY NANOCOMPOSITES

The nanocomposite constituting the SEC was developed for large scale applications. Given the prospective large quantities of composite materials used in SHM of mesosystems, research efforts have been conducted in developing more environmentally friendly materials. In particular, an alternative fabrication process, termed *melt mixing*, has been investigated to eliminate any solvents used in the process. Also, a novel vegetable oil–based nanocomposite has been developed. These two studies are discussed in the below subsections.

Melt Mixing Process

The melt mixing fabrication process for a dielectric is illustrated in Figure 1.7. A 60 ml capacity advanced torque rheometer plasti-corder heated shear mixer (C.W. Brabender Instruments, South Hackensack, NJ, USA) is used to melt and mix the SEBS particles with titania at 160°C and 50 rpm (Figure 1.7a). The residence time in the barrel is limited to 10 min to avoid thermal degradation of the mix. The extruded SEBS-TiO$_2$ is then compressed into 200 μm thick films at 160°C for 5 min (Figure 1.7b). The sample is finally left at room temperature to anneal before releasing the compression plates to create the pressed dielectric film (Figure 1.7c).

This alternative fabrication process was evaluated by comparing different dielectric samples under slight variations of the nanocomposite mix. Four types of samples were prepared: (1) pure SEBS; (2) SEBS filled with uncoated titania; (3) SEBS filled with uncoated titania and Si-69; and (4) SEBS filled with polydimethylsiloxane (PDMS) oil-coated titania. All samples with fillers were fabricated at a 15 vol% concentration.

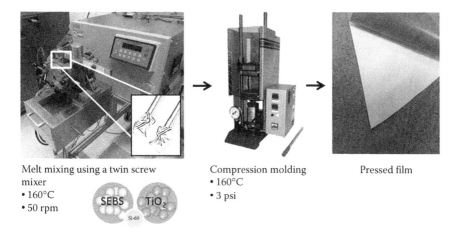

Melt mixing using a twin screw mixer
• 160°C
• 50 rpm

SEBS TiO$_2$
Si-69

Compression molding
• 160°C
• 3 psi

Pressed film

FIGURE 1.7 Melt mixing fabrication process. (Reprinted from *Polymer*, 55, Saleem, H. et al., Interfacial treatment effects on behavior of soft nano-composites for highly stretchable dielectrics, 4531–4537, Copyright 2014, with permission from Elsevier.)

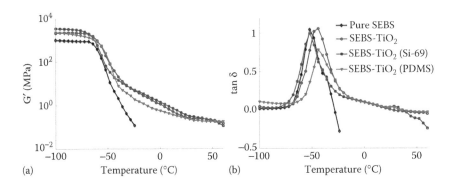

FIGURE 1.8 Dynamic mechanical analysis results: (a) storage modulus and (b) loss factor recorded at different temperatures. (Reprinted from *Polymer*, 55, Saleem, H. et al., Interfacial treatment effects on behavior of soft nano-composites for highly stretchable dielectrics, 4531–4537, Copyright 2014, with permission from Elsevier.)

Figure 1.8 shows the results of the dynamic mechanical analysis conducted on the four samples. An increase in the nominal storage modulus (G') and stability of the loss factor (tan δ) over larger temperatures can be observed for the titania-doped specimens. However, the modification of titania particles and the addition of Si-69 do not result in significant differences in both G' and the glass transition temperature (T_g), except for a loss of stability of the sample with Si-69 at high temperatures.

Figure 1.9 is a plot of the stress–strain curves for all four samples. The sample with uncoated titania shows superior performance by exhibiting a linear behavior over approximately 500% tensile strain (five times the original length).

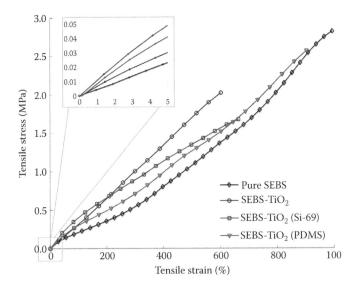

FIGURE 1.9 Stress–strain curves. (Reprinted from *Struct Health Monitor*, 14, Kharroub, S. et al., Bio-based soft elastomeric capacitor for structural health monitoring applications, 158–167, Copyright 2014, with permission from Elsevier.)

Results from this investigation show the promise of the melt mixing fabrication process, in particular for an SEBS matrix mixing with uncoated titania particles (the solution cast process uses PDMS-coated titania particles).

Castor Oil–Based SEC

Castor oil–based nanocomposites were investigated as a possible replacement to the pretroleum-based SEBS [19]. The polymer matrix is also filled with titania, which is considered as an environmentally friendly inorganic material [42]. In this case, uncoated particles are used in the solution cast process as PDMS-coated particles do not disperse well in this particular matrix.

To fabricate a sensor, a waterborne castor oil–based polyurethane dispersion (PUD) is first synthesized by a reaction of isophorone diisocyanate, castor oil, and dimethylol propionic acid (DMPA) as internal surfactant. The DMPA is used to incorporate carboxylic functionality in the prepolymer backbone. Tertiary amine (e.g., triethylamine and TEA) is added to neutralize the carboxylic groups and produce ionic centers, which stabilizes the polymer particles in water. Titania particles are dispersed in the PUD to increase the permittivity and mechanical robustness. An SEC is finally fabricated by sandwiching the dielectric with two conductive electrodes fabricated from a carbon black and SEBS mix. The replacement of electrodes by castor oil–based PUDs is currently being investigated by the authors. Figure 1.10 shows the SEM images of three samples fabricated with various titania content. The uncoated titania particles tend to sink within the matrix, but a closer look at the dense titania region (Figure 1.10b) shows that the particles are well dispersed. This nonuniformity of the dispersion along the thickness of the dielectric

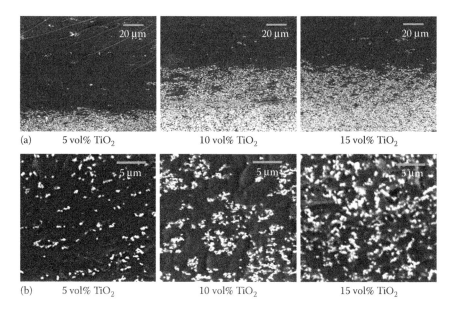

(a)   5 vol% TiO_2 10 vol% TiO_2 15 vol% TiO_2

(b)   5 vol% TiO_2 10 vol% TiO_2 15 vol% TiO_2

FIGURE 1.10 SEM images of castor oil samples taken (a) over the thickness of the membrane and (b) within the dense titania region.

does not affect the sensing capabilities of the SEC due to its planar applications. Current research is being conducted by the authors to ameliorate the dispersion of titania along the entire volume of the sensor.

Figure 1.11 shows test results of the resulting sensor subjected to a quasi-static increasing step load for three specimens with various titania vol% content. Strain response has been computed using the sensor's electromechanical model derived in the next section. The sensor performs well at tracking the strain time history, but shows a drift after each step load likely attributed to its viscoelastic behavior and significantly higher stiffness compared against the SEBS-based SEC. The gauge factors λ for the 5, 10, and 15 vol% samples were computed to be $\lambda = 1.415$, $\lambda = 1.376$, and $\lambda = 1.463$, respectively. These values compare very well against the theoretical value of the gauge factor $1.34 < \lambda < 1.49$. These test results demonstrate the promise of castor oil to fabricate environmentally friendly SEC skins.

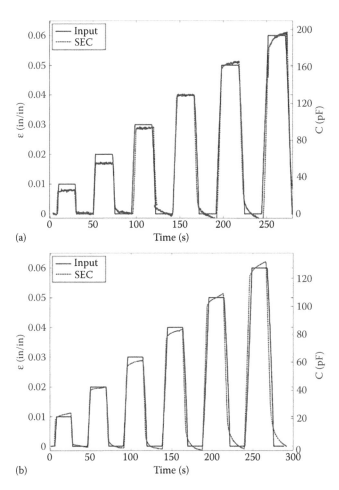

FIGURE 1.11 Time series results for the sensor subjected to an increasing step load: (a) 5% and (b) 10%. (*Continued*)

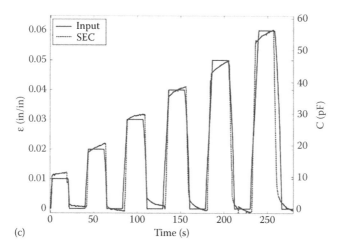

(c)

FIGURE 1.11 (Continued) Time series results for the sensor subjected to an increasing step load: (c) 15 vol% titania.

ELECTROMECHANICAL MODEL

Figure 1.12 is a schematic representation of an SEC. The sensor is designed to be deployed in planar mode. An epoxy is used to adhere the sensor onto a monitored surface along the x–y plane. Thus, the SEC measures strain ε along the x- and y-axes. Two conductive copper tapes are used to connect the SEC to the data acquisition system. The sensing principle consists of measuring strain (e.g., bending or crack) via a relative change in capacitance $\Delta C/C$.

Using Hooke's law under plane stress assumption, the out-of-plane stress ε_z can be written as a function of ε_x and ε_y:

$$\varepsilon_z = -\frac{\nu}{1-\nu}(\varepsilon_x + \varepsilon_y) \tag{1.1}$$

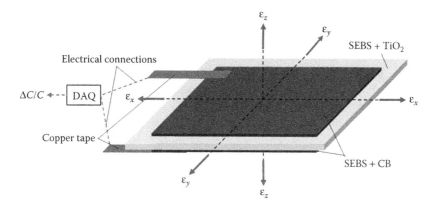

FIGURE 1.12 Schematic representation of an SEC.

where v is the SEC's Poisson's ratio. For pure SEBS $v \approx 0.49$ [41]. Assuming the SEC can be modeled as a nonlossy capacitor at low measurement frequencies (<1,000 Hz), its capacitance C can be written as

$$C = e_0 e_r \frac{A}{h} \qquad (1.2)$$

where $e_0 = 9.854$ pF/m is the vacuum permittivity, e_r the dimensionless polymer relative permittivity, $A = w \times l$ the sensor area with width w and length l, and h is the thickness of the dielectric. For small changes in C, the differential of Equation 1.2 leads to

$$\frac{\Delta C}{C} = \left(\frac{\Delta l}{l} + \frac{\Delta w}{w} - \frac{\Delta h}{h} \right) \qquad (1.3)$$

$$= \varepsilon_x + \varepsilon_y - \varepsilon_z$$

Substituting Equation 1.1 into Equation 1.3, yields

$$\frac{\Delta C}{C} = \lambda(\varepsilon_x + \varepsilon_y) \qquad (1.4)$$

with

$$\lambda = \frac{1}{1-v} \qquad (1.5)$$

Approximating $v \approx 0.5$, one obtains a gauge factor $\lambda = 2$, or a sensitivity $\Delta C / (\varepsilon_x + \varepsilon_y) = 2C$. Equation 1.4 can be specialized for uniaxial strain where $\varepsilon_y = -v_m \varepsilon_x$

$$\frac{\Delta C}{C} = \frac{1-v_m}{1-v} \varepsilon_x \qquad (1.6)$$

where v_m is the monitored material's Poisson's ratio, and the monitored material is assumed to be significantly stiffer than the SEC. A feature of the electromechanical model (Equation 1.4) is that the sensor measures the additive in-plane strain. While this feature is a strong advantage in monitoring of vibration signatures as it will be discussed later, the information on principal strain components (magnitude and direction) is hidden in the signal. To decompose the signal into values of interest, an algorithm has been developed that reconstitutes strain maps by leveraging the utilization of the sensor into an array configuration. This algorithm is described in the below subsection.

ALGORITHM FOR STRAIN DECOMPOSITION

The signal of the SEC can be decomposed into principal strain components by leveraging the network configuration of the SEC in mesoscale applications. The algorithm consists of assuming a shape deformation consistent with boundary

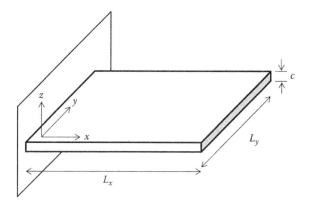

FIGURE 1.13 Schematic of a monitored cantilever plate.

conditions, deriving two-dimensional strain functions directly from the assumed shapes, and finding the coefficients of the strain functions using a least-squares estimator (LSE). For instance, consider the monitored cantilever plate shown in Figure 1.13. A polynomial displacement function $z(x, y)$ of the fourth degree can be used as a displacement shape [26]:

$$z(x,y) = a_1xy + a_2x^2 + a_3y^2 + a_4xy^2 + a_5x^2y + a_6x^3 + a_7y^3$$
$$+ a_8xy^3 + a_9x^2y^2 + a_{10}x^3y + a_{11}x^4 + a_{12}y^4 \tag{1.7}$$

The two-dimensional strain functions are derived directly from Equation 1.7 under the assumption of Kirchoff thin plates:

$$\varepsilon_x(x, y) = -\frac{c}{2}\frac{\partial^2 z}{\partial x^2} = -\frac{c}{2}\left(2a_2 + 2a_5y + 6a_6x + 2a_9y^2 + 6a_{10}xy + 12a_{11}x^2\right)$$
$$\varepsilon_y(x, y) = -\frac{c}{2}\frac{\partial^2 z}{\partial y^2} = -\frac{c}{2}\left(2a_3 + 2a_4x + 6a_7y + 6a + 8xy + 2a_9x^2 + 12a_{12}y^2\right) \tag{1.8}$$

or, in a more general form:

$$\varepsilon_x(x, y) = b_1 + b_2x + b_3y + b_4x^2 + b_5xy + b_6y^2$$
$$\varepsilon_y(x, y) = b_7 + b_8x + b_9y + b_{10}x^2 + b_{11}xy + b_{12}y^2 \tag{1.9}$$

Using Equation 1.4, a sensor signal s_i located at coordinates $\{x_i, y_i\}$ can be written as

$$s_i = \frac{\Delta C_{x_i,y_i}}{C_{x_i,y_i}} = \lambda\left(b_1 + b_2x_i + b_3y_i + b_4x_i^2 + b_5x_iy_i + b_6y_i^2 + b_7 + b_8x_i + b_9y_i\right.$$
$$\left. + b_{10}x_i^2 + b_{11}x_iy_i + b_{12}y_i^2\right) \tag{1.10}$$

or, in matrix form:

$$\mathbf{S} = \lambda \mathbf{H} \mathbf{B} \qquad (1.11)$$

with

$$\mathbf{S} = \begin{bmatrix} s_1 \\ s_2 \\ \vdots \\ s_n \end{bmatrix} \qquad \mathbf{B} = \begin{bmatrix} b_1 \\ b_2 \\ \vdots \\ b_{12} \end{bmatrix}$$

$$\mathbf{H} = \begin{bmatrix} 1 & x_1 & y_1 & x_1^2 & x_1 y_1 & y_1^2 & 1 & x_1 & y_1 & x_1^2 & x_1 y_1 & y_1^2 \\ 1 & x_2 & y_2 & x_2^2 & x_2 y_2 & y_2^2 & 1 & x_2 & y_2 & x_2^2 & x_2 y_2 & y_2^2 \\ \vdots & \vdots & \vdots & \vdots & \vdots & \vdots & \vdots & \vdots & \vdots & \vdots & \vdots & \vdots \\ 1 & x_n & y_n & x_n^2 & x_n y_n & y_n^2 & 1 & x_n & y_n & x_n^2 & x_n y_n & y_n^2 \end{bmatrix}$$

where:
   $\mathbf{S}$ is the vector of sensor signals for n sensors
   $\mathbf{H}$ is the sensor placement matrix
   $\mathbf{B}$ is the vector of parameters

Parameters $\mathbf{B}$ are estimated using LSE:

$$\hat{\mathbf{B}} = \frac{1}{\lambda}(\mathbf{H}^T\mathbf{H})^{-1}\mathbf{H}^T\mathbf{S} \qquad (1.12)$$

where the hat denotes an estimation. In its current form, $\mathbf{H}$ is multicollinear and $\mathbf{H}^T\mathbf{H}$ is singular. The system's initial conditions must be enforced in the matrix $\mathbf{H}$ to have $\mathbf{H}^T\mathbf{H}$ invertible. For instance, these initial conditions for a fixed plate model are: $\varepsilon_x(L_x, \alpha_y \leq y \leq L_y - \alpha_y) = -v_m\varepsilon_y$, $\varepsilon_y(\alpha_x \leq x \leq L_x - \alpha, 0) = \varepsilon_y(\alpha_x \leq x \leq L_x - \alpha_x, L_y) = -v_m\varepsilon_x$, and $\varepsilon_y(0, \alpha_y \leq y \leq L_y - \alpha_y) = 0$, where v_m is the Poisson's ratio of the plate, and α_x and α_y are constants such that $0 \leq \alpha_x \leq L_x$ and $0 \leq \alpha_y \leq L_y$ to account for changes in boundary conditions around the corners. Using these assumptions, the left or right entries of $\mathbf{H}$ associated with a particular boundary conditions are multiplied by a constant (0 or $1 - v_m$ with the assumptions listed above). The strain measurements of each sensor signal s_i can then be decomposed using the following expressions:

$$\hat{\varepsilon}_x(s_i) = \hat{b}_1 + \hat{b}_2 x_i + \hat{b}_3 y_i + \hat{b}_4 x_i^2 + \hat{b}_5 x_i y_i + \hat{b}_6 y_i^2$$
$$\hat{\varepsilon}_y(s_i) = \hat{b}_7 + \hat{b}_8 x_i + \hat{b}_9 y_i + \hat{b}_{10} x_i^2 + \hat{b}_{11} x_i y_i + \hat{b}_{12} y_i^2 \qquad (1.13)$$

EXPERIMENTAL RESULTS AND SIMULATIONS

STATIC AND DYNAMIC BEHAVIOR

The linearity of the SEC has been discussed above (Figure 1.3), thus validating the electromechanical model (Equation 1.4). That test was conducted under large strain. The resolution of the sensor has been evaluated by comparing the performance of the SEC at tracking a low strain time history versus an off-the-shelf resistive strain gauge (RSG) [24]. In this test, both the SEC and RSG were adhered onto the bottom surface of a simply supported aluminum beam, within the constant moment region obtained by a four-point load setup. The RSB was a 1 με resolution Vishay CEA-06-500UW-120. Data from the SEC were acquired at 48 Hz using an ACAM PCap01 data acquisition system, while data from the RSG were acquired at 1.7 Hz using an HP-3852 data acquisition system. The load configuration consisted of an increasing step load. Figure 1.14 plots the results from both time series, where the strain from the SEC was calculated using Equation 1.4. Results show that the SEC has an approximate resolution of 25 με. This resolution is currently limited by the electronics, and could be improved by developing a dedicated data acquisition system.

The dynamic response of the SEC has been evaluated over the frequency range 1–40 Hz. The sensor was installed on a cantilever beam subjected to a harmonic point load. The load frequency swept from 1 to 40 Hz with 1 Hz step increment, with data sampled at 145 Hz. The strain measured by the SEC was computed using Equation 1.4. Figure 1.15a compares the Fourier transform of both experimental and analytical strain signals. Figure 1.15b shows the frequency response function (FRF) obtained from results in Figure 1.15. Results show that the sensor can detect

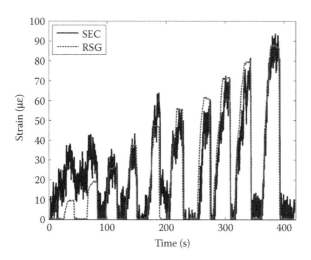

FIGURE 1.14 Strain tracking time history: SEC versus RSG.

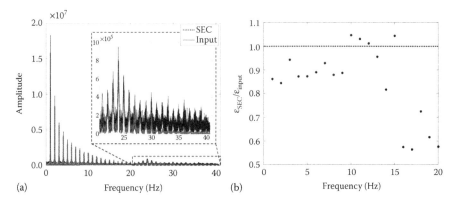

(a) Frequency (Hz)   (b) Frequency (Hz)

FIGURE 1.15 Dynamic response of the SEC: (a) comparison of Fourier transforms of the analytical and measured strain signals and (b) FRF. (Reprinted from Laflamme, S. et al., *J. Struct. Eng.*, 141, 04014186, 2014, with permission from ASCE.)

all frequencies under 40 Hz. However, the FRF shows that the electromechanical model proposed in the previous section can only be used under 15 Hz, as the sensor's response function is close to unity. The sensor's behavior beyond 15 Hz can be modeled by a frequency dependence of its permittivity as a function of frequency.

CRACK SENSING

One of the main advantages of the SEC over off-the-shelf solutions is its capacity to be deployed over very large areas. It is thus a great candidate for crack detection and localization. This subsection shows a test result for the SEC evaluated as a crack sensor. The laboratory setup (Figure 1.16b and 1.16d) consisted of deploying the sensor under the bottom surface of a simply supported wood specimen subjected to a three-point load. The signal of the SEC was compared against a second SEC located

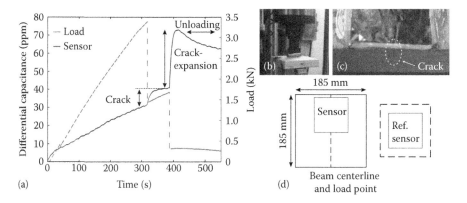

FIGURE 1.16 Crack sensing: (a) time series measurements; (b) test setup; (c) blow up on crack; and (d) setup schematic. (Reprinted from Laflamme, S. et al., *J. Eng. Mech.*, 139, 879–885, 2012, with permission from ASCE.)

externally to amplify the change in signal. The load was applied at a constant rate until cracking (Figure 1.16c) and failure of the specimen. Tests were repeated on five specimens. Figure 1.16a shows the time series of the load versus the differential signal of the sensor. Before the crack first forms, the signal of the SEC has a 98.5% correlation with the load (average value for all five tests). The crack in the specimen results in a jump in the capacitance measurement. This result validates that the SEC can be used as a crack sensor.

Vibration Monitoring

Given the performance of the SEC at detecting frequencies in the range 1–40 Hz, tests were conducted in References [25,39] to investigate the sensor's performance at monitoring vibration signatures. That way, networks of SECs could be deployed over mesosurfaces to localize changes in dynamic behaviors. Tests were conducted on a large-scale steel beam, and on a large-scale concrete beam. This subsection summarizes test results on the steel beam.

Two SECs were deployed onto the top surface of a 5.5 m HP10 × 42 simply supported steel beam, at a distance of 280 mm from each other, 1.8 m from the left support. One sensor was used as a reference, analogous to the crack tests described above, and data recorded at 200 Hz over 35 s using an ACAP PICOAMP PS021 data acquisition system. Note that newer generations of ACAM capacitance data acquisition systems have a reference capacitor located internally. The beam was excited using a 4,000 rpm capacity shaker, installed on the top flange of the beam, 2.85 m from the left support. The eccentric load of the shaker was directed along the strong axis of the beam, and consisted of a hand-induced chirp signal from up to approximately 50 Hz. Figure 1.17a is a picture of the test setup.

Figure 1.17b shows the normalized power spectral density plot of the SEC outputs, with six peaks identified as the six first natural frequencies. The Fourier domain of the signal includes energy above the excitation of 50 Hz, because the beam resonated around 24 Hz and bounced on its supports which excited higher frequencies. The identified frequencies were compared against a finite element model (FEM) of the beam created in SAP2000. The comparison is listed in Table 1.1 and confirms the ability of the SEC to identify vibration signatures. Small differences between experimental and numerical results are ascribed to modeling uncertainties. Figure 1.17c shows the wavelet transform of the signal. The chirp input is visible, along with the resonance of the beam around 24 Hz. Another feature in the signal is the ramping signal that is visible in the lower middle of the plot. This signal is attributed to secondary excitation from the shaker that occurred along the perpendicular direction.

Network Applications

Studies on the SEC have been extended to network applications, for deflection shapes and strain maps reconstruction. Two summary test results are described in the below subsections.

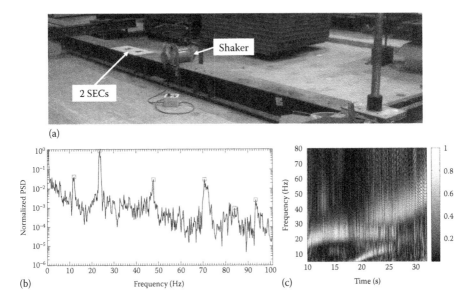

FIGURE 1.17 Vibration monitoring test: (a) test setup; (b) normalized PSD; and (c) wavelet transform. (Reprinted from Laflamme, S. et al., *J. Struct. Eng.*, 141, 04014186, 2014, with permission from ASCE.)

TABLE 1.1
Modal Frequency Identification

	Mode Number (Mode Axis)					
	1	**2**	**3**	**4**	**5**	**6**
	(Weak)	**(Torsion)**	**(Strong)**	**(Weak)**	**(Torsion)**	**(Strong)**
FEM (Hz)	12.3	26.5	42.9	65.8	91.3	117.0
SEC (Hz)	11.8	23.9	47.4	71.1	83.5	94
Difference (%)	−4.07	−9.81	10.5	8.05	−8.54	−19.7

Deflection Shape Reconstruction

Deflection shape reconstruction was investigated on an aluminum cantilever beam subjected to a point load at its free end. This test is described in Reference [23]. Briefly, the beam's unsupported dimensions was $406.4 \times 101.6 \times 6.35$ mm^3. Four SECs were deployed onto the bottom surface of the beam in a 1×4 array, covering an area of 280×70 mm^2, and compared against four RSG deployed at the same location. The test setup is shown in Figure 1.18. RSGs were 1 με resolution Vishay CEA-06-500UW-120, and an increasing step load applied using a hand-operated hydraulic (Enerpac). Data from the SECs were acquired at 48.0 Hz using an ACAM PCap01 data acquisition system, while data from the RSGs were acquired at 1.7 Hz using an HP-3852 data acquisition system. Tests were repeated three times.

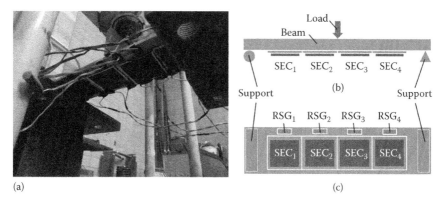

(a)   (c)

FIGURE 1.18   Test setup: (a) picture; (b) schematic (elevation view); and (c) schematic (bottom view).

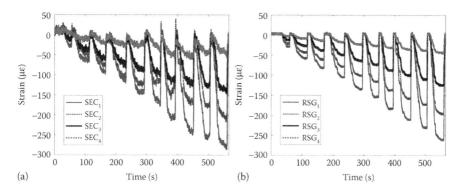

FIGURE 1.19   Strain measurements: (a) SECs and (b) RSGs.

Figure 1.19 is a plot of the time history of strain measurements for the SECs (Figure 1.19a) and the RSGs (Figure 1.19b). The SECs slightly overestimate strain with respect to RSGs. Figure 1.20 shows the absolute relative error between SEC_1 and RSG_1, corresponding to the largest error. The level of error remains generally under the 25 με threshold.

One of the possible monitoring applications of SEC networks is condition assessment of mesosystems. The sensing method can be used to extract physics-based features from the signal to provide insights on structural performance. For example, deflection shapes and strain magnitudes can be extracted and used as inputs for a time-based reliability condition assessment model. Figure 1.21 is an example of shape reconstruction using test data from the cantilever beam, for a typical result (here taken at $t = 400$ s). Strain data from both the sensors were fitted using a third-order polynomial, and the function integrated twice to obtain the deflection shape. Figure 1.21a is a plot of the nonnormalized deflection shapes obtained from the SECs and RSGs data, compared against the analytical deflection shape obtained using Euler–Bernoulli beam theory. The overall absolute performance from both the SECs

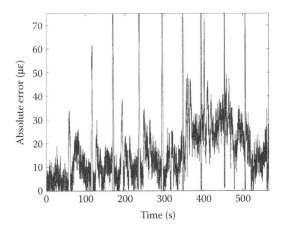

FIGURE 1.20 Strain measurement error.

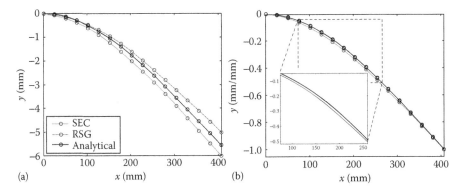

FIGURE 1.21 Deflection shape at $t = 500$ s: (a) nonnormalized and (b) normalized.

and the RSGs is similar. Figure 1.21b is a plot of the deflection shapes normalized at the highest deflection. The curvature obtained from all sensors is similar, except for a slight underperformance of the SECs showed in the blow up.

Strain Map Reconstruction

The utilization of the SEC networks for feature extraction has been extended to two-dimensional strain map reconstruction in Reference [26]. Numerical simulations were used to model a large number of SEC installed on various types of cantilever plates. The algorithm for strain decomposition discussed above was used to obtain principal strain components and reconstruct the strain map. Below is a summary result from numerical simulations on an asymmetric plate mimicking a 9 m wind turbine blade described in Reference [3]. The numerical model is shown in Figure 1.22. The plate is constituted with laminated sections of orthotropic materials, and subjected to two time varying wind pressures on two separated areas (over the entire height of the first 180 cm and the last 720 cm). Red filled circles

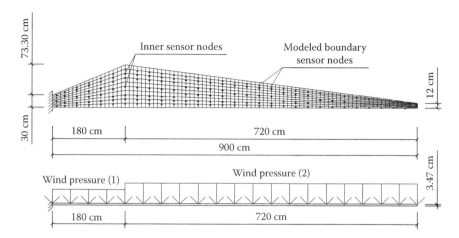

FIGURE 1.22 Strain map reconstruction on a wind turbine blade: simulation model.

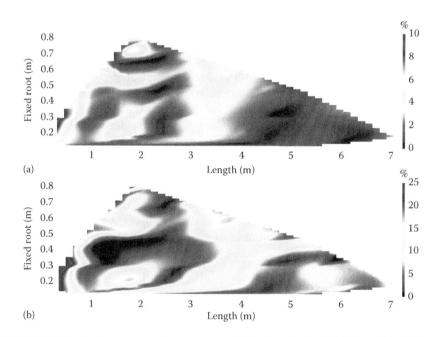

FIGURE 1.23 Strain map reconstruction on a wind turbine blade: (a) APE ε_x and (b) APE ε_y.

in Figure 1.22 denote sensors subjected to boundary conditions, while blue circles denote the sensors not subjected to boundary conditions.

Figure 1.23a and 1.23b are plots of the absolute percentage error (APE) for the strain along the x-axis (longitudinal in Figure 1.22) and y-axis (along the height in Figure 1.22), respectively, and constitute a typical result for the strain map reconstruction taken at a specific time step. Given that the plate is asymmetric and the material orthotropic, the boundary conditions around the plate are harder to estimate, leading to weaker assumptions. It follows that the error on the strain reconstruction is higher at the root of the blade, as shown in Figure 1.23. Nevertheless, the APE still remains small—mostly under 5% for ε_x and under 15% for ε_y. Note that the mean APE on the reconstructed deflection shape (not shown) and the maximum APE are 5.24% and 14.2%, respectively.

CONCLUSION

Recent progresses in LAEs have enabled skin-like sensing technologies for SHM of mesosystems. This chapter has presented a novel sensor developed by the authors. The sensor, termed SEC, can be deployed in a network configuration to monitor local deformations over global surfaces. It can be used for condition assessment by facilitating the reconstruction of mesosurface strain maps and deflection shapes, or by localizing defects such as fatigue cracks. A critical advantage of the SEC is its scalability. This scalability is a direct result of a simple fabrication process and utilization of inexpensive materials. Most of results presented in this chapter were obtained using sensors measuring 3×3 in², but the size of an SEC can be easily scaled up depending on the engineering application. The technology presented in this chapter constitutes one of the very first scalable skin-type sensing solution for monitoring of mesosystems.

Fabrication processes and in-depth characterization of the sensing material have been presented, including recent advances in developing an environmentally friendly version of the SEC using castor oil. An electromechanical model of the sensor was derived, as well as a dedicated algorithm for decomposing additive strain into principal components. This model and algorithm were later validated via laboratory experiments and numerical simulations. Experiments covered a wide range of loading conditions, including static and dynamic loadings, and showed that the sensor compares well against more mature technologies (e.g., off-the-shelf RSG and accelerometers). These results demonstrated the applicability of the proposed sensing solution. Numerical simulations on a plate mimicking a wind turbine blade showed the promise of the SHM method at condition assessment of mesosystems, by reconstructing strain maps and deflection shapes of a complex structure.

REFERENCES

1. Alvandi, A., and Cremona, C. Assessment of vibration-based damage identification techniques. *Journal of Sound and Vibration 292*, 1–2 (2006), 179–202.
2. Arshak, K., McDonagh, D., and Durcan, M. Development of new capacitive strain sensors based on thick film polymer and cermet technologies. *Sensors and Actuators A: Physical 79*, 2 (2000), 102–114.
3. Berry, D., and Ashwill, T. *Design of 9-Meter Carbon-Fiberglass Prototype Blades: Cx-100 and tx-100. SAND2007-0201*, Sandia National Laboratories, Albuquerque, NM (2007).

4. Brownjohn, J. Structural health monitoring of civil infrastructure. *Philosophical Transactions of the Royal Society A: Mathematical, Physical and Engineering Sciences 365*, 1851 (2007), 589–622.

5. Carlson, J., English, J., and Coe, D. A flexible, self-healing sensor skin. *Smart Materials and Structures 15* (2006), N129.

6. Chan, T., Yu, L., Tam, H., Ni, Y., Liu, S., Chung, W., and Cheng, L. Fiber bragg grating sensors for structural health monitoring of tsing ma bridge: Background and experimental observation. *Engineering Structures 28*, 5 (2006), 648–659.

7. Cross, E., Manson, G., Worden, K., and Pierce, S. Features for damage detection with insensitivity to environmental and operational variations. *Proceedings of the Royal Society A 468* (2012), 4098–4122.

8. Diani, J., Fayolle, B., and Gilormini, P. A review on the Mullins effect. *European Polymer Journal 45*, 3 (2009), 601–612.

9. Dobrzynska, J. A., and Gijs, M. Polymer-based flexible capacitive sensor for three-axial force measurements. *Journal of Micromechanics and Microengineering 23*, 1 (2013), 015009.

10. Gangopadhyay, A., and De, A. Conducting polymer nanocomposites: A brief overview. *Chemistry of Materials 12*, 3 (2000), 608–622.

11. Gao, L., Thostenson, E., Zhang, Z., Byun, J., and Chou, T. Damage monitoring in fiber-reinforced composites under fatigue loading using carbon nanotube networks. *Philosophical Magazine 90*, 31–32 (2010), 4085–4099.

12. Glisic, B., and Yao, Y. Fiber optic method for health assessment of pipelines subjected to earthquake-induced ground movement. *Structural Health Monitoring 11*, 6 (2012), 696–711.

13. Harrey, P., Ramsey, B., Evans, P., and Harrison, D. Capacitive-type humidity sensors fabricated using the offset lithographic printing process. *Sensors and Actuators B: Chemical 87*, 2 (2002), 226–232.

14. Heeder, N., Shukla, A., Chalivendra, V., and Yang, S. Sensitivity and dynamic electrical response of cnt-reinforced nanocomposites. *Journal of Materials Science 47* (2012), 3808–3816.

15. Hong, H. P., Jung, K. H., Min, N. K., Rhee, Y. H., and Park, C. W. A highly fast capacitive-type humidity sensor using percolating carbon nanotube films as a porous electrode material. In *Proceedings of IEEE Sensors* (2012), IEEE, pp. 1–4.

16. Hsieh, K., Halling, M., and Barr, P. Overview of vibrational structural health monitoring with representative case studies. *ASCE Journal of Bridge Engineering 11*, 6 (2006), 707–715.

17. Hu, Y., Rieutort-Louis, W. S., Sanz-Robinson, J., Huang, L., Glisic, B., Sturm, J. C., Wagner, S., and Verma, N. Large-scale sensing system combining large-area electronics and cmos ics for structural-health monitoring. *IEEE Journal of Solid-State Circuits 49*, 2 (2014), 513–523.

18. Hurlebaus, S., and Lothar, G. Smart layer for damage diagnostics. *Journal of Intelligent Material Systems and Structures 15*, 9 (2004), 729–736.

19. Kharroub, S., Laflamme, S., Madbouly, S., and Ubertini, F. Bio-based soft elastomeric capacitor for structural health monitoring applications. *Structural Health Monitoring 14* (2014), 158–167.

20. Ko, J., and Ni, Y. Technology developments in structural health monitoring of large-scale bridges. *Engineering Structures 27* (2005), 1715–1725.

21. Laflamme, S., Kollosche, M., Connor, J., and Kofod, G. Soft capacitive sensor for structural health monitoring of large-scale systems. *Structural Control and Health Monitoring 19*, 1 (2012), 70–81.

22. Laflamme, S., Kollosche, M., Connor, J. J., and Kofod, G. Robust flexible capacitive surface sensor for structural health monitoring applications. *Journal of Engineering Mechanics 139*, 7 (2012), 879–885.

23. Laflamme, S., Saleem, H., Song, C., Vasan, B., Geiger, R., Chen, D., Kessler, M., Bower, N., and Rajan, K. Sensing skin for condition assessment of civil infrastructure. In *Proceedings of the International Workshop on Structural Health Monitoring*, Stanford, CA (2013).

24. Laflamme, S., Saleem, H. S., Vasan, B. K., Geiger, R. L., Chen, D., Kessler, M. R., and Rajan, K. Soft elastomeric capacitor network for strain sensing over large surfaces. *IEEE/ASME Transactions on Mechatronics 18*, 6 (2013), 1647–1654.

25. Laflamme, S., Ubertini, F., Saleem, H., D' Alessandro, A., Downey, A., Ceylan, H., and Materazzi, A. L. Dynamic characterization of a soft elastomeric capacitor for structural health monitoring. *Journal of Structural Engineering 141* (2014), 04014186.

26. Laflamme, S., Wu, J., Song, C., Saleem, H., and Ubertini, F. Study of the bidirectional sensing characteristic of a thin film strain gauge for monitoring of wind turbine blades. In *Proceedings of the 6th World Conference on Structural Control and Monitoring*, Barcelona, Spain (2014).

27. Li, H., Li, D., and Song, G. Recent applications of fiber optic sensors to health monitoring in civil engineering. *Engineering Structures 26* (2004), 1647–1657.

28. Lipomi, D., Vosgueritchian, M., Tee, B., Hellstrom, S., Lee, J., Fox, C., and Bao, Z. Skin-like pressure and strain sensors based on transparent elastic films of carbon nanotubes. *Nature Nanotechnology 6*, 12 (2011), 788–792.

29. Loh, K., Hou, T., Lynch, J., and Kotov, N. Carbon nanotube sensing skins for spatial strain and impact damage identification. *Journal of Nondestructive Evaluation 28*, 1 (2009), 9–25.

30. Lopez-Higuera, J., Rodriguez Cobo, L., Quintela Incera, A., and Cobo, A. Fiber optic sensors in structural health monitoring. *Journal of Lightwave Technology 29*, 4 (2011), 587–608.

31. Magalhaes, F., Cunha, A., and Caetano, E. Vibration based structural health monitoring of an arch bridge: From automated oma to damage detection. *Mechanical Systems and Signal Processing 28* (2012), 212–228.

32. Materazzi, A., Ubertini, F., and D'Alessandro, A. Carbon nanotube cement-based transducers for dynamic sensing of strain. *Cement and Concrete Composites 37* (2013), 2–11.

33. Peeters, B., and De Roeck, G. One-year monitoring of the Z24-bridge: Environmental effects versus damage events. *Earthquake Engineering and Structural Dynamics 30* (2001), 149–171.

34. Saleem, H., Laflamme, S., Zhang, H., Geiger, R., Kessler, M., and Rajan, K. Static characterization of a soft elastomeric capacitor for non destructive evaluation applications. In *Proceedings of the 40th Annual Review of Progress in Quantitative Nondestructive Evaluation: Incorporating the 10th International Conference on Barkhausen Noise and Micromagnetic Testing* (2014), vol. 1581, AIP Publishing, Baltimore, MD, pp. 1729–1736.

35. Saleem, H., Thunga, M., Kollosche, M., Kessler, M., and Laflamme, S. Interfacial treatment effects on behavior of soft nano-composites for highly stretchable dielectrics. *Polymer 55*, 17 (2014), 4531–4537.

36. Stoyanov, H., Kollosche, M., McCarthy, D. N., and Kofod, G. Molecular composites with enhanced energy density for electroactive polymers. *Journal of Materials Chemistry 20* (2010), 7558–7564.

37. Suster, M., Guo, J., Chaimanonart, N., Ko, W. H., and Young, D. J. A high-performance mems capacitive strain sensing system. *Journal of Microelectromechanical Systems 15*, 5 (2006), 1069–1077.

38. Tata, U., Deshmukh, S., Chiao, J., Carter, R., and Huang, H. Bio-inspired sensor skins for structural health monitoring. *Smart Materials and Structures 18* (2009), 104026.

39. Ubertini, F., Laflamme, S., Ceylan, H., Materazzi, A. L., Cerni, G., Saleem, H., D'Alessandro, A., and Corradini, A. Novel nanocomposite technologies for dynamic monitoring of structures: A comparison between cement-based embeddable and soft elastomeric surface sensors. *Smart Materials and Structures 23*, 4 (2014), 045023.

40. Van der Auweraer, H., and Peeters, B. International research projects on structural health monitoring: An overview. *Structural Health Monitoring 2* (2003), 341–358.

41. Wilkinson, A., Clemens, M., and Harding, V. The effects of sebs-g-maleic anhydride reaction on the morphology and properties of polypropylene/pa6/sebs ternary blends. *Polymer 45*, 15 (2004), 5239–5249.

42. Yang, D., Tian, M., Dong, Y., Kang, H., Gong, D., and Zhang, L. A high-performance dielectric elastomer consisting of bio-based polyester elastomer and titanium dioxide powder. *Journal of Applied Physics 114*, 15 (2013), 154104.

43. Yoda, R. Elastomers for biomedical applications. *Journal of Biomaterials Science, Polymer Edition 9*, 6 (1998), 561–626.

44. Zhou, Z., Zhang, B., Xia, K., Li, X., Yan, G., and Zhang, K. Smart film for crack monitoring of concrete bridges. *Structural Health Monitoring 10*, 3 (2011), 275–289.

2 Use of Distributed Sensor Networks with Optical Fibers (Brillouin Scattering) for SHM of Composite Structures

Nozomi Saito, Takashi Yari, Kazuo Hotate,
Masato Kishi, Yoshihiro Kumagai,
Satoshi Matsuura, and Kiyoshi Enomoto

CONTENTS

INTRODUCTION

Structural health monitoring (SHM) contributes to reducing maintenance costs, improving flight safety, and fuel efficiency. SHM will change the maintenance scheme from conventional schedule based to condition based. That means hidden area detail inspection will be conducted only when SHM system indicates initiation of structural damages without disassembly and reassembly. Damage detection by SHM system during flights will be effective to improving flight safety.[1] Furthermore, innovative structural design by taking SHM into consideration has a capability of weight saving in the aircraft design phase. Conventional structural design includes damage tolerance such as fail-safe for flight safety that is because of heavy structural weight. SHM will be able to cover the damage tolerance design.[2]

Optical fiber sensor (OFS), which is one of the attractive sensor using SHM technologies, has great characteristics such as lightweight, high corrosion durability, high electromagnetic durability, and capability to be embedded into composites. Thus, there are many SHM technologies using OFS. The authors have been developing Brillouin optical correlation domain analysis (BOCDA)[3] because BOCDA has more superior characteristics than the other OFS technologies for wide-area monitoring as shown in Table 2.1. In Table 2.1, BGS (Brillouin gain spectrum) and DGS (dynamic gain spectrum) are BOCDA measurement results that mainly affect strain and temperature measurement values, respectively. BOCDA measures strain and temperature simultaneously using an OFS with polarization maintaining property utilizing the Brillouin scattering and birefringence phenomena utilizing Equation 2.1.[4] The most important characteristic of BOCDA is distributed strain and temperature sensing with high spatial resolution. The second is dynamic strain and temperature sensing at arbitrary points in an OFS.

$$
\begin{pmatrix} \Delta\varepsilon \\ \Delta T \end{pmatrix} = \frac{1}{C_v^\varepsilon \cdot C_f^T - C_v^T \cdot C_f^\varepsilon} \begin{pmatrix} C_f^T & -C_v^T \\ -C_f^\varepsilon & C_v^\varepsilon \end{pmatrix} \begin{pmatrix} \Delta v_B \\ \Delta f_{yx} \end{pmatrix}
\tag{2.1}
$$

TABLE 2.1

Characteristics of BOCDA

Characteristics	Value	Unit
Measurement range	1000	m
Spatial resolution of BGS	1.6	mm
Spatial resolution of DGS	100	mm
Sampling rate	1000	Hz

where:

$\Delta \varepsilon$ is the strain change

ΔT is the temperature change

C_v^ε and C_v^T are Brillouin frequency shift (BFS) coefficients affect to strain and
    temperature

C_f^ε and C_f^T are DGS coefficients affect to strain and temperature

Δv_B is the BFS change

Δf_{yx} is the DGS change

Application scenarios of the BOCDA-SHM system are distributed sensing along an OFS and real-time dynamic sensing at arbitrary points of an OFS during flights. In distributed sensing, structural damages will be detected by measuring distributed strain changes. In real-time dynamic sensing, not only structural damages will be detected, but fatigue life will be also predicted.

The authors have been improving the technology readiness level (TRL)[5] of the BOCDA-SHM system to realize these scenarios. In this chapter, the achievement of the BOCDA development[6–9] was explained such as principle and basic functions of BOCDA, measurement equipment developed to date, advantages of BOCDA for aircraft SHM, and application scenarios.

PRINCIPLE AND BASIC FUNCTIONS OF BOCDA

FIBER BRILLOUIN SCATTERING AS A SENSING MECHANISM
FOR STRAIN AND TEMPERATURE

In an optical fiber, back scattering takes place at any points along it. The scattering has three components, namely Rayleigh, Raman, and Brillouin scattering. These three scatterings change the properties, such as intensity, frequency, and/or spectrum, with temperature and/or strain variation. Thus, by getting these properties as a function of the position along the optical fiber, we can realize fully distributed sensing function for temperature and/or strain. The optical fiber distributed sensing system used in the SHM described in this chapter utilizes Brillouin scattering.

Figure 2.1 shows the mechanism to cause the Brillouin scattering in an optical fiber.[10] The optical fiber material, SiO_2, vibrates thermally at any positions, which makes acoustic wave along the fiber, as schematically illustrated in Figure 2.1a. The refractive-index grating induced by the acoustic wave acts as the Bragg grating, when the grating period is just equal to half the wavelength of the input lightwave. In the silica optical fiber, the Bragg condition is satisfied with about 10 GHz acoustic wave frequency for 1.55 μm input (pump) wavelength of the laser source. In this mechanism, the Bragg grating moves along the fiber, then the reflected lightwave, or the Brillouin scattering, has downshifted frequency compared with the input one due to the shift phenomenon. This is called BFS, f_B, which is equal to the acoustic wave frequency satisfying the Bragg condition. This scattering is called the spontaneous Brillouin scattering (SpBS), whose intensity is quite small.

When the other lightwave, which has the frequency difference of $-f_B$ from the pump, is launched into the fiber at another end as shown in Figure 2.1b, the

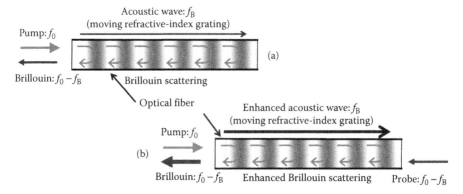

FIGURE 2.1   (a) Spontaneous and (b) stimulated Brillouin scattering in an optical fiber.

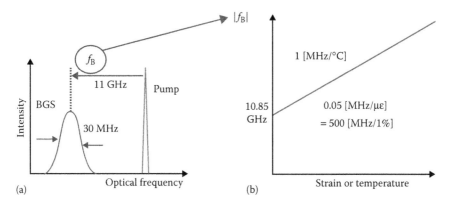

FIGURE 2.2   (a) Brillouin scattering spectrum with downshifted frequency from pump lightwave and (b) strain and temperature dependence of BFS.

refractive-index grating can be enhanced through the electrostriction property of the optical fiber. Then, the Brillouin scattering can also be enhanced, which is called the stimulated Brillouin scattering (SBS).

Figure 2.2a shows schematically the spectral relation between the pump and the Brillouin scattering. This spectrum (BGS) has the line width of about 30 MHz, which is due to the lifetime of the acoustic wave or the related phonon. The velocity of the acoustic wave is changed linearly with strain and/or temperature variation, as schematically shown in Figure 2.2b. The coefficients are 500 MHz/% (0.05 MHz/$\mu\varepsilon$) and 1[MHz/°C] for strain and temperature variation, respectively. Therefore, the Brillouin scattering can act as strain and/or temperature sensing mechanism through measuring BFS.

As for the ways to have distributed information along the optical fiber, time domain techniques with pulsed lightwave have been developed.[11–13] These are called Brillouin optical time domain analysis (BOTDA) or Brillouin optical time domain reflectometry (BOTDR). Former uses the SBS and the latter uses the SpBS. In the time domain technologies, the spatial resolution for the distributed measurement is given by the pulse width. When the pulse width is 1m, corresponding to the 1m spatial resolution, the spectrum is broadened to have about 100 MHz width. This means that the time domain technologies have the trade-off relation between the spatial resolution and the strain or temperature accuracy.[10,13] Then, in the basic time domain systems, spatial resolution is limited around 1 m. On the contrary, single pulses do not have enough energy. Therefore, time averaging is required in the time domain systems, which may reduce basically the measurement speed.

Recently, various improvements in the time domain systems have been proposed and demonstrated.[10] Then, 10 or 1 cm spatial resolution has been realized. However, these systems have still difficulties, such as requirement of specialty optical fibers and quite a high-speed data-acquisition electronics.

On the contrary, optical correlation domain techniques with utilizing continuous lightwaves have already overcome the difficulties.[10] The correlation domain systems are also classified into two types:[10] (i) BOCDA uses the SBS[14] and (ii) BOCDR uses the SpBS.[15,16] In these systems, interference nature of the continuous lightwave is manipulated by modulating the lightwave frequency in an appropriate waveform to form the specific one point along the optical fiber, at which the Brillouin characteristic is obtained position selectively.[10,14] The specific position can easily be scanned to realize the distributed measurement. In BOCDA, 1.6 mm spatial resolution, kHz-order sampling rate, and random accessibility to arbitrary multiple points along the fiber have already been demonstrated.[10] Simultaneous distributed measurement of strain and temperature has also been demonstrated based on the BOCDA system.[10]

PRINCIPLE OF BOCDA FOR DISTRIBUTED MEASUREMENT

Figure 2.3 shows a BOCDA system.[3,10,14] A distributed feedback laser diode (DFB-LD) is used as a light source, whose output lightwave is divided into two ways by a fiber

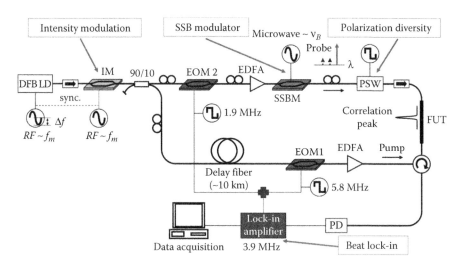

FIGURE 2.3 BOCDA system. Several schemes for increasing the functions are also shown. (Data from K. Y. Song, Z. He, and K. Hotate, *OSA Opt. Lett.*, 31, no. 17, 2006:2526–2528; K. Hotate and T. Hasegawa, *IEICE Trans. Electron.*, E83-C, no. 3, 2000:405–412.)

coupler with 90 versus 10 coupling ratio. In one arm, the lightwave is put into a single side-band modulator (SSBM) to shift the frequency by around BFS (v_B), which acts as the probe wave. In the other arm, the lightwave is chopped by an intensity modulator (electro-optic modulator 1 [EOM1]) for lock-in detection, amplified through an erbium-doped-fiber amplifier (EDFA), and launched into the fiber under test (FUT) as the pump lightwave. In the FUT, the SBS might take place. In the BOCDA system, the laser output is modulated in its frequency with a sinusoidal waveform as shown in Figure 2.3,[10,14] which realizes a position-selective excitation of the SBS.

Because of the frequency modulation, the frequency difference between the pump and probe fluctuates at almost all the points along the optical fiber, as shown in the left-hand side inset in Figure 2.4. Therefore, the stimulated scattering is prohibited at these points. On the contrary, specific points are realized along the optical fiber, where the frequency modulation is synchronized between the pump and probe to have the constant frequency difference between these. At the specific points, which are called correlation peak points, the SBS takes place position selectively, as shown in Figure 2.4.[10,14] The BOCDA system output shows the spectrum shape illustrated in the left-hand side of Figure 2.4. The peak frequency shown in the output corresponds to that in the BGS at the correlation peak point. This is the principle for the position selective measurement of BGS in BOCDA.

At the correlation peak position, the frequency modulation is just synchronized. Then, the spectral broadening does not take place at this position in BOCDA. This means that the higher spatial resolution can easily be realized. The spatial resolution is inversely proportional to the FM frequency and the FM amplitude.[10,14] In BOCDA, 1.6 mm spatial resolution has already been demonstrated. Because of the continuous wave, BOCDA can achieve a fast measurement. BGS measurement speed of 1,000 spectra/s has been demonstrated in a BOCDA system. As was explained, the measurement point is selected along the optical fiber in BOCDA. Then, the point can be set arbitrarily along the optical fiber, which provides us with the random accessibility to arbitrarily selected multiple points along the optical fiber. The random accessibility is the feature that cannot be realized by the time domain technologies.

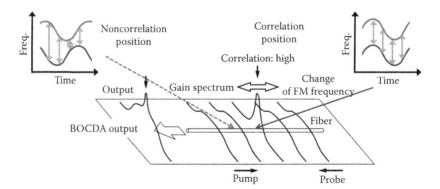

FIGURE 2.4 Position-selective excitation of SBS in BOCDA system. (Data from Y. Kumagai, S. Matsuura, and T. Yari et al., *Yokogawa Technical Report*, 56, no. 2, 2013:83–86; K. Hotate and T. Hasegawa, *IEICE Trans. Electron.*, E83-C, no. 3, 2000:405–412.)

In Figure 2.3, several schemes are also proposed and introduced to enhance the performance of the BOCDA. Intensity modulation synchronized with the frequency modulation can modify the time-averaged lightwave power spectrum shape, which can reduce the background spectrum shown in Figure 2.4. With this scheme, spatial resolution and strain dynamic range can be enhanced.[10] When an ordinary single mode fiber (SMF) is used as FUT, the state of the polarization fluctuates, which brings the fluctuation of the SBS gain. To avoid the fluctuation, time-division polarization-diversity scheme is introduced in Figure 2.3.[10] When Fresnel reflections take place in the system, the BOCDA output includes the noise. To avoid the noise, both of the pump and probe are intensity modulated to have a synchronous detection with beat frequency of the two modulations, which is called beat lock-in scheme.[10]

BASIC FUNCTIONS REALIZED BY BOCDA

Measurement range of the BOCDA is decided by the correlation peak position interval.[10,14] The typical range is several tens of meters. To enhance the range, schemes have been proposed and demonstrated. One is temporal gating scheme, in which rather long pulse is additionally used. By the pulse modulation with gated detection of the Brillouin scattering, we can select a portion with the pulse length along the optical fiber, then only one correlation point can be selected even when the total FUT length is longer than the correlation peak interval. Figure 2.5 shows a distributed measurement of BFS along the optical fiber obtained in the BOCDA system with the temporal gating scheme.[17] The spatial resolution of 7 cm with the measurement range of 1,030 m has been achieved in this experiment.

By using the beat lock-in scheme and the intensity modulation scheme, 1.6 mm spatial resolution has also been demonstrated, as shown in Figure 2.6.[3] The FUT is

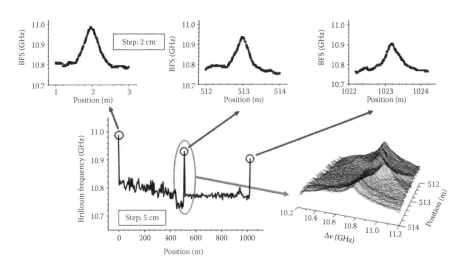

FIGURE 2.5 Distributed measurement with measurement range of 1,030 m and spatial resolution of 7 cm. (Data from K. Hotate et al., *SICE J Control Meas. Syst. Integr.*, 1, no. 4, 2008:271–274.)

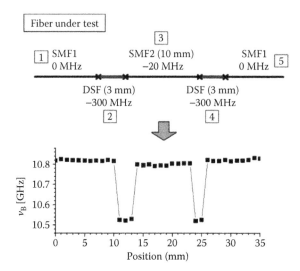

FIGURE 2.6 Demonstration of 1.6 mm spatial resolution in a BOCDA system. (Data from K. Y. Song et al., *OSA Opt. Lett.*, 31, no. 17, 2006:2526–2528.)

composed of an ordinary SMF with 10.83 GHz BFS and a dispersion shifted fiber (DSF) with 10.53 GHz BFS. DSFs of 3 mm length are spliced between the SMFs by a fusion splicer. In Figure 2.6, stepwise BFS changes are clearly observed with 1.6 mm spatial resolution.[3] The BFS difference of 300 MHz corresponds to 6000 με. Such a large strain can be measured by the BOCDA, because of the background spectrum suppression by the intensity modulation scheme.

BOCDA utilizes a continuous lightwave. Therefore, energy efficiency is better compared with the time domain technologies. Fast measurement of the BGS has been demonstrated as shown in Figure 2.7.[18] In the experiments, 1,000 spectra/s has been realized. The dynamic strain of 100 Hz sinusoidal waveform applied to the measurement point along the fiber is successfully measured.

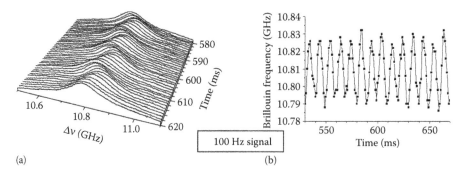

FIGURE 2.7 Demonstration of 1,000 spectra/s high-speed measurement for dynamic strain by a BOCDA system. (a) Brillouin spectra versus time and (b) Brillouin frequency shifts versus time. (Data from K. Y. Song and K. Hotate, *IEEE Photonics Technol. Lett.*, 19, no. 23, 2007:1928–1930.)

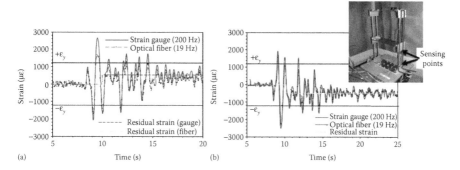

FIGURE 2.8   Demonstration of the random accessibility of BOCDA system. (a) and (b) show the dynamic strain behavior at two different sensing points. (Data from S. S. L. Ong et al., *Proc. 16th Int. Conf. Opt. Fiber Sensors*, Nara, Japan, 2003, We3-2.)

As explained above, one of the features of the BOCDA system is the random accessibility to arbitrarily multiple points along the fiber. Both with this feature and with the high-speed measurement capability shown above, dynamic strain measurement at multiple selected points can be realized. This function is similar to that realized by the multiplexed optical fiber sensing systems, for example, using fiber Bragg gratings (FBGs). The advantage brought by the BOCDA is, however, that the multiple points to be measured can be selected arbitrarily at any points along the optical fiber. Figure 2.8 shows the demonstration of the dynamic strain measurement at arbitrarily selected multiple points along the optical fiber.[19] A tall building model is shaken with an earthquake waveform, and the dynamic strain is measured simultaneously at the two points. Under the earthquake, the measured strain has exceeded the yielding strain, which means the structure had already been broken. The total measurement speed of BGS was about 30 spectra/s in this experiment. In other experiments, 200 spectra/s measurement has been demonstrated.[20]

Dynamic strain measurement at arbitrary selected multiple points in a tall building model.

BRILLOUIN DYNAMIC GRATING AND ITS APPLICATION TO DISCRIMINATIVE AND DISTRIBUTED MEASUREMENT OF STRAIN AND TEMPERATURE

A discriminative and distributed sensing of strain and temperature has recently been demonstrated with only using a single optical fiber. In the measurement, a polarization maintaining fiber (PMF) is used, such as polarization-maintaining and absorption-reducing (PANDA) fiber. Besides the Brillouin scattering nature, birefringence property of the PMF is also measured. When we launch the linearly polarized pump and probe lightwave into the PMF through the *x*-polarization axis as shown in Figure 2.9a, of course, the SBS takes place. In this process, the acoustic wave is also enhanced along the PMF, which is called Brillouin dynamic grating (BDG).[21] The acoustic wave is longitudinal wave, so it can reflect another lightwave with the orthogonal linear polarization. As shown in Figure 2.9b, the Bragg

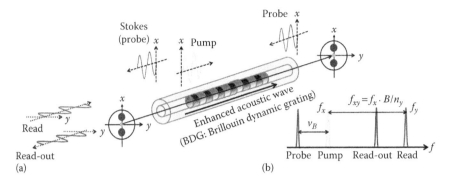

FIGURE 2.9   (a) BDG generation in a polarization maintaining optical fiber and (b) relation among pump, probe, stokes, read, and read-reflection lightwave.

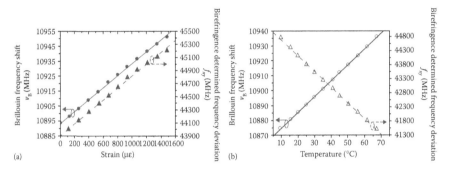

FIGURE 2.10   Dependence of BFS and BDG Bragg-reflection-spectrum peak on strain (a) and temperature (b). (Data from W. Zou et al., *OSA Opt. Express*, 17, no. 3, 2009:1248–1255.)

reflection for the y-polarized lightwave appears at the other lightwave frequency with the difference of f_{xy}. This is due to the birefringence of the PMF.

The Brillouin scattering and the birefringence of the PANDA fiber are different physical phenomena with each other. The strain and temperature dependence of the two parameters are studied experimentally, which are shown in Figure 2.10.[22] Strain dependence shows the same tendency between the two, but the temperature dependence shows the opposite slope, which means that the discriminative measurement can be performed in a superior accuracy by simply calculating the two data, v_B and f_{xy}.

Figure 2.11 shows a system configuration for the discriminative and distributed measurement.[23] This system is based on the BOCDA scheme. The FUT is a PANDA fiber. A DFB-LD output is modulated in intensity with IM2 to suppress the carrier component. The lower side band is used to make the SBS in x-polarization in the FUT. The BDG is enhanced in this process also in the FUT. The upper side band is used as the read-out light to be launched into the y-polarization axis. By changing the radio frequency to modulate IM2, BDG reflection spectrum is measured, and f_{xy} is obtained.

Figure 2.12 shows the measured distribution of v_B and f_{xy}.[4] Two experiments have been performed. In the blue experiment, portions B, D, E, F, and H are heated, but no strain are applied at any portions. In the red experiment, strain is

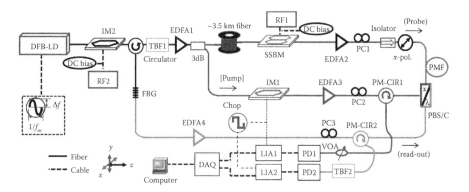

FIGURE 2.11 BOCDA system for discriminative and distributed measurement of strain and temperature using Brillouin scattering and BDG. (Data from W. Zou et al., *OSA Opt. Express*, 19, no. 3, 2011:2363–2370.)

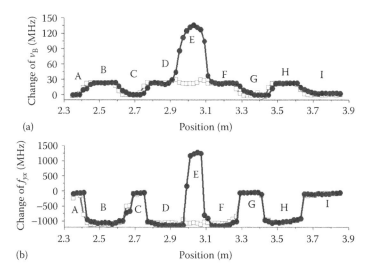

FIGURE 2.12 (a) Measured distribution of BFS and (b) BDG Bragg-reflection-spectrum peak. (Data from W. Zou et al., *IEEE Photonics Technol. Lett.*, 22, no. 8, 2010:526–528.)

applied additionally only at E portion. By using the data shown in Figure 2.12, strain and temperature distribution have been calculated, and the results are shown in Figure 2.13.[4] In Figure 2.13a, temperature distribution are clearly measured both in the blue and red experiment. In Figure 2.13b, strain applied only at E portion in the red experiment are also clearly shown. The discriminative and distributed measurement of strain and temperature has been successfully demonstrated by using BOCDA and BDG phenomenon. By the temporal gating scheme, the measurement range for the discriminative and distributed measurement has been enhanced to be 500 m.[24]

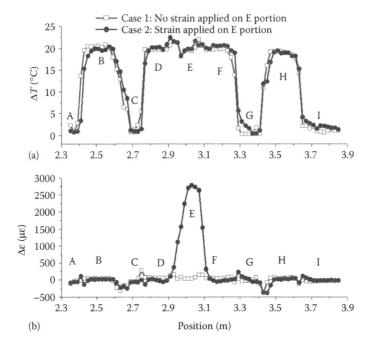

FIGURE 2.13 Discriminative and distributed measurement of strain and temperature obtained from the data in Figure 2.12. (a) Temperature distribution and (b) strain distribution. (Data from W. Zou et al., *IEEE Photonics Technol. Lett.*, 22, no. 8, 2010:526–528.)

MEASUREMENT EQUIPMENT DEVELOPED TO DATE

Two on-board prototype BOCDA devices for aircraft SHM are under development. Both devices are confirmed that it is able to apply both of them to aircraft SHM. The first prototype BOCDA device measures BGS only. Because the BGS is affected by both strain and temperature, some temperature compensating method is needed for accurate strain measurement. On the other hand, it is simple in construction, and cost effective, not only the measurement device but also the sensing fiber, because commercial available SMF can be applied. The second prototype BOCDA device can measure strain accurately because it can discriminate between strain and temperature by measuring BGS and DGS with a PMF. But it is more complex in construction than first prototype BOCDA device. If the system measures BGS only, it behaves as first prototype BOCDA device. Details of the each prototype will be discussed in the below subsections.

First Prototype BOCDA System[25]

Overview

The first prototype BOCDA device[14] can measure the BGS over a length of 500 m SMF with a spatial resolution of 30 mm, a sampling rate of 70 Hz, and strain accuracy of ±13 με. This device consists of an optical unit, a power unit, and a measurement unit.

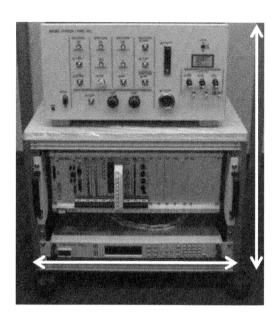

FIGURE 2.14   Outside view of the first prototype BOCDA device.

The outside view of the device is shown in Figure 2.14. The system is 430 mm in width, 460 mm in height, 550 mm in depth, and 45 kg in weight.

The first prototype BOCDA device is equipped with its performance using the temporal gating scheme, the polarization diversity technique, and the automatic bias-adjustment function of an optical modulator. These techniques contribute to improve the measurement reliability, the measurement range, and the setup time. Details of the temporal gating scheme and the polarization diversity technique will be discussed in following sections.

The outline of the optical circuit of the first prototype BOCDA device is shown in Figure 2.15. To keep the linearly polarization state of the probe and pump wave, the optical fibers from the DFB laser to the polarization switch (Pol. SW), also from the DFB laser to the EOM1 consist of PANDA-type PMF. The measuring unit consist of four function generators (FGs), an analog to digital converter (A/D), and a CPU module. All of them are PXI (PCI eXtensions for Instrumentation) modules and installed in a PXI mainframe to synchronize and control.

Temporal Gating Scheme[26]

In the basic BOCDA technology, the peak interval and the spatial resolution are in trade-off relation with the measurement range. The measurement range is restricted by the correlation peak interval, because it appears periodically along a fiber. The temporal gating scheme is one of the strong methods to avoid this restriction. Figure 2.16 shows the operation of the temporal gating scheme. In Figure 2.16, provided that two or more correlation peaks stand along an FUT. Here, we introduce "pulsed pump" wave with a narrower pulse width than the correlation peak interval. The correlation signal in the FUT is generated at the point of overlap of a probe wave

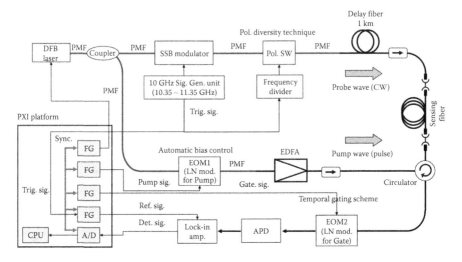

FIGURE 2.15   Outline of the optical circuit.

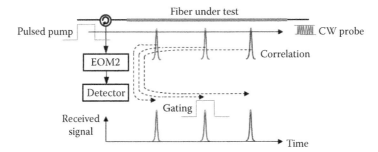

FIGURE 2.16   Temporal gating operation to select a specified correlation peak.

(CW: continuous wave) and a pulsed-pump wave. That means the correlation peak signals along the FUT can be expressed on the time domain. When the signal from specified correlation peak is propagated in the FUT and reaches the detector, the EOM2 pass the probe wave. On the other hand, when the signals from other correlation peaks reach the detector, the EOM2 blocks the probe wave. Thus, the information only on specified correlation peak is obtained from the probe wave. Here, the pulse width of a pulsed pump wave (t_{pw}) is restricted by the correlation peak interval so that one correlation peak can be extracted from two or more correlation peaks in time. Equation 2.2 is derived as follows:

$$t_{pw} = \frac{2d_m}{V_g} = \frac{1}{f_m} \tag{2.2}$$

where:
    d_m is the interval of the correlation peaks
    V_g is the group velocity of light in the FUT
    f_m is the modulation frequency

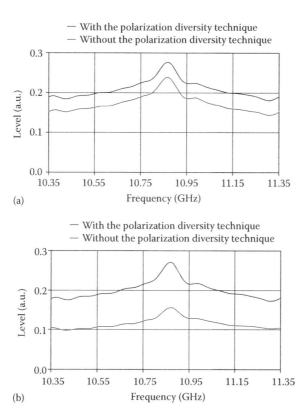

FIGURE 2.17   Result of BGS measurement from a portion of an FUT: (a) polarization state 1 and (b) polarization state 2.

Polarization Diversity Technique[27]

The measurement accuracy of the BOCDA device is affected by polarization-dependent Brillouin gain fluctuation. The polarization diversity technique is introduced to suppress the fluctuation.

To confirm availability of the polarization diversity technique, BGS in two different polarization states is measured. Figure 2.17 shows the results of BGS measurement from a portion of an FUT. The red colored traces in Figure 2.17 are measured without the polarization diversity technique, and the black colored traces are measured with the polarization diversity technique. The red colored trace in Figure 2.17a shows a BGS by a polarization state that makes high Brillouin gain, and the red colored trace in Figure 2.17b shows a BGS by another polarization state that makes low Brillouin gain. By employing the polarization diversity technique, the black colored traces in the Figures 2.17a and b become almost the same level. It is confirmed that this technique compensates for the effect of the polarization state change of pump and probe waves.

Demonstration on Spatial Resolution

The configuration of the FUT to evaluate the spatial resolution of the BOCDA device is shown in Figure 2.18. The FUT was composed of a series of two kinds of SMF,

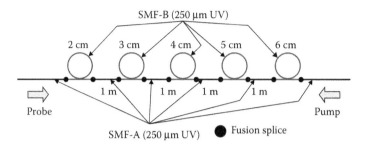

FIGURE 2.18 Configuration of the FUT.

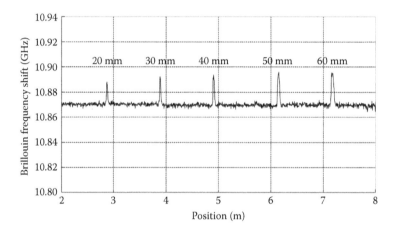

FIGURE 2.19 Result of spatial resolution measurement.

SMF-A and SMF-B, which have different BFSs, ν_B. The ν_Bs of the SMF-A and the SMF-B are 10.870 and 10.895 GHz, respectively. The magnitude of FM modulation Δf is set to 3.2 GHz, the FM modulation frequency f_m is controlled within 9.8–10 MHz according to sensing position, makes the correlation peak period to about 10 m. Figure 2.19 shows the result of spatial resolution measurement. Under these conditions, we confirmed the BGS distribution measurement with a spatial resolution of 30 mm.

SECOND PROTOTYPE BOCDA DEVICE

Overview

The second prototype BOCDA device that can discriminate between strain and temperature is under development. It measures BGS and DGS to discriminate between strain and temperature. A strain measurement accuracy of ± 60 μs train, a temperature measurement accuracy of $\pm 1°C$, a spatial resolution of 30 cm, and measurement range of over 500 m has achieved.

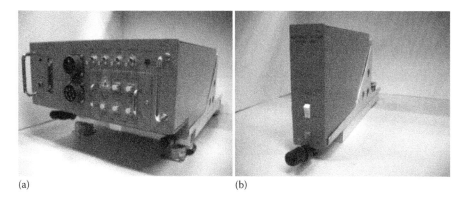

(a)   (b)

FIGURE 2.20   Outside view of the second prototype BOCDA device: (a) optical unit and (b) power unit.

The outside view of a second prototype BOCDA device is shown in Figure 2.20. It consists of an optical unit, a power unit, and a measurement unit. The form factor of the optical unit and the power unit are based on ARINC 600 series standards, which define the size and weight of boxes and trays used in aircrafts called the avionics modular concept unit (MCU). The size of optical unit is 12MCU and that of power unit is 2MCU.

The configuration of the second prototype BOCDA device is very similar to the first prototype BOCDA device. Because it measures not only BGS but also DGS, some components are used to measure DGS. The block diagram of the second prototype BOCDA device is shown in Figure 2.21. The power unit supplies ±15 V and +5 V of DC to the optical unit from +28 V of DC that is commonly used in aircrafts. The measurement unit is consist of two arbitrary waveform generators (AWG), two FGs, an A/D, a digital to analog converter (D/A), a switch, a digital multi meter (DMM), and a controller. All of them are PXI modules and installed in a PXI mainframe. The outline of the optical unit is shown in Figure 2.22. To measure BGS and DGS, all optical components and optical fibers use PANDA-type PMF. The AWG2, DFB-LD2, EDFA3, Circulator 2, and PBSC (polarization beam splitter cube) are used to measure DGS. The DC biases of the EOM and the SSB (single side band) modulator are controlled by automatic bias adjustment functions. Figure 2.23 shows GUI screen of the BOCDA system. All parameters such as measurement mode, distance range, and spatial resolution can be set through in the GUI screen. The measurement results such as measured BGS, DGS, and calculated strain and temperature are also displayed in the GUI screen.

Demonstration on Discrimination between Strain and Temperature

The experimental setup and the configuration of the FUT to demonstrate the discrimination between strain and temperature is shown in Figure 2.24. The FUT of a 500 m length is composed of a PANDA-type PMF. A part in the FUT at a length of 1.4 m is loosely inserted into a temperature controlled water bath. Another part in the

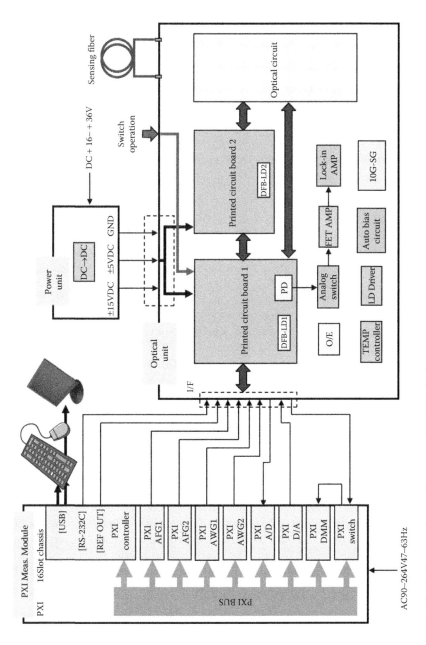

FIGURE 2.21 Block diagram of second prototype BOCDA device.

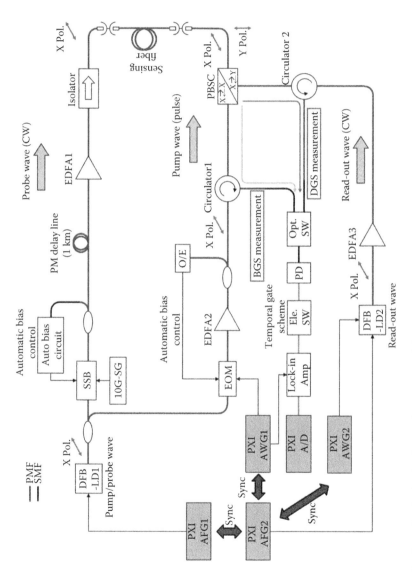

FIGURE 2.22 Outline of optical unit of second prototype BOCDA device.

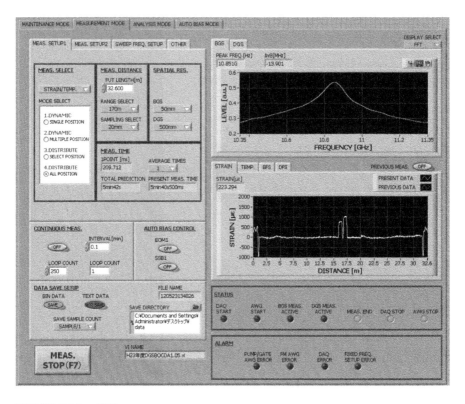

FIGURE 2.23   GUI screen.

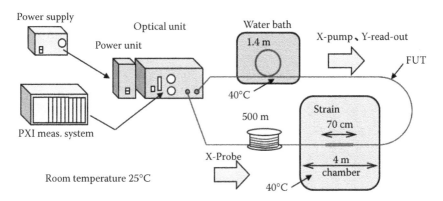

FIGURE 2.24   Experimental setup and configuration of the FUT.

FUT at length of 4 m is inserted into a temperature controlled chamber and more-over pulled to load strain at length of 70 cm in the temperature chamber. The spatial resolution of BGS and the DGS are set to 3 and 30 cm, respectively. The measured distributed strain and temperature along the FUT is shown in Figures 2.25 and 2.26. These results show good agreement with the applied strain and temperature.

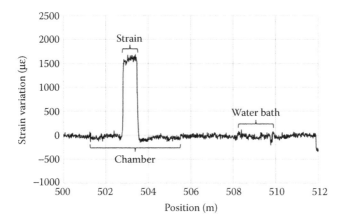

FIGURE 2.25   Discriminative distributed measurements of strain.

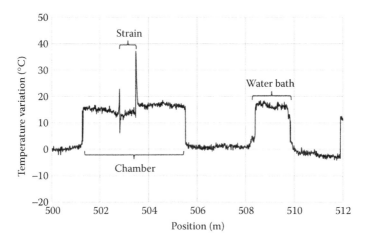

FIGURE 2.26   Discriminative distributed measurements of temperature.

APPLICATION SCENARIOS AND CASE STUDIES OF BOCDA-SHM SYSTEM

APPLICATION SCENARIOS OF BOCDA-SHM SYSTEM

Utilizing the superior characteristics of BOCDA, application scenarios of the BOCDA-SHM system, which includes OFS and BOCDA device, are distributed sensing along an OFS and real-time dynamic sensing at arbitrary points of an OFS during flights (see Figure 2.27). In distributed sensing, the system monitors distributed strain and temperature changes. When strain distribution changes occur due to structural damages, the system detects changes of distributed strain states. In real-time dynamic sensing, the system monitors dynamic strain changes at structural critical portions such as wing–body junctions and pressure bulkheads.

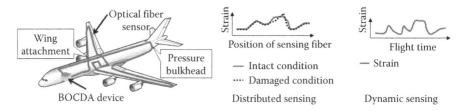

FIGURE 2.27   Application scenarios of BOCDA-SHM system.

The dynamic sensing will be able not only to detect structural damages, but also to predict fatigue life of these portions.

CASE STUDIES OF BOCDA-SHM SYSTEM

The authors have raised the TRL[5] of the BOCDA-SHM system to realize these scenarios. The prototype of BOCDA device has been already manufactured on trial,[6] durability evaluation of the system involving attached sensors and measurement devices, probability of detection (PoD) evaluation have been also conducted.[6–8,28] The TRL of the system is approximately 6 established by these developments. In this section, our development results are shown as case studies.

Probability of Detection

Verification of PoD is needed for the system to real use. SHM system using OFS must have the same level of PoD with the conventional nondestructive inspections such as ultrasonic inspections for composite structures. Measurement targets of the BOCDA-SHM system are strain and temperature as shown in section "Application Scenarios of BOCDA-SHM System." Therefore, PoD verification tests were conducted using Al plate that is usually used on aircraft structures. An OFS were attached on the surface of the specimens with no prestrained and with pre-tensile strain. Strain gauges and a thermocouple were also attached near the OFS to compare the measurement results with BOCDA-SHM system. Figures 2.28 and 2.29 show the test specimen and the

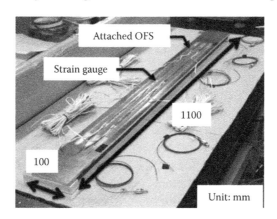

FIGURE 2.28   Test specimen in PoD test.

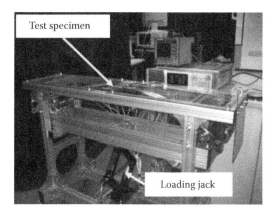

FIGURE 2.29 Test setup in PoD test.

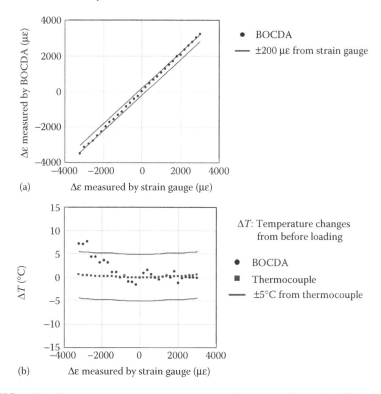

FIGURE 2.30 Temperature measurement results with pre-tensile strain OFS: (a) strain measurement results and (b) temperature measurement results.

setup of PoD verification test, respectively. Four-point bending test apparatus was used and tension/compression strains were loaded to each surface by bending loads. Figure 2.30 shows the test results using no prestrain OFS. Strain measurement results by BOCDA-SHM system were coincided with strain measured by strain gauges closely. Temperature measured by BOCDA with no prestrained OFS has lager error

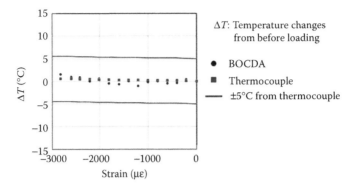

FIGURE 2.31 Temperature measurement results with pre-tensile strain OFS.

as OFS compression strain increased. The causes of these errors might be that coefficients shown in Equation 2.1 are adapted only when OFS was tensioned. When OFS was compressed, the coefficients will depend on compression strain. To eliminate these errors, the authors propose a new OFS installation method. In this new method, OFS shall be attached on the specimen surface with pretensile strain. Figure 2.31 shows the measurement test results with pretensile strain OFS when OFS was compressed. Temperature measurement errors between by BOCDA-SHM system and the thermocouple were dramatically decreased. From the test results, strain and temperature measurement errors will be ± 200 $\mu\epsilon$ and $\pm 2°C$, respectively.

Durability of OFS

Application targets of the BOCDA-SHM system are whole aircraft structures such as main wing, fuselage, and so on. OFS will be attached on the surface of these component structures. Therefore, OFS will be under severe environmental conditions such as high/low temperature, high/low pressure, high humidity, and fuel. In our developments, durability verification categories that will be needed to put our sensor system on real use have been ensured based on RTCA/DO-160.[29] Table 2.2 shows a part of the durability verification categories for our OFS system. Our OFS system includes not only OFS but also sealant and top coat as OFS protection under severe conditions. Figure 2.32 shows the OFS system in our development. Almost all durability tests have been conducted and durability of OFS system have already been verified. Figure 2.33 shows an overview of the durability verification tests for OFS system.

Durability of BOCDA Device

The BOCDA device will be at an avionics bay on an aircraft in the application scenarios of the BOCDA-SHM system. Therefore, durability of the device has to be verified although environmental conditions will not be more severe than the OFS system. In our developments, durability verification categories for the devices have been also ensured based on RTCA/DO-160.[29] Table 2.3 shows a part of the durability verification categories for the BOCDA device. Almost all durability tests have been conducted and durability of the BOCDA device have already verified although

TABLE 2.2
Durability Verification Categories
for OFS System (Partly)

Durability Verification Category

Mechanical loading
High temperature
Low temperature
Temperature variation
Water immersion
Skydrol immersion
Fuel immersion
Humidity
Low pressure
High pressure
Decompression

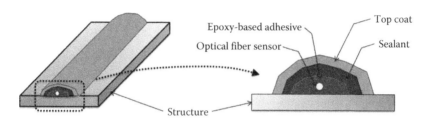

FIGURE 2.32 OFS system.

High temperature   Low temperature   Water immersion   Humidity

FIGURE 2.33 Overview of durability verification tests for OFS system.

measurement accuracy decreased under vibration and shock loading. Figure 2.34 shows overview of the durability verification tests for the BOCDA device.

Flight Demonstration Test

To apply our BOCDA-SHM system to real use during operation, it is needed to verify the operability and feasibility of the system. Therefore, flight demonstration tests were conducted with the airborne BOCDA-SHM system as described before using the 11-seater business jet, MU-300. An OFS was attached on the front spar of

TABLE 2.3

Durability Verification Categories for BOCDA Device (Partly)

Durability Verification Category

Vibration

Shock

Low temperature

High temperature

Temperature variation

Humidity

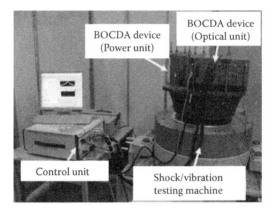

FIGURE 2.34   Overview of durability verification tests for BOCDA device.

the vertical tail and the BOCDA device was on the cabin floor of the jet. Figure 2.35 shows the test bed on the test.

Figure 2.36 shows dynamic measurement results and Figure 2.37 shows static distribution measurement results on the flight demonstration test. Dynamic strain changes during ascending, descending, and right–left turning flights were measured and those

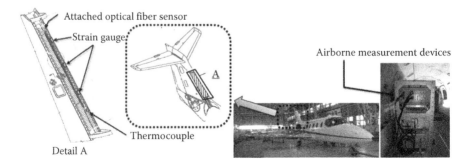

FIGURE 2.35   Test bed on the flight demonstration test.

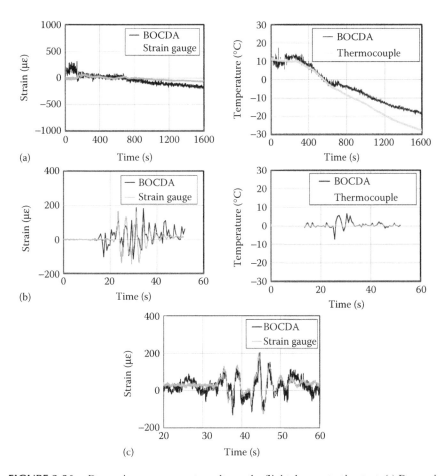

FIGURE 2.36 Dynamic measurement results on the flight demonstration test. (a) Dynamic strain and temperature changes during take-off and ascending. (b) Dynamic strain and temperature changes during right–left turning flights. (c) Dynamic strain changes during right–left turning flights.

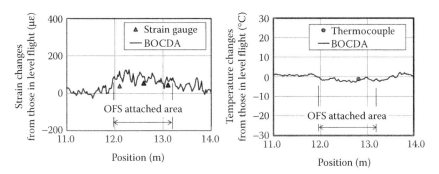

FIGURE 2.37 Distribution measurement results during steady banking flights.

values were close to those measured by strain gauges. There were, however, some errors on temperature measured by BOCDA compared with temperature measured by a thermocouple. The cause of these errors was considered when the coefficients related to the birefringence phenomena might be different from the coefficients shown in Equation 2.2 when OFS was compressed. The authors, therefore, conducted four-point bending tests to investigate the cause of these errors and to develop a new OFS installation method decreasing the temperature measurement errors as shown in section "Probability of Detection." Strain and temperature distributions measured by BOCDA during steady bank flights denoted the same tendency of those measured by the legacy measurement. The values of strain and temperature did not coincide with the reference values. The reason why there were margins of these values was that actual strain and temperature changes from during level flight were too small compared with measurement accuracies of the BOCDA device in this measurements, which were ± 50 $\mu\varepsilon$ and $\pm 2°C$.

Development for Practical Use

The application scenarios of the BOCDA-SHM system are whole structure monitoring utilizing the characteristics of the system such as distributed wide area monitoring, high spatial resolution, and high sampling rate. If the system could detect not only wide area damages but also small damages on narrow area, the system will be more attractive. Therefore, the authors developed a new damage detection technique for small damages, especially micro-damages due to bearing failure at bolted joints in carbon-fiber-reinforced polymer (CFRP). Furthermore, to monitor whole structure, installation of OFS on a wide area will be needed. Attaching OFS using the conventional epoxy-based adhesives on wide area will be a big time-consuming procedure. To overcome these problems, the authors developed an easy OFS installation method using thermoplastic-resin-based adhesives.

Bearing Damage-Detection Technique

Because of the high characteristics of CFRP such as light weight, high strength, and high corrosion durability, aircraft components with CFRP were employed into service recently. These components were usually assembled by bolted joints. Figure 2.38

FIGURE 2.38   Bolted joint portion on aircraft structures.

shows a bolted joint portion on aircraft structures. At bolted joints, however, stress concentrations will initiate bearing failures. That is why these joints portions are important portions at inspections.

Although it is important to inspect these portions, these portions are usually hidden by panels, so disassembly and reassembly are needed for inspection. Therefore, the authors developed a new damage-detection technique at CFRP bolted joints without disassembly and reassembly, utilizing BGS, which is one of the measurement outputs. It is well known that bearing failures on CFRP structures attribute to microdamages such as reinforced fiber buckling, matrix shear cracking, and delamination. To detect these microdamages, OFS were embedded in CFRP laminates and monitored strain status changes in CFRP laminates due to occurrences of these microdamages.

Figures 2.39 and 2.40 show a CFRP test specimen with embedded OFS and test data processing flow, respectively. Figure 2.41 shows the test results, in which BGS

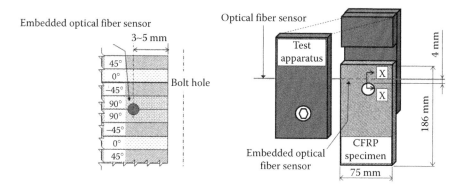

FIGURE 2.39 CFRP test specimen with embedded OFS.

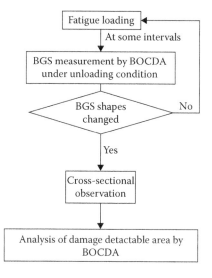

FIGURE 2.40 Test data processing flow.

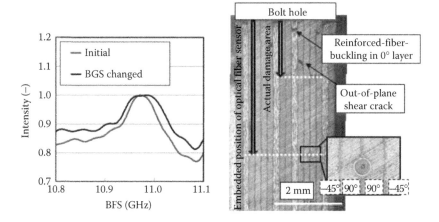

FIGURE 2.41   Bearing damage detection test result.

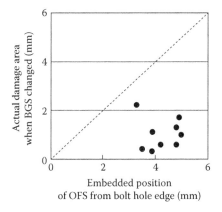

FIGURE 2.42   Relationship between OFS embedded position and actual damage area.

clearly changes before and after microdamages occurred and cross section of the test specimen. Then, OFS embedded position and actual damage areas were observed by cross-sectional observations using optical microscope. Figure 2.42 shows the relationship between OFS embedded position and actual damage area, which indicates that the BOCDA will be able to detect occurrences of microdamages even when microdamages do not reach the OFS.

Easy OFS Installation Method

OFS will be attached hardly on the structure surface to monitor strain changes properly. Epoxy-based adhesives are used conventionally to attach OFS. It is a time-consuming procedure to use these adhesives because curing time of the epoxy-based adhesives is usually long. To overcome these problems, the authors developed new OFS attaching procedures, which can be easily used as thermoplastic-resin-based adhesives. Thermoplastic-resin-based adhesives are known as short curing time.

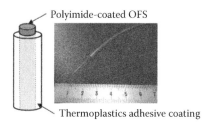

Polyimide-coated OFS

Thermoplastics adhesive coating

FIGURE 2.43   New OFS.

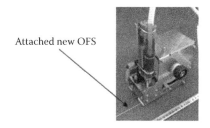

Attached new OFS

FIGURE 2.44   New installation system.

TABLE 2.4
Test Result of Installation Speed Validation Tests

Installation Procedure	Installation Speed (mm/min)
New system using thermoplastic adhesive	200
Conventional using epoxy adhesive	50

The authors manufactured the new OFS with thermoplastic-resin adhesive coating, which is shown in Figure 2.43. An installation system, furthermore, was developed to attach the new OFS on the surface using hot air. Figure 2.44 shows the installation system. Using the new OFS and installation system has a capability to improve OFS installation speed compared with using a conventional epoxy-based adhesive. The authors conducted validation tests by the new and conventional installation procedures. Table 2.4 shows the installation speed validation test results. The new installation system improves OFS installation speed about 3 times faster than conventional procedure.

Challenges for Practical Use

The BOCDA-SHM system has a big potential to reduce maintenance costs, improve flight safety, and fuel efficiency. The authors have been developing the system to practical use as noted above. However, there are some challenges for practical use of the system. For example, measurement accuracy of the BOCDA-SHM system has to be improved under shock/vibration conditions, OFS installation procedures have to be easier and big data processing procedures. Especially for the BOCDA-SHM system, lightwaves have to enter from both sides of the OFS, therefore OFS network with redundancy shall be developed because the BOCDA system will be impossible to measure when the OFS is broken.

Challenges as noted above are technical challenges. On the other hand, certification process and establishment of operation center are important challenges for practical use.

SUMMARY

In this chapter, the history of the BOCDA development such as principle and basic functions of BOCDA, measurement equipment developed to date, advantages of BOCDA for aircraft SHM, and application scenarios was explained. Although there are many sensing technologies utilizing Brillouin scattering such as BOTDA, BOTDR, BOCDA, and BOCDR, the authors have been developing the BOCDA on the view of technology level, measurement stability, and practical utility. BOCDA has more superior characteristics than the other OFS technologies for wide-area monitoring of aircraft structure such as distribution sensing, dynamic sensing, and discrimination of strain/temperature.

For practical use of the BOCDA-SHM system, prototype BOCDA systems were manufactured and many verification tests were conducted using the prototype systems such as the flight demonstration test. The BOCDA has a big probability to be applied as on-board SHM system of aircraft structures in years to come even though there are still some challenges for practical use.

ACKNOWLEDGMENTS

This work was conducted as a part of the project, "Advanced Materials and Process Development for Next-Generation Aircraft Structures" under the contract with SOKEIZAI Center, founded by Ministry of Economy, Trade and Industry (METI) of Japan.

REFERENCES

1. D. Roach, and S. Neidigk, "Does the maturity of structural health monitoring technology match user readiness?," *Proceedings of the 8th International Workshop on Structural Health Monitoring,* no. 1 (2011):39–52.
2. B. Beral, and H. Speckmann, "Structural health monitoring (SHM) for aircraft structures: A challenge for system developers and aircraft manufactures," *Proceedings of the 4th International Workshop on Structural Health Monitoring,* no. 1 (2003):12–29.
3. K. Y. Song, Z. He, and K. Hotate, "Distributed strain measurement with millimeter-order spatial resolution based on Brillouin optical correlation domain analysis," *OSA Optics Letters,* 31, no. 17 (2006):2526–2528.
4. W. Zou, Z. He, and K. Hotate, "Demonstration of Brillouin distributed discrimination of strain and temperature using a polarization-maintaining optical fiber," *IEEE Photonics Technology Letters,* 22, no. 8 (2010):526–528.
5. J. C. Mankins, "Technology Readiness Levels," *A white paper* (1995). https://www.hq .nasa.gov/office/codeq/trl/trl.pdf (last accessed 9 February, 2016).

6. T. Yari, M. Ishioka, K. Nagai, M. Ibaragi, K. Hotate, and Y. Koshioka, "Monitoring aircraft structural health using optical fiber sensors," *Mitsubishi Heavy Industries Technical Review*, 45, no. 4 (2008):5–8.

7. T. Yari, N. Saito, K. Nagai, K. Hotate, and K. Enomoto, "Development of a BOCDA-SHM system to reduce airplane operating costs," *Mitsubishi Heavy Industries Technical Review*, 48, no. 4 (2011):65–69.

8. N. Saito, T. Yari, K. Hotate, M. Kishi, S. Matsuura, Y. Kumagai, and K. Enomoto, "Developmental status of SHM applications for aircraft structures using distributed optical fiber," *Proceedings of the 9th International Workshop on Structural Health Monitoring*, no. 1 (2013):2011–2018.

9. Y. Kumagai, S. Matsuura, T. Yari, N. Saito, K. Hotate, M. Kishi, and M. Yoshida, "Fiber-optic distributed strain and temperature sensor using BOCDA technology at high speed and with high spatial resolution—Taking on aircraft structural health monitoring," *Yokogawa Technical Report*, 56, no. 2 (2013):83–86.

10. K. Hotate, "Fiber distributed Brillouin sensing with optical correlation domain techniques," *Optical Fiber Technology*, 19 (2013):700–719.

11. T. Horiguchi, T. Kurashima, and M. Tateda, "A technique to measure distributed strain in optical fibers," *IEEE Photon Technology Letters*, 2, no. 5 (1990):352–354.

12. X. Bao, D. J. Webb, and D. A. Jackson, "22-km distributed temperature sensor using Brillouin gain in an optical fiber," *Optics Letters*, 18, no. 7 (1993):552–554.

13. N. Marc, L. Tevenaz, and P. A. Robert, "Brillouin gain spectrum characterization in single mode optical fibers," *Journal of Lightwave Technology*, 15, no. 10 (1997):1842–1851.

14. K. Hotate, and T. Hasegawa, "Measurement of Brillouin gain spectrum distribution along an optical fiber using a correlation-based technique—Proposal, experiment and simulation," *IEICE Transactions on Electronics*, E83-C, no. 3 (2000):405–412.

15. Y. Mizuno, Z. He, and K. Hotate, "One-end-access high-speed distributed strain measurement with 13-mm spatial resolution based on Brillouin optical correlation-domain reflectometry," *IEEE Photonics Technology Letters*, 21, no. 7 (2009):474–476.

16. M. Sitthipong, M. Kishi, Z. He, and K. Hotate, "1-cm spatial resolution with large dynamic range in strain distributed sensing by Brillouin optical correlation domain reflectometry based on intensity modulation," *Proceedings of the 3rd Asia Pacific Optical Sensors Conference*, Th-C23 (2012):APO12–51.

17. K. Hotate, H. Arai, and K. Y. Song, "Range-enlargement of simplified Brillouin optical correlation domain analysis based on a temporal gating scheme," *SICE Journal of Control, Measurement, and System Integration*, 1, no. 4 (2008):271–274.

18. K. Y. Song and K. Hotate, "Distributed fiber strain sensor at 1 kHz sampling rate based on Brillouin optical correlation domain analysis," *IEEE Photonics Technology Letters*, 19, no. 23 (2007):1928–1930.

19. S. S. L. Ong, M. Imai, Y. Sako, Y. Miyamoto, S. Miura, and K. Hotate, "Dynamic strain measurement and damage assessment of a building model using a Brillouin optical correlation domain analysis based distributed strain sensor," *Proceedings of 16th International Conference on Optical Fiber Sensors*, Nara, Japan (2003):We3-2.

20. K. Hotate, M. Numasawa, M. Kishi, and Z. He, "High speed random accessibility of Brillouin optical correlation domain analysis with time division pump-probe generation scheme," *Proceedings of the 3rd Asia Pacific Optical Sensors Conference*, WB-3, Sydney, Australia (2012): APO12-99.

21. K. Y. Song, W. Zou, Z. He, and K. Hotate, "All-optical dynamic grating generation based on Brillouin scattering in polarization-maintaining fiber," *OSA Optics Letters*, 33, no. 9 (2008):926–938.

22. W. Zou, Z. He, and K. Hotate, "Complete discrimination of strain and temperature using Brillouin frequency shift and birefringence in a polarization-maintaining fiber," *OSA Optics Express*, 17, no. 3 (2009):1248–1255.

23. W. Zou, Z. He, and K. Hotate, "One-laser-based generation/detection of Brillouin dynamic grating and its application to distributed discrimination of strain and temperature," *OSA Optics Express*, 19, no. 3 (2011):2363–2370.

24. R. K. Yamashita, W. Zou, Z. He, and K. Hotate, "Measurement range elongation based on temporal gating in Brillouin optical correlation domain distributed simultaneous sensing of strain and temperature," *IEEE Photonics Technology Letters*, 24, no. 12 (2012):1006–1008.

25. Y. Kumagai, S. Matsuura, S. Adachi, and K. Hotate, "Enhancement of BOCDA system for aircraft health monitoring," *SICE Annual Conference*, 2B02-1 (2008).

26. M. Kannou, S. Adachi, and K. Hotate, "Temporal gating scheme for enlargement of measurement range of Brillouin optical correlation domain analysis for optical fiber distributed strain measurement," *Proceedings of the 16th International Conference on Optical Fiber Sensors*, Nara, Japan, 2003, Paper We3-1.

27. K. Hotate, K. Abe, and K. Y. Song, "Suppression of signal fluctuation in Brillouin optical correlation domain analysis system using polarization diversity scheme," *IEEE Photonics Technology Letters*, 18, no. 24 (2006):2653–2655.

28. Y. Kumagai, S. Matsuura, T. Yari et al., "Fiber-optic distributed strain and temperature sensor using BOCDA technology at high speed and with high spatial resolution—Taking on aircraft structural health monitoring," *Yokogawa Technical Report*, 56, no. 2 (2013):81–84.

29. RTCA DO-160G, Environmental Conditions and Test Procedures for Airborne Equipment (2010).

3 Development of Embedded FBG Sensor Networks for SHM Systems

Gayan Chanaka Kahandawa, Jayantha Ananda Epaarachchi, John Canning, Gang-Ding Peng, and Alan Lau

CONTENTS

INTRODUCTION

Fiber-reinforced polymer (FRP) composites being an engineering material for many decades. The main attraction of the FRP is its superior strength-to-weight ratio. In particular, aircraft and defense industries have been spending billions of dollars on investment in these composites to produce lightweight subsonic and supersonic aircrafts. Other desirable properties, such as the ease of fabrication of complex shapes and the ability to tailor desirable properties to suit different engineering applications, are enviable for an advanced material. Research and development work in past few decades FRP composites has made inroads to delicate aerospace and space industries. Therefore the use of FRP Composite in aerospace industry has been increased by a significant amount in recent years.

The weight-save or positive weight spiral in the aircraft industry is directly translated to the enhancement of the load carrying capacity of civil aircraft, whereas for the fighter jets, it will be translated to the performance enhancement (mainly on the fuel carrying capacity versus the flying speed). As composites are partially made from polymer-based materials, they possess very good damping and fatigue resistance properties as compared with traditional metallic materials.

The commercial aircraft industry is gradually replacing metallic parts with FRP composites as much as possible. Hence, the FRP composites are frequently applied to primary load-bearing structures in the newly developed aircraft such as Boeing 787 and Airbus A380. However, the main disadvantages of using FRP composites in the aircraft industry are their difficulty for repair, anisotropic behavior, high initial setup cost, and most importantly the complex failure criteria. Because of these undesirable properties, the FRP composite structures in the aircraft need to be closely monitored to prevent unexpected failure.

Aircraft structures have numerous stress-concentrated regions such as pin-loaded holes and other cut-outs. These stress concentrations easily induce damages such as concurrent splitting, transverse cracking, and delamination (Chang and Chang, 1987; Kortscho and Beaumont, 1990; Kamiya and Sekine, 1996). Unlike metals, the damage accumulation and the damage prediction of composites are very difficult to predict and can be catastrophic. Because of this reason, it is essential to monitor advance composites structures such as expensive aerospace structures regularly. As a consequence, structural health monitoring (SHM) technique has recently been developed for FRP composite structures majorly for aerospace structures (Zhou and Sim, 2002; Chang, 2003). During past few decades monitoring of structural health of composites began with damage detection techniques such as vibration and damping methods (Adams et al., 1978). Then sophisticated and expensive offline nondestructive testing (NDT) methods were developed for the safe operation of composite structures. However, with the increasing complexity of structures, offline NDT was criticized as insufficient and developments of proper SHM systems became vital.

There are many procedures and technologies need to be integrated to form intelligent SHM systems. The indicator for damage, one of the most important parameter of an SHM, is defined as changes to the material properties or changes to the structural response of the structure. The SHM process involves the observation of a system over time using periodically sampled dynamic response measurements from an array of sensors. The SHM system developed to monitor aircraft and space structures must be capable of identifying multiple failure criteria of FRP composites (Reveley et al., 2010). Since the behavior of composites is anisotropic, multiple numbers of sensors must be in service to monitor these structures under multidirectional complex loading conditions. The layered structure of the composites makes it difficult to predict the structural behavior only by using surface sensors.

An SHM system must be efficient, robust and accurate to detect complex and hidden the process of damage propagation (Reveley et al., 2010). Because of recent developments in the aerospace industry, utilization of FRP composites for primary aircraft structures, such as wing leading-edge surfaces and fuselage sections, has increased leading to a rapid growth in the field of SHM. Various causes such as impact, vibration, and loading can initiate damages, such as delamination and matrix cracking, to FRP composite structures. Moreover, the internal material damage can be undetectable using conventional techniques, making inspection of the structures for damage and clear insight into the structural integrity difficult using currently available evaluation methods.

Since the behavior of composites is anisotropic, multiple numbers of sensors must be in service to monitor these structures under multidirectional complex loading conditions. The layered structure of the composites makes it difficult to predict the structural behavior only by using surface sensors.

Many modern light aircraft are being increasingly designed to contain as much lightweight composite material as possible. For elevated-temperature applications carbon-fiber-reinforced composite is in use. Concord's disk brakes used this material, rocket nozzles and re-entry shields have been fashioned from it, and there are other possibilities for its use as static components in jet engines. Rocket motor casings and rocket launchers are also frequently made of reinforced plastics. A particularly interesting (and important) application of composites is in its development in Australia as a means of repairing battle damage (patching) in metal aircraft structures.

Space applications offer many opportunities for employing light-weight, high-rigidity structures for structural purposes. Many of the requirements are the same as those for aeronautical structures, since there is a need to have low weight and high stiffness to minimize loads and avoid the occurrence of buckling frequencies. Dimensional stability is at a premium, for stable antennae and optical platforms, for example, and materials need to be transparent to radio-frequency waves and stable toward both UV radiation and moisture. Progressive damage in laminated composite can be subdivided into two:

1. Micromechanics (matrix-fiber)
2. Macromechanics (lamina)

The evolution of a matrix crack as the initial stage of damage, followed by delamination, is also true for composites subjected to impact load (Marshall et al., 1985). In most of the damage models reported in literature are transverse matrix cracks, splitting, and delamination. The degradation of effective elastic module of damaged laminates used as the damage parameters. The observed damage parameters by online health monitoring, such parameters can further be interpreted to damage states. The final stage of damage progression identified as "Delamination," which is the failure of the interface between two plies, is known as the silent killer of the composite structures. It is caused by normal and shear tractions acting on the interface, which may be attributed to transverse loading, free edge effect, ply-drop-off, or local load introduction. Delamination can significantly reduce the structural stiffness and the load carrying capacity and, therefore, is considered as one of the critical failure modes in laminated composites.

SHM SYSTEMS FOR FRP COMPOSITES

The process of implementing a damage detection and characterization strategy for engineering structures is referred to as SHM. The SHM process involves the observation of a system over time using periodically sampled structural response measurements from an array of sensors. Most of offline NDT methods do not fall into SHM.

With the complex failure modes of FRP composites, the need for SHM of composite structures becomes critical. With the recent developments in the advanced composite applications, utilization of FRP composites for primary aircraft structures, such as wing leading-edge surfaces and fuselage sections, has increased. This is one of the major reasons for the rapid growth in the research fields related to SHM. Impact from flying objects, excessive vibration, and loading can cause damage such as delamination and matrix cracking to the FRP composite structures. Moreover, the internal material damage in the FRP composite structures can be invisible to the human eyes. In some cases, delaminations and cracks remain closed while the structure is under no loaded condition. As a consequence, inspection for damage and clear insight into the structural integrity become difficult using currently available evaluation methods.

The process of implementing a damage detection and characterization strategy for engineering structures is referred to as SHM. The SHM process involves the observation of a system over time using periodically sampled structural response measurements from an array of sensors. Most of offline NDT methods do not fall into SHM.

With the complex failure modes of FRP composites, the need for SHM of composite structures becomes critical. Impact from flying objects, excessive vibration, and loading can cause damage such as delamination and matrix cracking to the FRP composite structures. Moreover, the internal material damage in the FRP composite structures can be invisible to the human eyes. In some cases, delaminations and cracks remain closed while the structure is under no loaded condition. As a consequence, inspection for damage and clear insight into the structural integrity become difficult using currently available evaluation methods.

DEVELOPMENT OF FBG SENSORS FOR SHM SYSTEMS

The SHM system developed to monitor FRP composite structures must be capable of identifying the multiple failure criteria of composites (Reveley et al., 2010). Since the behavior of most composites is anisotropic, multiple numbers of sensors must be in service to monitor these structures under multidirectional complex loading conditions. The layered structure of the composites makes it difficult to predict the structural behavior by using surface mounted sensors only. To address this issue embedded sensors need to be used, and these sensors must be robust enough to service the structure's lifetime. It is impossible to replace embedded sensors after fabrication of the parts.

In particular, the fiber Bragg grating (FBG) sensor is one of the most suitable sensors for the SHM of aircraft FRP structures. The FBG sensors can be embedded into FRP composites during the manufacture of the composite part with no adverse effect on the strength of the part as the sensor is diminutive in size. Furthermore, this sensor is suitable for networking as it has a narrowband response with wide wavelength operating range, hence can be highly multiplexed. As it is a nonconductive sensor it can also operate in electromagnetically noisy environments without any interference. The FBG sensor is made up of glass which is environmentally more stable and with a long lifetime similar to that of FRP composites. Because of its low transmission loss, the sensor signal can be monitored from longer distances making it suitable for remote sensing (Kersey et al., 1997; Hill and Meltz, 1997).

The FBGs' capability of detecting stress gradients along its grating length can be used to identify the stress variations in the FRP composites by means of chirp in the reflected spectra of the FBG sensor (Hill and Eggleton, 1994; Le Blanc et al., 1994). This phenomenon can be used to detect damage in the composite structures (Okabe et al., 2000; Takeda et al., 2002). But, it has been reported that the chirp of the FBG spectrum is not only due to stress concentrations caused by damage accumulation in the composite structure (Wang et al., 2008). There are other reasons for chirping of spectra and eliminating such effects during the processing of spectra is necessary to identify damage accurately. The most recent development in the fiber optic sensor field is the pulse-pre-pump Brillouin optical time domain analysis (PPP-BOTDA) (Che-Hsien et al., 2008). The PPP-BOTDA is capable of achieving a 1 cm spatial resolution for strain measurements and the resolution is further improving. The PPP-BOTDA-based system has been successfully used in various industrial applications; however, PPP-BOTDA is so far only able to measure the static or quasistatic strain and soon will be developed for dynamic readings.

FIBER BRAGG GRATINGS

FBGs are formed by constructing periodic changes in the index of refraction in the core of a single mode optical fiber. This periodic change in index of refraction is typically created by exposing the fiber core to an intense interference pattern of UV radiation. The formation of permanent grating structures in optical fiber was first demonstrated by Hill and Meltz in 1978 at the Canadian Communications Research Centre (CRC) in Ottawa, Ontario, Canada. In ground breaking work, they launched high-intensity argon-ion laser radiation into germanium doped fiber and observed an increase in reflected light intensity. After exposing the fiber for a period of time it was found that the reflected light had a particular frequency. Subsequent spectral measurements were taken, and these measurements have confirmed that a permanent narrowband Bragg Grating filter had been created in the area of exposure.

The Bragg grating is named after William Lawrence Bragg who formulated the conditions for X-ray diffraction (Bragg's law). These concepts, which won him the Nobel Prize in 1915, related energy spectra to reflection spacing. In the case of FBGs, the Bragg condition is satisfied by the above-mentioned area of the modulated index of refraction in two possible ways based on the Grating's structure. The first is the Bragg reflection grating, which is used as a narrow optical filter or reflector. The second is the Bragg diffraction grating which is used in wavelength division multiplexing and demultiplexing of communication signals.

The gratings first written at CRC, initially referred to as "Hill gratings," were actually a result of research on the nonlinear properties of germanium-doped silica fiber. It established, at the time, a previously unknown photosensitivity of germanium-doped optical fiber, which led to further studies resulting in the formation of gratings, Bragg reflection, and an understanding of its dependence on the wavelength of the light used to form the gratings. Studies of the day suggested a two-photon process, with the grating strength increasing as a square of the light intensity (Lam and Garside, 1981). At this early stage, gratings were not written from the "side" (external to the fiber) as commonly practiced now, but were written by creating a standing wave of radiation (visible) interference within the fiber core introduced from the fiber's end.

After their appearance in late 1970s, the FBG sensors had been using for SHM of composite materials efficiently for more than two decades. Recent advances in FBG sensor technologies have provided great opportunities to develop more sophisticated *in situ* SHM systems. There have been a large number of research efforts on the health monitoring of composite structures using FBG sensors. The ability to embed them inside FRP material between different layers provides a better opportunity to receive valuable data inside the structure. The attractive properties such as small size, immunity to electromagnetic fields, and multiplexing ability are some of the advantages of FBG sensors. The lifetime of an FBG sensor is well above the lifetime of the FRP structures and also it provides the measuring of multiple parameters such as load/strain, vibration, and temperature (Kashyap, 1999).

In 1989, Meltz et al. showed that it was possible to write gratings from outside the fiber. This proved to be a significant achievement as it made possible future low-cost manufacturing methods of Bragg Gratings and enabled continuous writing or "writing-on-the-fly." With this method of writing gratings, it was discovered that a grating made to reflect any wavelength of light could be created by illuminating the fiber through the side of the cladding with two beams of coherent UV light. By using this method (holography), the interference pattern (and, therefore, the wavelength of reflected light from the grating) could be controlled by the angle between the two beams, something not possible with the internal writing method, as seen in Figure 3.1. The figure shows two methods of manufacturing a side-written grating. Figure 3.1b shows the light beam incident to the fiber with a phase mask. In Figure 3.1a, the two coherent beams of light form an interference pattern which creates a standing wave with variable intensity of light. The variable radiation intensity occurs within the fiber core. This variation in radiation intensity creates a modulated index of refraction profile within the fiber core. In Figure 3.1b, the modulation is created by using a single light beam and a phase mask. In areas where the mask allows light transmission, the index of refraction is changed within the core, creating the grating. This technique is particularly useful to write gratings quickly. Both of these methods allowed for "tuning" of the grating to whatever wavelength was desired. This, in itself was an important development, as it allowed gratings to be easily written at various wavelengths to follow the communications industry's changing source wavelengths. In addition, it was found at the time that this method was far more efficient.

The first method of fabricating gratings was internal writing through standing waves of radiation and the second method was the holographic side writing of gratings. Consequently, both of these methods have been surpassed by the use of the

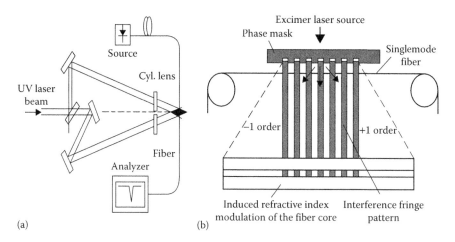

FIGURE 3.1   (a) Split beam interferometer and (b) phase mask technique.

phase mask (Anderson et al., 1993; Hill et al., 1993). The phase mask is a planar slide of silica glass or similar structure which is transparent to UV light. A periodic structure with the appropriate periodicity is etched onto the glass slide to approximate a square wave using photolithography (as viewed from the side). As shown in Figure 3.2, use of phase mask for fabrication of grating in use of the phase mask for fabrication of grating, the optical fiber is placed very close to the phase mask while the grating is written. UV light is introduced to the fiber and is diffracted by the periodic structure of the phase mask, creating the grating structure described above. The periodic structure created in the fiber is half that of the spacing of the periodic structure in the phase mask. In this manufacturing technique, the periodicity of the FBG is independent of the wavelength of the UV light source. The wavelength of the UV light source is selected based on the absorbance spectra of the doped optical fiber core, thereby maximizing the source's efficiency in writing gratings. Use of

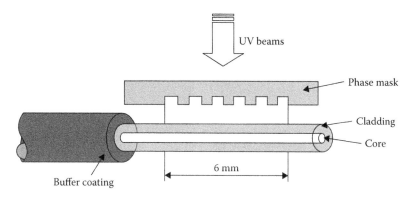

FIGURE 3.2   Use of phase mask for fabrication of grating.

phase masks made lower cost, and made greater precision Bragg gratings possible by simplifying the manufacturing process. In addition, the phase mask technique made it possible to automate grating writing, and to write multiple gratings on a fiber simultaneously. The phase mask procedure allowed for the efficient writing of other types of gratings such as chirped gratings which have nonconstant periodicities for a wider spectral response.

The process of writing a grating using the phase mask method is illustrated in Figure 3.3. Figure 3.3a and b show the setup for the grating writing process with the phase mask and mirror arrangements to direct the laser to the phase mask and the phase mask mounted just above the fiber, respectively. Figure 3.3c shows the writing in progress and Figure 3.3d shows the corresponding response from the created grating.

The reference spectrum shown in Figure 3.3d was used to maintain the consistent reflection power of the sensors. When the reflected spectrum from the grating reached the reference spectrum power, the writing process was terminated. Excessive exposure of the fiber to the laser will broaden the reflection spectra as shown in Figure 3.4, which makes it hard to track the peak of the spectrum.

The side lobes are one of the major drawbacks of using phase mask technique for FBG fabrication. As shown in Figure 3.5, side lobes of an FBG sensor response spectrum are inherent to a particular phase mask and will be written to all FBG sensors fabricated using that phase mask. "Apodization" technique can be used to get rid of those side lobes, but it will extinguish the uniformity of the grating length (λ) which drops the sensitivity of the sensor for SHM purposes.

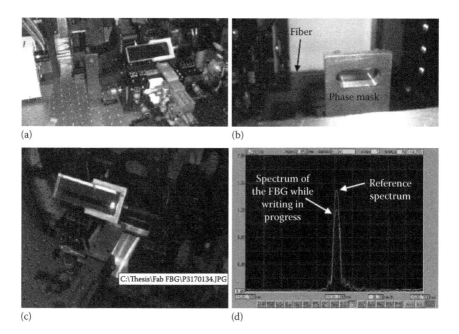

(a)

(b)

(c)

(d)

FIGURE 3.3 Fabrication of FBG sensor using phase mask method (a) setup for the writing, (b) phase mask, (c) writing in progress, and (d) response of the grating.

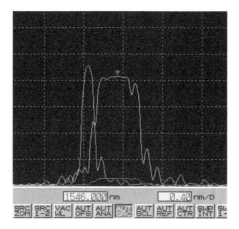

FIGURE 3.4   Broadening of the reflection spectra due to over exposure to laser light.

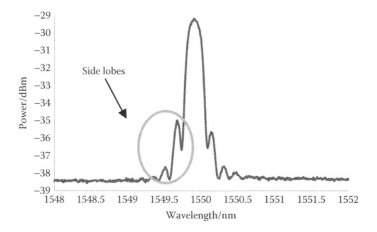

FIGURE 3.5   Side lobes of an FBG sensor response spectrum.

PRINCIPLE OF BRAGG GRATING

As described previously, FBG sensors are fabricated in the core region of specially fabricated single mode low-loss germanium doped silicate optical fibers. The grating is the laser-inscribed region which has a periodically varying refractive index. This region reflects only a narrow band of light corresponding to the Bragg wavelength λ_B, which is related to the grating period Λ_0:

$$\lambda_B = \frac{2n_0\Lambda_0}{k} \tag{3.1}$$

where:

  k is the order of the grating

  n_0 is the initial refractive index of the core material prior to any applied strain

Because of the applied strain, ε, there is a change in the wavelength, $\Delta\lambda_B$, for the isothermal condition,

$$\frac{\Delta\lambda_B}{\lambda_B} = \varepsilon P_e \tag{3.2}$$

where P_e is the strain optic coefficient and is calculated as 0.793.

The Bragg wavelength is also changing with the reflective index. Any physical change in the fiber profile will cause variation of the reflective index. The variation of the Bragg wave length λ_B, as a function of change in the refractive index $\Delta\delta n$ and the grating period $\delta\Lambda_0$ is given below.

$$\delta\lambda_{\text{Bragg}} = 2\Lambda_0\eta\Delta\delta n + 2n_{\text{eff}}\delta\Lambda_0 \tag{3.3}$$

where:

η is the core overlap factor of about 0.9 times the shift of the Bragg wavelength
n_{eff} is the mean refractive index change
Λ_0 is the grating period

For the Gaussian fit, the sensor reflectivity can be expressed as

$$s(\lambda, \lambda_s) = y_0 + S_0 \exp\left[-\alpha_s(\lambda - \lambda_s)^2\right] \tag{3.4}$$

where:

y_0 is the added offset to represent the dark noise
α_s is a parameter related to the full width at the half maximum (FWHM)
λ is the wavelength
λ_s is the central wavelength
S_0 is the initial reflectivity of the fiber

A major advantage of using FRP composites is the possibility of deciding on the number of layers and layup orientation based on the required structural behavior. In an FRP composite aerospace structure there are a number of layers with multiple orientations. The layers are placed one on top of the other and hence it is possible to embed FBG sensors in any layer during the manufacturing of the structure.

EMBEDDING FBG SENSORS IN FRP STRUCTURES

The process of embedding FBG sensors in FRP composites is quite complicated. The level of difficulty is largely dependent on the geometry of the part, lay-up configuration and embedding location of the sensors in the part. In general, FBG sensors will be placed closer to critical sections of the structure where high stress concentrations are predicted. However, in reality, locating FBG sensors in predicted locations are not always possible. On the other hand, reliance on a single sensor is not recommended as it is not possible to replace failed embedded sensors after manufacturing.

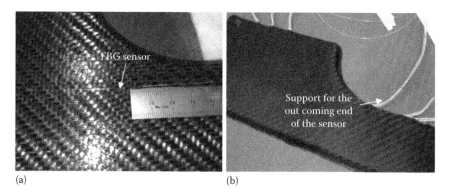

(a) (b)

FIGURE 3.6 (a) Embedding FBG sensors and the support for the out coming end before sending to the autoclave. (b) Cured sample from the autoclave.

As a result, many FBG sensors need to be embedded in the surrounding area closer to the critical locations of the structure to capture strain levels reliably.

As such, multiplexed FBG sensors play a critical role in the SHM of aerospace structures. Normally in FRP, the damage starts from stress concentrations. In the process of implementing SHM systems, identification of the locations that have potential for damage is essential. Finite element analysis techniques are being widely used to identify stress concentrations and, hence, to locate FBG sensors. It is less likely that FBG sensors are placed in simple planer structures in real applications, apart from where the requirement is mere strain rather than the damage detection.

Difficulties associated with the manufacturing of composite structures with embedded FBG sensors are the main problem with placing FBG sensors in a complicated location. The aerospace industry's advanced manufacturing technologies, (such as prepreg and autoclave process) creates hazards environments for brittle sensor. Every precaution needs to be taken to not apply loads on the sensor in the noncured resin matrix during the manufacturing process. With applied pressures as high as 700 kPa, even the egress ends of the sensors need to be supported to avoid breakage. It is essential to develop methods to protect FBG sensors during the FRP composite manufacturing processes. Since there is no way of replacing damaged FBG sensors after manufacturing of the component, a strict set of procedures must be developed to follow during manufacture.

Figure 3.6b shows a support given to the egress end of the sensor. Sometimes it is helpful to have an extra protective layer of rubber applied to the fiber to maximize the handling of samples without damage to the sensors.

DISTORTION OF EMBEDDED FBG SENSOR SPECTRA

FBG sensors are very good in strain measurements and the linear unidirectional sensitivity in the axial direction of the sensor is desirable for accurate and reliable strain readings. In such applications, the FBG sensor undergoes pure elongation or contraction and hence, the cross section always remains in a circular shape. In multidirectional loading cases, an FBG sensor may be subjected to torsional deformations

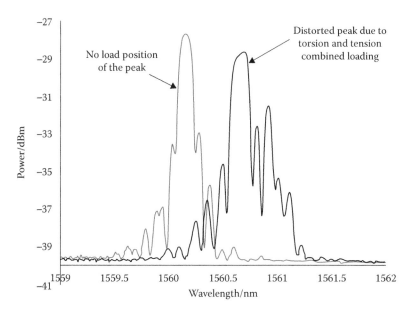

FIGURE 3.7   Distortion of the peak due to applied torsion and tension combined loading.

other than linear elongation or contraction. For example, when a torque is applied to a composite sample which has an embedded FBG sensor, it undergoes a twist which may cause changes to its cross section.

Another possibility of changed cross section of FBG sensors under torsional loading is due to microbending of the grating. The embedded sensor is not always laid on the matrix and there is a possibility of laying an FBG between reinforced fibers. In case, if a structure is under lateral pressure, the fiber sitting on the FBG sensor will press the FBG sensor against the fibers, causing the sensor to experience microbends. These changes of the cross section of the FBG lead to changes in the refractive index of the core material of the sensor. Because the changes are not uniform along the grating length, the refractive index of the sensor unevenly varies along the grating length of the sensor causing distortion in the FBG spectra.

It is obvious that the distortion of FBG sensors depends on the type of loading. The effect of the twist and microbending of FBG sensors under multiaxial loading has been identified as the causes for this discrepancy. The change of section geometry of the FBG sensor due to microbending and twisting, leads to a variation of the refractive index of the FBG core material which causes distortion of the FBG response spectra. Figure 3.7 illustrates a distorted FBG sensor response due to tension and torsion combined loading on the FRP panel which the FBG is embedded (Kahandawa et al., 2010a, b, 2011, 2013b).

The majority of research work on FBG sensors in SHM of composite structures has focused on investigation of the spectra of FBG sensor embedded in the vicinity of damage. Observations of the distorted sensor spectra due to stress concentrations caused by delaminations and cracks have been used to estimate the damage conditions. Many researchers have investigated purposely damaged axially loaded specimens, and the changes of FBG spectra were attributed to the damage and successfully identified the damage (Takeda et al., 2008). In real life situations, the applied loads are not limited to uniaxial loads and hence the performance of FBGs in multiaxial loading situation needs to be investigated for a complete understanding of damage status. The FBG spectral response is significantly complicated under multi-axial loading conditions (Sorensen et al., 2007; Kahandawa et al., 2010a, 2012a). The distortion of FBG spectra is not only due to the accumulated damage, but also the loading types. In the previous section, it was shown that embedding FBGs between nonparallel fiber layers and the application of torque caused substantial distortions to the FBG spectra.

Even through the distortion of the response spectra of an FBG sensor has been widely used in SHM applications to detect structural integrity, unfortunately there is still no definite method available to quantify the distortion. This has been a signifi-cant drawback in the development of SHM systems using embedded FBG sensors for decades. Quantification of distortion of the FBG sensor will allow referencing and comparison to monitor for the progressive damage status of a structure. An explicit method to quantify distortion to the sensor including self-distortion needs to be developed.

Kahandawa et al. (2013a) has proposed a "distortion index" (DI) to quantify dis-tortion in the FBG spectrum. The DI is calculated related to the original spectrum before the presence of any damage.

The DI is calculated related to the original spectrum before the presence of any damage (Kahandawa et al., 2013a). Distortion (D_s) is defined using the FWHM value of FBG spectra and the maximum power of the FBG response spectrum. FWHM is an expression of the extent of a function given by the difference between the two extreme values of the independent variable at which the dependent variable is equal to half of its maximum power (Figure 3.8).

Generally for an FBG response spectrum, the FWHM value increases with the distortion while the peak value decreases. FWHM $=(x_2 - x_1)$.

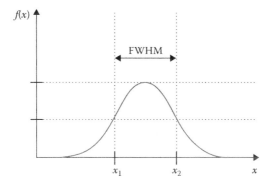

FIGURE 3.8   Full width at half maximum.

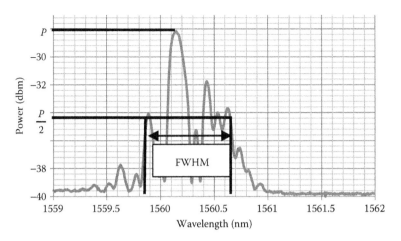

FIGURE 3.9   Peak value and the FWHM of FBG response spectra.

For the comparison, the distortion D_s and the DI, need to be calculated for the same load case. Distortion of the peak at a particular load case, D_s, can be expressed as

$$D_s = \frac{\text{FWHM}}{P} \tag{3.5}$$

where P is the peak strength of the FBG sensor reflection spectra in dB as shown in Figure 3.9. With the calculated distortion value, the DI can be calculated.

DI at the same load case:

$$\text{DI} = \frac{D_{si}}{D_{s0}} \tag{3.6}$$

where:

D_{si} is the current distortion

D_{s0} is the distortion at the original condition (no damage)

If the structure is undamaged, D_{si} is equal to D_{s0} for the same load case. In that case the DI is equal to unity. With the presence of damage, the response spectrum of an FBG sensor broadens (FWHM increases), while the peak power of the spectrum decreases. As a result D_{si} increases making the DI value above unity. This phenomenon can be used to identify the presence of damage in a structure.

To verify the DI, it is better to use several load cases to calculate distortion and corresponding DIs. At the initial stage, the structure can be subjected to several known load cases, and corresponding distortion, D_{s0} can be recorded. In future operations, those load cases can be used to calculate the DI to identify damage. In the following experiment, the DI has been calculated to investigate the relationship of DI to a growing defect.

DAMAGE PREDICTION

Previous discussions in this chapter have shown that multiple causes lead to the distortion to the FBG response spectra. Most of the intrinsic effects cause by embedded FBG sensor between parallel fiber laminates, parallel to the fiber orientation, and also multidirectional loading, could not be isolated from the spectra of FBG sensors. To identify damage from the distortions to the FBG response spectra, the individual effect from each effect needs to be identified and eliminated. To identify the pure effects from the damage, distinguished from the other effects, extensive computational power is required for postprocessing of the spectral data. Figure 3.10 shows FBG response spectra from an FBG embedded near a purposely created damage location with the part under the complex multidirectional loading.

In a laboratory condition a crack can be simulated and related FBG spectra can be obtained easily. In these purposely created experimental situations the response spectra (distortion) of embedded FBG can be explained. However, distorted response spectrum from an embedded FBG sensor is not easy to convert to exact damage condition of the structure. This incongruity made some of the SHM researchers disappointed and discouraged.

Kahandawa et al. (2012b) have proposed a novel statistical technique using artificial neural network (ANN)-based approach to handle a large data base of FBG sensor network for predict accurately the damage condition. The response of the FBG during the undamaged states of the structure is recorded, and this recorded data are used as a "reference" for the analysis. Therefore, the isolation of possible "reference" data from a distorted spectrum of any embedded FBG sensor will be possible and subsequent distortions to the spectra caused during the operational life of the structure/component.

The main difficulty for this approach is to develop a system to reference the FBG response spectra. Historically, statistical methods such as ANNs have been used to

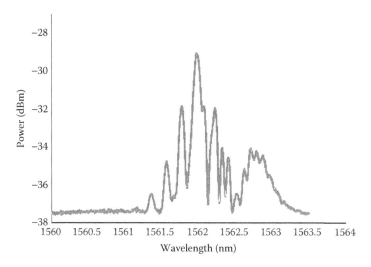

FIGURE 3.10   Distorted FBG spectra due to multiple effects.

analyze such complicated data associated with a large number of random variables. The main advantage is the ability to train an ANN with undamaged data, and the trained ANN can be used to distinguish new spectral variations of FBG sensor responses. To input spectral data to the ANN, decoding system needs to be developed. To address the above issues, the "fixed FBG filter decoding system" (FFFDS) was developed to capture the distortion to the FBG sensor response spectra.

DECODING DISTORTED FBG SENSOR SPECTRA

During the past decade, many systems for decoding FBG spectra using fixed FBG filters have been developed (Lewis et al., 2007; Nunes et al., 2007; Veiga et al., 2008; Zimmermann et al., 2008; Lopes et al., 2010). Figure 3.11 illustrates a general arrangement of a fixed FBG filter system. The system consists of a tunable laser (TLS), fixed FBG filter, optical couplers (CPs) and photo detector (PD). A high-frequency data acquisition system (DAQ) has been used to acquire the PD voltage values.

Figure 3.11 illustrates the simplest form of this system, using only one (1) FBG filter, which is the building block of the complete decoding system. TLS light, A, is transmitted to the FBG sensor and the reflected light, B, of the FBG sensor is fed to the FBG filter through an optical CP. The intersection of the wavelengths reflected from the sensor and the wavelengths' reflected light, C, by the filter (λ), or conversely, the wavelengths which are not transmitted through the filter, are reflected to the photodetector. L_1 and L_2 are the light transmitted through the FBG sensor and filter, respectively.

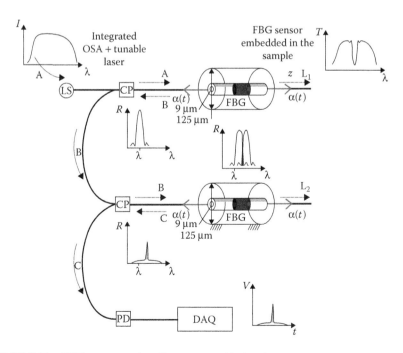

FIGURE 3.11   FBG spectrum decoding system $\alpha(t)$, timed response spectra.

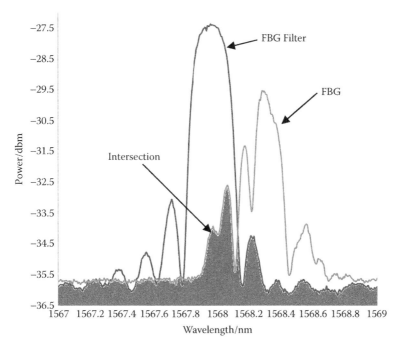

FIGURE 3.12   Intersection of the FBG spectra.

While the sensor receives the total wavelength range from the TLS source, the filter only receives the wavelengths reflected by the sensor. Hence, the filter can only reflect light (to the photodetector) if the wavelength from the sensor is within the filter's grating range (λ). The reflected light of the filter is captured using the photodetector, converted to a voltage, and recorded in the DAQ system. A system can be implemented with multiple FBG filters with λ_n wavelengths as required for a specific application.

Figure 3.12 shows the reflected spectra of the FBG sensor and the filter. The filter can only reflect light if the received wavelength from the sensor reflection is within the filter's grating range. Thus, the filter reflects the intersection as shown in Figure 3.12.

The reflected light of the FBG filter was captured using the PD and the voltage was recorded using the DAQ. Figure 3.13 shows the PD voltage in the time domain corresponding to the intersection of the spectra shown in Figure 3.12. TLS sweeping frequency allows transformation of voltage reading to time domain. Since the filter spectrum is fixed, the intersection of the two spectra only depends on the sensor spectrum position. Variation of the intersection can be used to identify the location of the peak, the strain at sensor, and the damage status of the structure. Any distortion to the spectrum is visible from the PD voltage–time plot (Figure 3.13). By matching the TLS swept frequency with the DAQ sampling frequency, it is possible

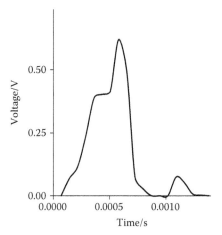

FIGURE 3.13   PD reading due to the intersection of the FBG spectra.

to transform voltages to respective wavelength values accurately. More filter readings will increase the accuracy, the operating range and robustness of the system.

There were several attempts to fit the FBG spectra using mathematical functions such as the commonly used Gaussian curve fit (Nunes et al., 2004). Sensor reflectivity can be expressed as

$$S(\lambda,\lambda_s) = y_0 + S_0 \exp\left[-\alpha_s(\lambda-\lambda_s)^2\right] \tag{3.7}$$

where:
y_0 is the added offset to represent the dark noise
α is a parameter related to FWHM
λ is the wave length

Unfortunately Gaussian fit always gives an error for a distorted spectrum as shown in Figure 3.14a. Realistically, a distorted spectrum must be considered as a piecewise continuous function, f_{pc} to capture the distortion. Consequently, optical power, P, of the distorted signal can be obtained using the following integral:

$$P = \beta \int_{t_a}^{t_b} f_{pc} dt \tag{3.8}$$

where:
β is a constant dependent on the power of the source
t_a and t_b are the integral limits in the time domain, respectively (Figure 3.14b)

The power integral at each point can be used to estimate the strain in the sensor by using an ANN. The sensitivity of the integrated data depends on the integral limits; larger integration limits reduce the sensitivity and very small limits cause data to scatter. Both cases make the algorithms inefficient. Optimum limit values have to be set to achieve better results.

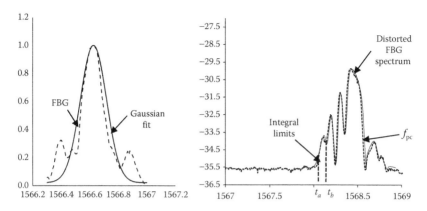

FIGURE 3.14   Gaussian fit and the piecewise continuous function.

DEVELOPMENT OF ANN-BASED DECISION-MAKING ALGORITHM

The SHM systems used in damage detection in FRP composites must be capable of identifying the complex failure modes of composite materials. The damage accumulation in each layer of a composite laminate is primarily dependent on the properties of the particular layer (McCartney, 1998; McCartney, 2002) and the loads which are imposed onto the layer. As such, the layered structure of the composite laminates makes it difficult to predict the structural behavior using only surface attached sensors. Over the past few years, this issue has been critically investigated by many researchers using embedded FBG sensors (Eric, 1995; Lee et al., 1999; Takeda et al., 2002, 2003, 2008).

The majority of the research works were focused on the investigation of the spectra of FBG sensors embedded in the vicinity of damage loaded with unidirectional loading. However, in real life situations, the applied loads are not limited to uniaxial loads and hence the performance of FBGs in multiaxial loading situation needs to be investigated for comprehensive damage characterization.

The FBG spectral response is significantly complicated by multiaxial loading conditions (Sorensen et al., 2007), fiber orientation, and the type of damage present in the structure (Kahandawa et al., 2010a, b). It has been shown that FBG's embedded between nonparallel fiber layers and subjected to torque create significant distortions in the spectra (Figure 3.15).

It is clear that the cause of the distortion of FBG spectra depends, not only on the consequences of accumulated damage, but also on loading types and the fiber orientation. Embedding FBGs in between nonparallel fiber layers and the application of torsional loading to the component have caused substantial distortions to FBG spectra. To identify damage using the response of the FBG sensor, the other effects imbedded in the response needs to be identified and eliminated. The introducing referencing technique for the FBG spectrum using fixed wavelength FBG filters, provides the capability of identifying the variations to the FBG spectrum and distinguish the other effects causing distortions. Consequently, elimination of distortions caused by other effects will permit identification of distortions of FBG

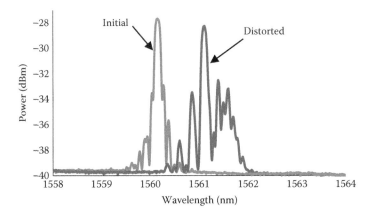

FIGURE 3.15 A typical distortion of FBG spectra.

spectra caused by the damage. The proposed system is used to capture the distortions of reflected spectra of an embedded FBG sensor inside a composite laminate, thus enabling a quantitative estimate of the damage size in the vicinity of the sensor.

The aforementioned effects on the FBG spectrum along with the accumulation of damage make the response of the FBG highly nonlinear. The nonlinearity of the response varies from structure to structure hence the estimation of transfer functions is extremely difficult. In this scenario, statistical methods provide promising results for data processing. Among the methods available, the ANN has provided proven results for nonlinear systems with high accuracy. Application of ANNs is an efficient method for modeling nonlinear characteristics of physical parameters and creates a system which is sensitive to wide range of noise.

The decoding of spectral data to feed in to ANN was addressed using a fixed filter FBG decoding system. The main objectives in this work are to decode the spectral data to determine the average strain at the embedded location. Furthermore, identification of damage also will be discussed. This method eliminates lengthy postprocessing of data, and bulky equipment for data acquisition (as shown in Figure 3.16).

ANN-Based Damage Detection

With the complex damage modes of composite materials and complex spectral responses of FBG sensors under complex operational loading, the damage detection in composite materials using FBG sensors becomes extremely difficult. Incorporation of multiple sensor readings is also a challenging task. Further, extraction of important date and the elimination of valueless data imbedded in the response spectra of an FBG is a challenging task. Even though it is possible to avoid these complications in the laboratory environment, in real applications these are not avoidable. To overcome these difficulties the introduced novel FFFDS with an ANN is being used.

The FFFDS uses the desirable characteristics of ANN to work and train with complications, to identify the real working environment. As a result of considering the working environment as the base (reference), the system's sensitivity to changes

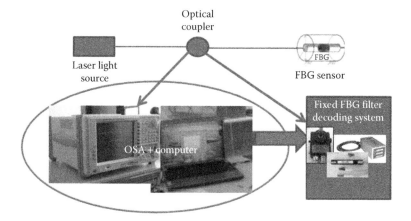

FIGURE 3.16 Replacement of OSA with fixed filter decoding system (FFFDS).

such as damage, was remarkably improved. The following section introduces the field of ANNs and the characteristics.

Introduction to Neural Networks

ANNs are commonly referred as "Neural Networks" (Haykin, 1998). This concept emerged while scientists were looking for a solution to replicate human brain. In some cases like identification and prediction, the human brain tracks the problem more efficiently than other controllers. That is because the human brain computes in an entirely different way from the conventional digital computer.

The human brain is a highly complex, nonlinear and parallel computer (information processing system) (Kartalopoulos, 1995). It has the capacity to organize its structure constituents, known as neurons, so as to perform certain computation many times faster than the fastest digital computer available today.

For an example, in human vision, the human routinely accomplishes perceptual recognition tasks such as recognizing a familiar face embedded in a familiar scene in approximately 100–200 ms, whereas tasks of much lesser complexity may take hours on a conventional computer (Freeman and Skapura, 2007). Hence, we can say that the brain processes information superquickly and superaccurately. It can also be trained to recognize patterns and to identify incomplete patterns. Moreover, the trained network works efficiently even if certain neurons (inputs) failed. The attraction of ANN, as an information processing system is due to the desirable characteristics presented in the next section.

The fixed wavelength filters and data capturing system is used to decode spectral data from an FBG sensor to a form which can be fed into an ANN to estimate strain and/or damage in a composite structure.

Figure 3.17 illustrates a data flow diagram of the structural strain and damage assessment process. First, reflected spectral data from the FBG sensor mounted on the structure is entered into an FBG filter. The reflected spectral data from the FBG filter (representing the spectral intersection of the reflected FBG sensor data and

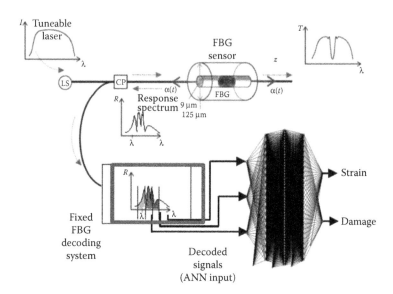

FIGURE 3.17   Decoding FBG spectrum using fixed FBGs and use of ANN.

the inherent filter characteristics) is entered into a photodetector. The voltage time domain output of the photodetector is entered into a data acquisition unit. The data are then processed and entered into a neural network. Finally, the neural network output is the state of strain and/or damage (Figure 3.18).

Figure 3.19 depicts a general arrangement of the proposed system. The system consists of a TLS, three (3) fixed FBG filters (Filters 1–3), optical CP and three (3) photodetectors (PD). A high-frequency DAQ was used to record the photodetector output voltages which are subsequently fed into the processor to determine the state of strain and damage.

Three (3) filters are used to increase the accuracy, operating range and robustness of the system, and the composite signal created by the photodetectors represents a unique signature of the sensor spectrum. Furthermore, the output of the photodetectors contains information relative to the sensor spectra in a form which can be used with an ANN. The number of filters and FBG sensors is not limited to the example shown in this figure. Any one filter in the system is capable of covering an approximate range of 500 microstrain/1 nm movement of the peak of the sensor spectrum. More filters can be used with the same system to cover a wider operating range of the embedded FBG sensors and to obtain more precise data.

POSTPROCESSING OF **FFFDS** DATA USING **ANN**

As discussed in Section "Distortion of Embedded FBG Sensor Spectra," under complex load conditions, the spectrum of the FBG distorts. Figure 3.20 shows the FBG sensor response of an FBG embedded in a composite structure during operation under several load cases. The complicated response makes it extremely difficult to

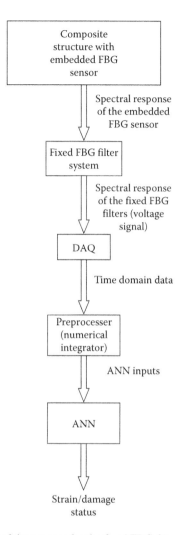

FIGURE 3.18   Flowchart of the process for the fixed FBG filter system.

model using mathematical transfer functions to estimate strain as well as damage. In such cases, an ANN provides promising functional approximations to the system.

One of the commonly used neural network architectures for function approximation is the multilayer perceptron (MLP). Backpropagation (BP) algorithms are used in a wide range of applications to design an MLP successfully (Lopes and Ribeiro, 2001).

The ANN Model

Figure 3.21 shows a general arrangement of the ANN used in this study. The ANN consists of three input neurons, which accommodate three FBG fixed filters, three hidden layers, and an output layer.

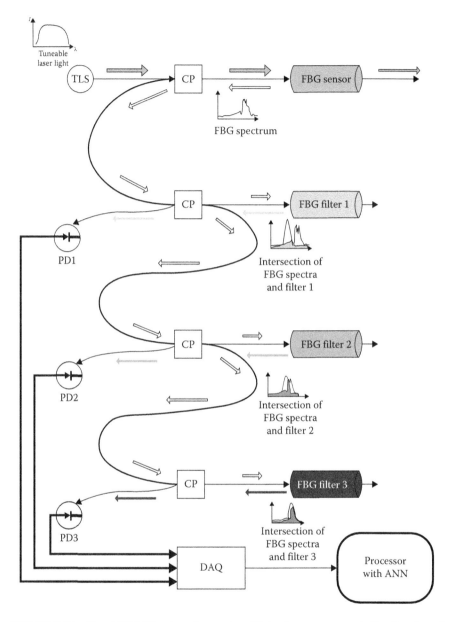

FIGURE 3.19 Fixed FBG filter decoding system (Note: Arrows represents the direction of the light signal).

The neurons of the hidden layers are with Gaussian activation functions, and the k value used is 1, as shown in Figure 3.22a. It takes a parameter that determines the center (mean) value of the function used as a desired value. The neuron of the output layer is with sigmoid activation function with $k = 1$, as shown in Figure 3.22b. This function is especially advantageous for use in neural networks trained by BP

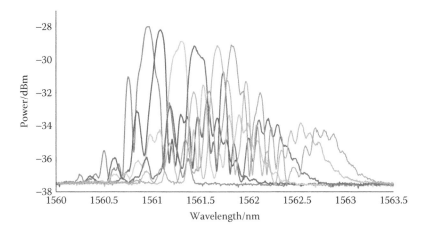

FIGURE 3.20   Distorted FBG response.

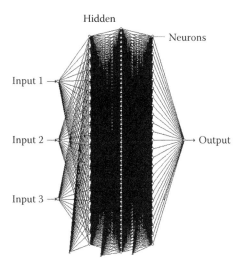

FIGURE 3.21   ANN developed to estimate strain.

algorithms because it is easy to distinguish, and can minimize the computational capacity of training (Karlik and Olgac, 2010). Initial weights (at the start of the training process) of the neurons were randomly places between −1 and 1.

Three different composite specimens, with embedded FBG sensors, were investigated using the developed system to evaluate the system performance for estimation of strain and/or damage.

CONCLUSION

The superior performances and the unique advantages of the FBG sensors have strongly established their place for the SHM of FRP composite structures. At this stage, the success of the SHM with FBG sensors is expanded from to the laboratory

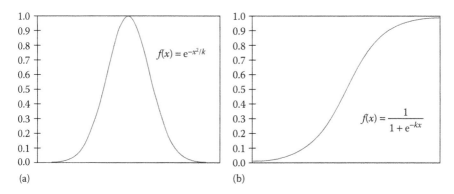

FIGURE 3.22 Activation functions of the neurons of ANN: (a) hidden layer neurons and (b) output neuron.

environment to the real structures. However, to make this technology reliable and readily accessible, further research is warranted. The embedding technology, robustness of the sensors and FBG interrogation techniques must be critically addressed. The postprocessing of FBG spectral data needs to be developed with the recent advancements of statistical data analysis algorithms.

REFERENCES

Adams, R. D., Cawley, P., Pye, C. J. and Stone, B. J. 1978. A vibration technique for non-destructively assessing the integrity of structures. *Journal of Mechanical Engineering Science*, 20, 93–100.

Anderson, D. Z., Mizrahi, V., Erdogan, T. and White, A. E. 1993. Production of in-fiber gratings using a diffractive optical element. *Electronic Letters*, 29, 566–568.

Chang, F. and Chang, K. 1987. A progressive damage model for laminated composites containing stress concentrations. *Journal of Composite Materials*, 21, 834–855.

Chang, F. K. 2003. *Structural Health Monitoring*. DESTechnol Publications, Lancaster, PA.

Che-Hsien, L. I., Nishiguti, K. and Miyatake, M. 2008. PPP-BOTDA method to achieve 2cm spatial resolution in Brillouin distributed measuring technique. *OFT*, Tokyo, Japan.

Eric, U. 1995. *Fibre Optic Smart Structures*. Wiley, New York.

Freeman, J. A. and Skapura, D. M. 2007. *Neural Networks, Algorithms, Applications, and Programming Techniques*. Pearson Education, Boston, MA.

Haykin, S. 1998. *Neural Networks: A Comprehensive Foundation*. Prentice Hall, NJ.

Hill, K. O., Malo, B., Bilodeau, F., Johnson, D. C. and Albert, J. 1993. Bragg gratings fabricated in monomode photosensitive optical fiber by UV exposure through a phase mask. *Applied Physics Letters*, 62, 1035–1037.

Hill, K. O. and Meltz, G. 1997. Fiber Bragg grating technology fundamentals and overview. *Journal of Lightwave Technology*, 15, 1263–1276.

Hill, P. C. and Eggleton, B. J. 1994. Strain gradient chirp of fibre Bragg gratings. *Electronics Letters*, 30, 1172–1174.

Kahandawa, G., Epaarachchi, J. and Wang, H. 2011. Identification of distortions to FBG spectrum using FBG fixed filters. *Proceedings of the 18th International Conference on Composite Materials*, Republic of Korea, pp. 1–6.

Kahandawa, G., Epaarachchi, J., Wang, H. and Lau, K. T. 2012a. Use of FBG sensors for SHM in aerospace structures. *Photonic Sensors*, 2, 203–214.

Kahandawa, G. C., Epaarachchi, J. A. and Lau, K. T. 2013a. Indexing damage using distortion of embedded FBG sensor response spectra. *Proceedings of 4th International Conference on Smart Materials and Nanotechnology in Engineering*, SPIE, Gold Coast, Australia, 87930K–87930K-6.

Kahandawa, G. C., Epaarachchi, J. A., Wang, H. and Canning, J. 2010a. Effects of the self distortions of embedded FBG sensors on spectral response due to torsional and combined loads. APWSHM3, Tokyo, Japan.

Kahandawa, G. C., Epaarachchi, J. A., Wang, H., Followell, D. and Birt, P. 2013b. Use of fixed wavelength fibre-Bragg grating (FBG) filters to capture time domain data from the distorted spectrum of an embedded FBG sensor to estimate strain with an artificial neural network. *Sensors and Actuators A: Physical*, 194, 1–7.

Kahandawa, G. C., Epaarachchi, J. A., Wang, H., Followell, D., Birt, P., Canning, J. and Stevenson, M. 2010b. An investigation of spectral response of embedded fibre Bragg grating (FBG) sensors in a hollow composite cylindrical beam under pure torsion and combined loading. ACAM 6, Perth, Australia.

Kahandawa, G. C. A., Wang, H. and Epaarachchi, J. 2012b. Signal monitoring system and methods of operating same. US Patent App. 13/441,064.

Kamiya, S. and Sekine, H. 1996. Prediction of the fracture strength of notched continuous fiber-reinforced laminates by interlaminar crack extension analysis. *Composites Science and Technology*, 56, 11–21.

Karlik, B. and Olgac, A. V. 2010. Performance analysis of various activation functions in generalized MLP architectures of neural networks. *International Journal of Artificial Intelligence and Expert Systems (IJAE)*, 1, 111–122.

Kartalopoulos, S. V. 1995. *Understanding Neural Networks and Fuzzy Logic: Basic Concepts and Applications*. IEEE Press, US.

Kashyap, R. 1999. *Fiber Bragg Gratings*. Academic Press, San Diego, CA.

Kersey, A. D., Davis, M. A., Patrick, H. J., Leblanc, M., Koo, K. P., Askins, C. G., Putnam, M. A. and Friebele, E. J. 1997. Fiber grating sensors. *Journal of Lightwave Technology*, 15, 1442–1463.

Kortscho, M. T. and Beaumont, P. W. R. 1990. Damage mechanics of composite materials: I—Measurements of damage and strength. *Composites Science and Technology*, 39, 289–301.

Lam, D. K. W. and Garside, B. K. 1981. Characterization of single-mode optical fiber filters. *Applied Optics*, 20, 440–445.

Le Blanc, M., Huang, S. Y., Ohn, M. M. and Measures, R. M. 1994. Tunable chirping of a fibre Bragg grating using a tapered cantilever beam. *Electronics Letters*, 30, 2163–2165.

Lee, C. S., Hwang, W., Park, H. C. and Han, K. S. 1999. Failure of carbon/epoxy composite tubes under combined axial and torsional loading 1. Experimental results and prediction of biaxial strength by the use of neural networks. *Composites Science and Technology*, 59, 1779–1788.

Lewis, E., Sheridan, C., O'Farrell, M., King, D., Flanagan, C., Lyons, W. B. and Fitzpatrick, C. 2007. Principal component analysis and artificial neural network based approach to analysing optical fibre sensors signals. *Sensors and Actuators A: Physical*, 136, 28–38.

Lopes, N. and Ribeiro, B. 2001. Hybrid learning in a multi-neural network architecture. *Proceedings of International Joint Conference on Neural Networks*, Washington DC, Vol. 4 pp. 2788–2793.

Lopes, P. A. M., Gomes, H. M. and Awruch, A. M. 2010. Reliability analysis of laminated composite structures using finite elements and neural networks. *Composite Structures*, 92, 1603–1613.

Marshall, D. B., Cox, B. N. and Evans, A. G. 1985. The mechanics of matrix cracking in brittle-matrix fiber composites. *Acta Metallurgica*, 33, 2013–2021.

McCartney, L. N. 1998. Prediction transverse crack formation in cross-ply laminates. *Composites Science and Technology*, 58, 1069–1081.

McCartney, L. N. 2002. Prediction of ply crack formation and failure in laminates. *Composites Science and Technology*, 62, 1619–1631.

Meltz, G., Morey, W. W. and Glenn, W. H. 1989. Formation of Bragg gratings in optical fibers by a transverse holographic method. *Optics Letters*, 14, 823–825.

Nunes, L. C. S., Olivieri, B. S., Kato, C. C., Valente, L. C. G. and Braga, A. M. B. 2007. FBG sensor multiplexing system based on the TDM and fixed filters approach. *Sensors and Actuators A: Physical*, 138, 341–349.

Nunes, L. C. S., Valente, L. C. G. and Braga, A. M. B. 2004. Analysis of a demodulation system for Fiber Bragg Grating sensors using two fixed filters. *Optics and Lasers in Engineering*, 42, 529–542.

Okabe, Y., Yashiro, S., Kosaka, T. and Takeda, N. 2000. Detection of transverse cracks in CFRP composites using embedded fiber Bragg grating sensors. *Smart Materials and Structures*, 9, 832.

Reveley, M. S., Kurtoglu, T., Leone, K. M., Briggs, J. L. and Withrow, C. A. 2010. Assessment of the state of the art of integrated vehicle health management technologies as applicable to damage conditions. *NASA/TM-2010-216911*.

Sorensen, L., Botsis, J., Gmur, T. and Cugnoni, J. 2007. Delamination detection and characterisation of bridging tractions using long FBG optical sensors. *Composites Part A: Applied Science and Manufacturing*, 38, 2087–2096.

Takeda, S., Okabe, Y. and Takeda, N. 2002. Detection of delamination in composite laminates using small-diameter FBG sensors. SPIE, San Diego, CA, pp. 138–148.

Takeda, S., Okabe, Y. and Takeda, N. 2003. Application of Chirped FBG Sensors for Detection of Local Delamination in Composite Laminates. SPIE, San Diego, CA, pp. 171–178.

Takeda, S., Okabe, Y. and Takeda, N. 2008. Monitoring of delamination growth in CFRP laminates using chirped FBG sensors. *Journal of Intelligent Material Systems and Structures*, 19, 437–444.

Veiga, C. L. N., Encinas, L. S. and Zimmermann, A. C. 2008. Neural networks improving robustness on fiber Bragg gratings interrogation systems under optical power variations. In: DAVID, D. S. (Ed.). *Proceedings of the 19th International Conference on Optical Fibre Sensors*. SPIE, Australia, pp. 700462.

Wang, Y., Bartelt, H., Ecke, W., Willsch, R., Kobelke, J., Kautz, M., Brueckner, S. and Rothhardt, M. November 7–9, 2008. Fiber Bragg gratings in small-core Ge-doped photonic crystal fibers. *1st Asia-Pacific Optical Fiber Sensors Conference*. China, pp. 1–4.

Zhou, G. and Sim, L. M. 2002. Damage detection and assessment in fibre-reinforced composite structures with embedded fibre optic sensors-review. *Smart Materials and Structures*, 11, 925–939.

Zimmermann, A. C., Veiga, C. and Encinas, L. S. 2008. Unambiguous signal processing and measuring range extension for Fiber Bragg Gratings sensors using artificial neural networks—A temperature case. *Sensors Journal*, IEEE, 8, 1229–1235.

4 Internet of Things for Structural Health Monitoring

Aravinda S. Rao, Jayavardhana Gubbi, Tuan Ngo,
Priyan Mendis, and Marimuthu Palaniswami

CONTENTS

SMART CITIES

Australia is an urbanized society. Happily, Australian cities are regularly rated as world's most liveable. Like most developed countries, major challenges to urban landscape are emerging: higher (and aging) populations, environmental and resource stresses, infrastructure capacities, security issues, and so on [36]. These challenges need to be addressed now to prepare our cities for tomorrow: future cities have to be "smart"—digitally smart—if they are to be sustainable. Present urban universe is neatly partitioned into autonomous spheres: (a) real—where people, systems, services, and environments lie and (b) virtual—where information and communication lie. In smart cities, this divide does not exist; the real and the virtual are seamlessly integrated. Technologically, this integration is predicated on the creation of some very new disruptive paradigms on the data–knowledge–action axis of real–virtual symbiosis. The Internet revolution led to the interconnection between people at an unprecedented scale and pace. The next revolution will be the interconnection between objects to create a smart environment. Currently, there are 9 billion interconnected devices, and it is expected to reach 24 billion devices by [59]. According to the Groupe Speciale Mobile Association, this amounts to $1.3 trillion revenue opportunities for mobile network operators alone spanning vertical segments such as transport, infrastructure, health, and entertainment [20]. The Digital Europe strategy of the European Union [12] points to some of these new paradigms: the Internet of things (IoT) [50], Big Data [4], and Cloud computing [6]. The first integrates the information highways and the physical world through ubiquitous sensing and actuation, the second is to extract actionable knowledge from constantly generated massive data from diverse sources (human and machine), and the third is to manage the required heterogeneity of computational platforms and goals in pervasive environments.

The emerging technology is applicable in almost all areas of city management. Some of the key application areas include

Environmental monitoring—The large-scale high-resolution environmental data will enable continuous monitoring of the city environment.

Video monitoring—Video sensor networks that integrate image processing, computer vision, and networking frameworks will help develop a new challenging scientific research area at the intersection of video, infrared, microphone, and network technologies. Surveillance, the most widely used camera network application, helps track a person, identify suspicious activities, detect left luggage, and monitor unauthorized access.

Structural health monitoring—As cities age, it is important that the critical infrastructure are regularly monitored and appropriate measures taken to extend the longevity. This includes measuring strains in structural components using advanced sensor technologies.

Traffic congestion and impact monitoring—Urban traffic is the main contributor to traffic noise pollution and a major contributor to urban air quality degradation and greenhouse gas emissions. Improved dynamic traffic information will improve freight movement, allow better planning and improve scheduling. The applications areas within traffic include dynamic traffic management and control strategies involving interactions among vehicles; vehicles and infrastructure; and different modes of transport.

In this chapter, we narrow our focus on structural health monitoring (SHM), particularly on critical infrastructure monitoring. Critical infrastructures are those that are essential for everyday living. Examples include physical structures such as long bridges connecting major locations for people to commute, tall buildings where region's workforce is mostly located; communication, electricity and water supply networks; information technologies; food and supply chains, and so on. Here, the focus is on civil infrastructure that mainly includes bridges and buildings. Integrity of the civil infrastructures is very important to maintain safety of everyone in and around such structures. To ensure that the structural integrity is maintained, health monitoring of these structures are carried regularly based on approximate estimates.

STRUCTURAL HEALTH MONITORING

According to Sohn et al. [46], "The process of implementing a damage identification strategy for aerospace, civil and mechanical engineering infrastructure is referred to as *Structural Health Monitoring (SHM)*." SHM implies monitoring of the state of the structures through sensor networks in an online mode and are pertinent to aircraft and buildings [55]. SHM can be further divided into two categories [7]: (1) global health monitoring and (2) local health monitoring. Global health monitoring is the process where the damage to the structural integrity is determined by identifying damage to the structure; whereas, local health monitoring deals with identifying the specific locations in the structures where the damage has occurred.

Damage to a structure can be defined as altering the true state of the structure that may hamper the performance of the structure. The structural damage can be due to the changes in material composition, physical constraints such as load, and also due to environmental factors and aging. Detection and evaluation of damages to structures are part of most of the engineering systems. The five key areas where damage detection is primarily focused are: (1) SHM [55], (2) condition monitoring (CM [5]), (3) nondestructive evaluation (NDE [44]), (4) statistical process control (SPC [35]), and (5) damage prognosis (DP [15]).

SHM is performed by using acoustic sensors, ultrasonic sensors, strain gauges, optical fibers, and so on. CM is assessment of the damage to the rotating machinery, where vibration, temperature, and accelerations provide insight about the damages to the structure. NDE is used to assess the local characteristics of the structure once the damage has been localized. Tests such as ultrasonic, shear stress, and magnetic particle inspection are some of the NDE methods. SPC evaluates the quality of the process where statistical methods are used. The monitoring process provides stability and minimum quality standards to the structures depending on the materials and the process used. Once the damage to the structure has been detected, DP is typically used to assess and predict the end time of the structure. SHM and CM are typically performed online, whereas NDE, SPC, and DP are usually offline. Figure 4.1 shows the schematics of SHM using traditional sensing techniques.

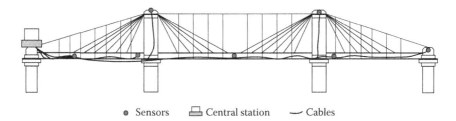

● Sensors   ⊟ Central station   ⌒ Cables

FIGURE 4.1 SHM using traditional sensing modalities. The central station collects the data from all the sensors using long cables.

SENSORS AND SENSOR NETWORKS

The development of advanced sensors, miniaturizations in integrated circuits, advancements in design engineering, and reduction of power consumption and cost acted as catalysts to the development of sensors during the late 1990s. Further impetus was provided the much-needed communication network, the Internet. The integration of these two technologies spawned a new widespread technology commonly known as sensor networks. The ability of the sensor networks to send data over the Internet further enhanced the scope and usage of the sensor networks. The capability of wireless communication by the sensor nodes along with computation power and sensing proved to be a benign technology. Sensor networks with wireless communications are referred to as wireless sensor networks (WSNs). "Smart Dust" [25,39], the pioneering node with 5 square millimeters [58] in size included micro electromechanical systems sensor, wireless communication, and processing.

TRADITIONAL SENSORS AND USAGE

Sensor networks can use wide range of sensors. Typically, the sensors are chosen based on the application and end-user requirements. Power consumption and cost are also other important factors that affect the selecting of the sensors. Some of the most commonly used sensors are [2,11] temperature, pressure, humidity, soil condition, movement of vehicles, noise levels, light levels, detection of specific objects, and mechanical stress; characteristics of an object such as speed, direction, and size can also be monitored; event detection, continuous monitoring, and remotely controlling actuators make the sensor networks attractive solutions toward added smartness to the human interactions. The application areas of sensors and sensor networks are numerous. These can be broadly categorized as military applications, environmental applications, healthcare, home automation, and other industrial applications [2]. The focus here will be on the use of WSN on civil engineering and infrastructure applications, in particular, on SHM.

LIMITATIONS OF TRADITIONAL MONITORING

SHM allows us to assess and estimate the damage to the structures. This may be due to structural aging, environmental effects, or the damage due to the structural material. The key factors required in assessing the damage are (1) temporal variations in structural integrity and (2) degree of damage to structural integrity [27]. Visual inspection, X-ray, and accelerators are some of the commonly used traditional methods. Specifically, accelerators and strain gauges are connected to a computer with appropriate software to collect the sensor data. Many times, the structures are too long and unreachable. Additionally, tall buildings put lives of many people at risk in collecting and monitoring the sensor data. There is also unease of carrying the computer and connecting the long wires. It is also important to note that the length of senor cables connecting to the computer affects the sensor readings, and sometimes become corrupted. Some of the major drawbacks are (1) high cost of measurement due to long cables and sophisticated protection equipments, (2) high cost of maintenance as both experts and nonexperts from sensors to structures are required, (3) only localized sensor data are obtained (i.e., the long-term behavior of the structural changes are not known), and (4) as a consequence, the estimation of lifetime of the structure becomes rather difficult with sparse data. There is a need for new sensors, sensing methodologies, and also methods in constructing the structures.

CARBON NANOTUBES FOR STRUCTURES

Carbon nanotubes have been in popularity since 1991, after the synthesis of needle-like structures that produce fullerenes [23], although initial developments were made in 1976 by growing carbon fibers [38]. The applications of carbon nanotubes are widespread due to their excellent properties. The applications of nanotubes include [24] electronics, energy applications, mechanical applications, sensors, fields emission and lighting applications, and biological applications. Nanotubes have Young's modulus higher than that of diamond [42,52]. Theoretical [57] and experimental [13] evidence show that nanotubes can undergo huge deformations and yet recover elastically. With a 63 GPa tensile strength, they are *two* magnitude order greater than steel [61]. Undoubtedly, nanotubes can be used as materials for formation of civil structures. For instance, Li et al. [31] found that 20 mm long ropes of single-walled carbon nanotube (SWNT) ropes had the tensile strength of 3.6 ± 0.4 GPa, which is comparable to carbon fibers. They concluded that for high-performance composite materials, SWNT make an ideal reinforcement candidate. With their low-density (1.3 g/cm^3), they have the highest specific strength $>46,000$ kN-m/kg [8]. Reference [8] provides the study of tensile strength of kilometer long bridges made of carbon nanotubes. Carbon nanotubes can be used to new sensors as well as building strong structures with their excellent properties. Thus, they form the core of the future structures.

CONTINUOUS MONITORING USING WSNs

However, as mentioned before, there are several drawbacks in conventional way of performing SHM. WSNs offer a wealth of opportunities to conduct the monitoring of structural health. They have the ability to continuously monitor the over long periods of time without human intervention. WSNs provide an important platform for long-term and continuous monitoring. The temporal variations of structural integrity and the damage severity can be measured in high resolutions (i.e., data can be collected in seconds). WSNs also eliminate the need for the personnel to be present during sensing, effectively reducing the life risks of the workers. WSNs are low-cost and low-power devices that would reduce the installation and maintenance costs of the SHM. They can form long networks by wirelessly communicating with neighboring/designated nodes. This reduces the need for long cables running over the bridges and tall buildings. By continuous monitoring of the structures, the integrity of the structures can be accurately estimated. Figure 4.2 shows the schematics of SHM using WSN.

However, WSNs face several challenges in continuous monitoring. Energy available to the sensor nodes is limited. They must operate at a very low power. Computation must also be reduced to reduce power. However, some of the computations such as processing sensor data and wireless transmissions are a must. The distance of communication also poses a severe challenge. If the distance is increased, then the power of transmitter must be increased or the packets will fail. Any distance beyond communication range will never be reached. Placing sensor nodes close to each other would overcome the communication problem to some extent, but the cost increases because the number of nodes in a given area increases with reduced distance. Furthermore, sometimes the packets get corrupted and the data will become void. Also, there are times when packet loss can be noticed due to environmental factors. Because the sensor nodes are made of low-cost hardware, the clocks on the devices have low-end oscillators that cause the clocks to drift. This introduces the problem of time synchronization. This is regarded as one of the important problems in WSNs. Because of time error, the analysis of the sensor data may deceive the assessment to falsely predict the integrity of the structures. Some of the core challenges in WSNs are power management, low-power design, programming limitations, computational power imitations, real-time communication, low bandwidth, network failure, security and privacy of WSN data, data fusion, interpolation of data, Big Data analysis, visualization of data, policymaking, and citizen engagement.

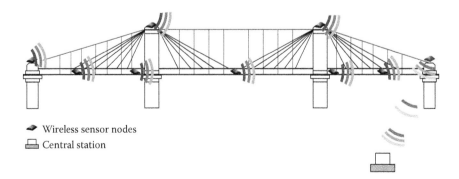

◄ Wireless sensor nodes
🖳 Central station

FIGURE 4.2   SHM using WSN. The central station collects the data from all the nodes through gateway router node.

Factors that influence the design of WSN are [2]

- *Reliability*—Sensor nodes fail due to physical damage, environmental effects, or power outage.
- *Scalability*—Algorithms for network creation and communication must be scalable. Algorithms must be designed so that ad hoc sensor nodes can still work with existing networks and network does not break down because of large number of sensor nodes.
- *Cost*—Hardware design and production of sensor nodes should be such that cost of single nodes are reduced. Sophisticated sensor nodes might cost high compared with tractional sensor nodes deployment. Maintenance of senor nodes and network for continuous monitoring is another major challenge.

TIME SYNCHRONIZATION IN WSN

Time synchronization is an inherent problem in any communication network; for instance, two computers connected to a common local area network will have a difference in their clock timings if both are left untethered to the Internet for a significant amount of time (a few days). This may be attributed to physical, mechanical, and electrical properties of the quartz crystal oscillators residing in each of the computers. However, computers connected to the Internet maintain their timings equal, as they are synchronized to higher level stratum time server using network time protocol [34]. To maintain accurate timings with moveable devices, global positioning system (GPS) can be used. GPS consists of a constellation of 27 satellites positioned in geosynchronous orbits such that at least three satellites are visible for a device on any geographical position on the Earth. These satellites use atomic clocks to maintain accuracy of 10 ns [3]. In case of low-cost and low-power networks such as WSN, sensor nodes use lower end quartz crystal oscillators as their clock source. As a consequence, the clocks experience time drifts and also skewness.

To address time synchronization problem in WSNs, numerous algorithms and protocols have been proposed by the research community. An early comprehensive survey can be found in Reference [49]. Reference broadcast synchronization (RBS) [9], timing-sync protocol for sensor networks (TPSNs) [16], flooding time synchronization protocol (FTSP) [33], lightweight time synchronization for sensor networks (LTS) [19], tiny-sync [45], and mini-sync [45] are few of the most commonly known protocols. In this chapter, we are presenting a new algorithm to compensate the time error between nodes in an energy-efficient way. In this manner, we would be able to use sensor networks for different applications and also with the inclusion of error compensation (time synchronization), it will provide a twofold benefits to the system as a whole. The tests were conducted using Castalia-3.2 simulator. A hierarchical network structure was created, and the algorithm was compared against TPSN and lightweight tree-based synchronization (LTS) methods (centralized LTS). The simulation was conducted for a period of s with two resynchronization intervals: 300 s and 1,000 s.

LITERATURE

Physical time plays a crucial role for WSN application [10,28] and is important that the deployed nodes have identical time stamps. Because of the absence of real time clock on the WSN motes, a software scheme is required to synchronize their on-board clock. The importance of physical time with respect to WSNs is detailed in Reference [41], where three scenarios are discussed: (1) interaction between the sensor network and the observer, (2) interaction between nodes, and (3) interface between nodes and the real world. Another problem with time synchronization is that once a time is set for a node, the node-time drifts over a period of time due to frequency drift. Hence, there is a need for clocks to be reset at prespecified intervals or dynamic intervals such that the resynchronization interval does not contribute to the inaccurate timings of the collected data. So far, many solutions are available in literature to tackle the problem: (a) devising software clocks, (b) adopting unidirectional synchronization method, (c) estimation of round trip synchronization time, and (d) synchronization using reference signals. All of these have their own merits and demerits for a given scenario, network, accuracy, and constraints. There exist other schemes to reduce error by collecting samples at higher data rate [41].

Clock drift is shown to be a major source of time synchronization error [29]. As a result, frequent synchronization of the clock is necessary. However, frequent synchronization of the clocks comes at the cost of energy. Krishnamurthy et al. [29] have also reported the synchronization accuracy requirement for modal analysis of about 0.6–9 μs. Wang et al. [54] have reported a drift of up to 5 ms in 6 s period, which definitely leads to a significant timing error if a network decides to synchronize less frequently to save energy. As pointed out by Krishnamurthy et al. [29], there is a need to synchronize the clock more frequently than originally intended, and this will have a bearing on the lifetime of the network. In this context, a few time synchronization methods proposed in literature have been identified to be very useful and are outlined below.

RBS [9] is a simple and effective way to eliminate the nondeterministic delay between the sender and the receiver. In this technique, a reference node sends a reference message and the receivers exchange their messages to calculate the offset. Elson et al. conducted the experiment under laboratory conditions and achieved the clock precision of 1.85 ± 1.28 μs between sender and the receivers. Receivers reported their received time by triggering a general purpose input–output pin, which is in turn connected to an external logic analyzer to record the time from each of the associated receivers. Elson et al. have also reported a precision of 3.68 ± 2.57 μs for 4 hops. RBS is mainly devised on measuring the receiver-to-receiver synchronization error. Although authors claim that as simply as a periodic reference provider, a closer look into RBS reveals that it is equivalent to external reference time provider that provides signal periodically such as GPS. Furthermore, a reference node alone is incapable of providing timing signals to all the nodes because of limited wireless range. To little extent, communication can be achieved by increasing radio range up to maximum radio limits while sacrificing its energy. Additionally, if the number of reference nodes is increased to support the entire network coverage, there exists a problem of providing timing information to all the reference nodes themselves.

Moreover, as the number of nodes increases, the computation and time required to calculate relative error increases: A node has to estimate the relative error among all the other nodes. Because of the increased number of nodes, it takes a fair amount of time to exchange timing information between nodes. Indeed, each node has to wait until the currently active node completes its message exchange. Therefore, each node has to keep a large random back-off time to allow sufficient time with currently active nodes. An inherent problem with this methodology is the resynchronization interval has to be determined externally and therefore lacks system dynamism.

TPSN [16] is a hierarchical structure-based algorithm that aims to provide network-wide synchronization. It is built on the notion of "always-on" model. TPSN works in a two-step process termed as "level discovery phase" and "synchronization phase," to synchronize all network nodes to reference node. First, it creates a hierarchical structure of sensor nodes and later, synchronizing between each pair along the edges. Hierarchical structure is created by assigning node levels to each of the nodes. "Root node" is always at level 0, and node at level i is ensured to have communication ability with at least a node belonging to level $i-1$. Next, root node initiates the synchronization phase in which level at level i synchronizes to a node at level $i-1$. Ganeriwal et al. have reported an average accuracy of 20 μs between neighboring nodes tested on Berkerly motes. Figure 4.3 depicts the typical two-way message exchange between two nodes. TPSN estimates the drift between clocks and the propagation delay as given by Equations 4.1 and 4.2, respectively.

$$\Delta = \frac{(T_2 - T_1) - (T_4 - T_3)}{2} \tag{4.1}$$

$$d = \frac{(T_2 - T_1) + (T_4 - T_3)}{2} \tag{4.2}$$

The experiment setup included programming of two motes to toggle a particular pin (connected to the Digital Analyzer) every 8 ms after two-way message exchange. Both the motes ran unsynchronized clocks. The error between the motes was measured by observing the phase shift between two waveforms produced by the motes. For 200 independent trials on both TPSN and RBS techniques, authors [16] report TPSN as roughly twofold better performing candidate than RBS. However, in a

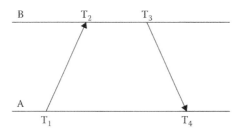

FIGURE 4.3 Two-way message communication between two nodes. (From Ganeriwal, S., Kumar, R., and Srivastava, M. B. *Proc. 1st Int. Conf. Embedded Networked Sensor Systems*, pp. 138–149. ACM, 2003.)

typical sensor network scenario, "always-on" model is collectively unattainable because of limited power reserve: sensor nodes run out of battery after specified amount of energy is expended. Additionally, due to unforeseen circumstances, a glitch that causes a sensor node to reboot might hinder the process of executing the algorithm successfully. In case of multihop scenario, sender synchronizes to receiver at every hop before data are transported. A careful analysis reveals that if the receiver at level $i-1$ is not synchronized to receiver at level $i-2$, then a datum sent from level i will lose its true time because of time differences between consecutive receivers at levels $i-1$ and $i-2$, and further up to root node. In other words, first, the immediate root node neighbors must send the data and next, the nodes at subsequent levels must send the data in an increasing hierarchical order. Therefore, a node at level i must be aware of synchronizing times of nodes hierarchically above it. This may be lead to data latency by a leaf node as compared with data from higher ups.

LTS [19] is similar to TPSN where in, with the aid of spanning tree algorithms, a spanning tree is created and later pair-wise synchronization is performed along the edges of the spanning tree for multihop networks. The scheme requires three messages to perform a pair-wise synchronization, and error properties of the synchronization are modeled as Gaussian. The method requires $n-1$ pair-wise synchronization be performed for a set of nodes. It has both centralized and distributed synchronizing schemes. Greunen et al. have shown that, for a multihop network, the communication complexity and accuracy depends on construction and related depth of the spanning tree; clock drift and accuracy are attributed to resynchronization interval. The work carried out was aimed at reduced computation and complexity at the cost of relaxed accuracy. One major assumption of Greunen et al. is that the drift of the clock is bounded. Pair-wise synchronization is performed as shown in Figure 4.4.

Two nodes j and k are synchronized upon exchanging packets 1 and 2 and calculating the offset. A third message is used to convey the offset calculated by node j relative to node k. The offset d is calculated using unknown transmission time D calculated as given by Equation 4.4

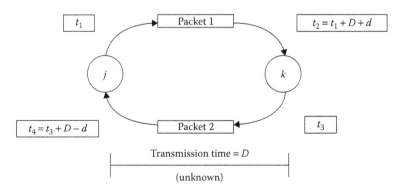

FIGURE 4.4   Pair-wise synchronization between two nodes. (From Greunen, J. V., and Rabaey, J. *Proc. 2nd ACM Int. Conf. Wirel. Sensor Netw. Appl.*, pp. 11–19. ACM, 2003.)

$$t_2 - t_4 = t_1 - t_3 - D + D + 2d \qquad (4.3)$$

$$d = \frac{\left(t_2 - t_4 - t_1 + t_3\right)}{2} \qquad (4.4)$$

FTSP floods the network with synchronization packets to the neighbors [33]. A global root node is elected and all other nodes synchronize their clocks to this node. The root node is dynamically reelected. A sender transmits its time by using byte alignment so as to reduce errors. In FTSP, a node is synchronized only upon having enough data (eight former data from the sender) to perform linear regression. A shortcoming with this approach is that if a node receives the data at less intervals, then it will not be in a position to complete the linear regression analyzes. Thereby, it deters any progress toward time synchronization: a situation may arise where in a node will never be able to synchronize.

In the recent past, very limited work has been focused on both time synchronization and energy efficiency simultaneously. More recently, Kim et al. [26] have developed an algorithm to increase efficiency by reducing the number of packets exchanged thereby conserving power. This again is a combination of RBS, TPSN, and LTS methods but reduces the number of packets exchanged. Again, the topology used is identical to TPSN and the best achievable error rate is affected by the number of hops. Of all the methods discussed above, less importance has been devoted toward energy efficiency. Shahzad et al. [43] proposed energy-efficient time synchronization protocol for WSNs algorithm that combines RBS and TPSN for achieving better efficiency in energy. Although these methods improve energy efficiency, however, in large multihop network applications—such as real-time traffic management, health monitoring of a structure—with 8–10 multihop levels, the synchronization error is higher than required for critical infrastructure monitoring.

CLOCK MODEL

Clock can be modeled as a function of linearly increasing function with time, that is, $f(x) = m \times x + c$. When we say it is linear, it is assumed that the clock function will not fluctuate randomly rather maintains linear slope. This property was also observed during simulations. Let the subscript I represent the ideal clock (i.e., having slope $m_I = 1$). Then the clock function $\left(f(X_I)\right)$ of an ideal clock is given by Equation 4.5:

$$f(X_I) = m_I \times x_I + c_I \qquad (4.5)$$

An imperfect clock has the slope $m > 1$ or $m < 1$. This implies that the clock is either faster or slower as compared with ideal clock.

The synchronization error is the time difference between two clocks at a given time t. The error (also offset) is caused by the skewness (first-order derivative of clock with respect to real time t). Therefore, the offset $f(O)$ between ideal clock $f(X_I)$ and an imperfect clock $f(X_C)$ can modeled as

$$f(O) = f(X_I) - f(X_C) \tag{4.6}$$

$$= \left[m_I \times x_I + c_I \right] - \left[m_C \times x_C + c_C \right] \tag{4.7}$$

$$= \left[(m_I - m_C) \times x_I \right] + \left[c_I - c_C \right] \tag{4.8}$$

because $(x_I = x_C = t; \; \forall x_I \in t, \; \forall x_C \in t)$ is true w.r.t. time t. The subscript C refers to a child node. The skewness is caused by the factor $m_C \neq 1$. c_I represents the offset at the starting of the timer. If c_I or $c_C > 0$ implies that the clock had a residual time. However, in general, $c_I = 0$ and $c_C = 0$. The following conditions are true for two clocks:

1. *Condition 1*—If $m_C > 1$, then $f(O) < 0$, indicating the child clock has a skew that is greater than m_I
2. *Condition 2*—If $m_C < 1$, then $f(O) > 1$, highlighting the fact that the child node has a clock that is running slower than the clock y_I measured at time t
3. *Condition 3*—If $c_I = 0$ and $c_C > 0$ results in a negative offset $(f(O) < 0)$ at time $t = 0$ when measured with respect to ideal clock, inferring that the child clock has started before ideal clock
4. *Condition 4*—If $c_I = 0$ and $c_C < 0$, then this condition results in a positive offset with respect to ideal clock $(f(O) > 0)$ pointing that child clock is slower. This also directs that the child clock will have a positive value when $m_C * t > c_C$

A sensor node clock can be of any one of the three types (ideal, fast, or slow). Therefore, the relative offset between clocks may be positive $(d/dt \; C_i(t) > d/dt \; C_j(t))$ or negative $(d/dt \; C_i(t) < d/dt \; C_j(t))$ for a node i with respect to node j depending on whether i is faster or slower than j. Figure 4.5 shows the linear variation of fast, ideal (perfect) and slow clocks.

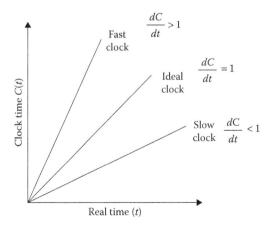

FIGURE 4.5   Fast, ideal and slow clocks. (Reprinted from Sundararaman, B., Buy, U., and Kshemkayani, A. D., *Ad Hoc Networks*, 3:281–323, 2005.)

It is very important to note that because of skewness of the clock, a fast clock samples faster than a slow clock. Thus, when a node is allowed to collect data, if it is not ideal, then both the collection time and collected data represent different timings.

VIRTUAL CLOCK–BASED TIME SYNCHRONIZATION

Our aim of time synchronization is to maintain a global time across the network through external synchronization (i.e., to maintain a precise time with respect to external source). For the purpose of synchronization, we make use of implicit synchronization technique and adopt hierarchical structure-based network. The network is dynamically formed and nodes can be added at any time. The communication mode is broadcast and communication distance is multihop. Our objective is to minimize the synchronization error while reducing the energy consumption.

Consider N nodes deployed as a group of sensor networks for monitoring certain activities. It is assumed that each of the N is in communication range from at least one other node, and there exists one node among them that acts a root node (connected to external time source). A root node is a node that maintains the real time through external synchronization or through any other means. Every node has its own unique identification tag and is aware of its own tag. The sensor network time synchronization can be achieved by having a hierarchical structure followed by data transportation from leaf nodes, intermediate nodes to root node. This network creation is accomplished in two stage process: network creation and time synchronization. The flowchart for network creation and time synchronization are shown in Figure 4.6 and the steps involved are described as follows.

NETWORK CREATION

1. The sensor network deployed creates its hierarchical structure being root node at level i and its immediate neighbors being at level $i+1$ (similar to TPSN, but not entirely). The immediate neighbors of level $i+1$ become level $i+2$ peers and so on. The process of network creation continues dynamically until all the nodes in the network have their levels determined—each node knows its parent (except root node) and each node knows its child (except leaf nodes).
2. Parent node sends a broadcast signal to determine neighbors.
3. Parent node waits for t_w time units to receive "request" signals from child nodes.
4. After t_w time units, parent node estimates its children count and sends "level" signal for the child node. This "level" is sent only once upon determining a child for the lifetime (until child node requests "parent" as mentioned in step 6. Child simply receives the parent level and increments it by one and stores. The message also contains the current time of the parent. This child node now acts as parent and repeats steps from 1. Parent node processes each child according to ascending (or descending) order of identification number. After a child node is processed (by sending "level"), it processes the next child until all the child nodes are processed.
5. A node can sense a physical activity only after having a "level" signal from its parent and can send datum to its parent whenever a datum is available.

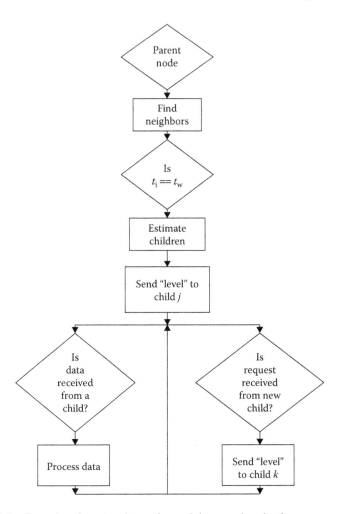

FIGURE 4.6   Procedure for network creation and time synchronization.

6. If a new node is added to (or change in topology is made—node displaced, node power cycled) the network, then it has to broadcast a request "parent" signal so that nodes nearer to it will respond. Each of the nodes will respond with their level numbers. The new child has the ability to choose the node with lowest level (closet to the parent) number as its parent node. After this an acknowledgment is sent to the parent. The parent will send a "level" signal to the new node member. Here, acting-parent nodes responding to the new node respond only if they think that they can handle additional load from this newly added node and hence the new network. In this way, nodes can avoid additional onus being experienced.

7. Every time a datum is received form a child, time synchronization is performed. Parent node updates the datum received time as described in section "Time Synchronization."

TIME SYNCHRONIZATION

Let t_i represent the local clock time of each the node and $i = 0,1,2, \ldots, N$. For simplicity, the algorithm described below is between any two nodes; that is, a parent and child node. This can be extended to single-parent multichild situations. Let i represent a parent node and j represent child node among N set of nodes. Let t_{ij} represent the relative clock time difference (offset) between nodes t_i and t_j, where $t_{ij} = t_i - t_j$ w.r.t. to its parent i. In this paper, all the time difference conventions are from parent's perspective.

As the time elapses, there will be time differences between parent and child: if the derivative of clock source of the child is fast clock ($d/dt\ C_j(t) > 1$), then $t_{ij} < 0$, and $t_{ij} > 0$, for $d/dt\ C_j(t) < 1$. Therefore there is a need for manipulation of child's time individually by the parent. The relative synchronization error at any point in time between two nodes is given by

$$t_{re} = t_{ij} \tag{4.9}$$

$$= t_i - t_j \tag{4.10}$$

The above set of equations are devoid of adaptively corrective mathematical terms that foster the sensor nodes to keep track of relative time. The relative time from parent's perspective can be obtained as

$$t_{j_{re}} = t_i - \left[(t_i - t_{i-1}) - (t_j - t_{j-1}) + t_{j_{te}} \right] \tag{4.11}$$

$$= t_i - \left[(t_p) - (t_c) + t_{j_{te}} \right] \tag{4.12}$$

where $t_{j_{re}}$ symbolizes relative error, $t_{j_{te}}$ represents total relative error measured by parent i for child j up to time t_i. t_{i-1} represents the parent time when a message was previously received from the child node with the child time as t_{j-1}, which was time stamped along with the message. t_p corresponds to the time elapsed because the current and previous message as observed by parent by its self-timing $t_p = t_i - t_{i-1}$, for child with timings $t_c = t_j - t_{j-1}$. The relative error is updated every time a new message is received by using Equation 4.13:

$$t_{j_{te}} = t_{j_{te}} + t_{j_{re}} \tag{4.13}$$

Note that $t_{j_{te}}$ could be positive or negative. It is observed that, when there is momentary variations in child or parent clock, $t_{j_{te}}$ swings drastically, but maintains a steady state after 2–3 successive measurements. In physical terms, $t_{j_{te}}$ is the sum of all relative time differences between a child and a parent until recent observation. In other words, parent node i is maintaining a cumulative account of relative offsets for child node j. The relative offset error is a combination of both clock offsets due to drift and nondeterministic delays. In Equation 4.12, parent i is correcting child's clock at each message communication instances. Therefore, relative error between parent i and child j remains extremely precise. Using this method, the relative offset reaches a steady-state error value for all t as shown in Figure 4.7, for a parent and child node. $t_i \pm \beta$ represents the relative offset between parent and child. Here, parent assumes

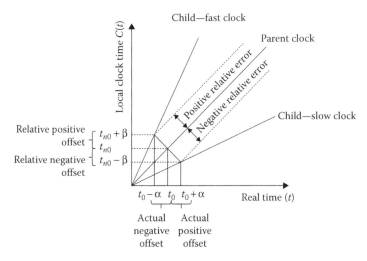

FIGURE 4.7   Relative error stabilized using Equation 4.12.

itself as ideal clock (for the purpose of calculation). $t_0 \pm \alpha$ would provide actual time (real time) between parent and child. It is to be noted from Figure 4.7 that for a fast clock, relative to parent $(t_i + \beta)$, the actual time is $(t_0 - \alpha)$, and for a slow clock $(t_i - \beta)$, the actual time is $(t_0 + \alpha)$, measured at actual time t_0. $(t_p - t_c)$ corresponds to $t_i \pm \beta$ shown in Figure 4.7. $(t_p - t_c) > 0$ for slow child clock, whereas $(t_p - t_c) < 0$ for fast child clock. $(t_p - t_c)$ provides an estimate of child clock skew for the time period $t_i - t_{i-1}$. Our algorithm implements Equations 4.12 and 4.13. The following sections provide improvements that can be undertaken to further improve the accuracy: (a) the section "Algorithmic Improvements" is about improving the accuracy taking packet size in to consideration, (b) section "Time Prediction and Transformation" is for predicting approximate time to estimate synchronization interval and time transformation as postfacto synchronization, and (c) section "Computing Resynchronization Intervals" highlights about short-term and long-term synchronization techniques.

ALGORITHMIC IMPROVEMENTS

Algorithmic improvements can be implemented to further lessen the synchroniza-tion error by considering the notion of time of reception that it takes more time for a sensor node to receive large-sized packet data, as against small-sized message packet. Given two nodes i and j, there is time associated with the length of the message. Let t_i denote the reception time of data at node i and let t_j represent the sending time at node j. Under ideal conditions, if node j is commanded to send k number of $n1$ bytes of data to node i at regular intervals t_s and also to send k number of $n2$ bytes of data, where $n2 > n1$, then the reception time difference t_d at node i is given as

$$t_d(i) = t_s(j) - t_s(j-1)$$
$$t_d(i+1) = t_s(j+1) - t_s(j)$$

$$t_d(i+2)=t_s(j+2)-t_s(j+1)$$

$$\cdot$$
$$\cdot$$
$$\cdot$$

$$t_d(i+k)=t_s(j+k)-t_s(j+k-1)$$

If we subscript i by i_{n1} and i_{n2}, then, because t_s is constant for all messages, it is found that $t_d(i_{n1}+k) \neq t_d(i_{n2}+k)$, for $n2 > n1$. Therefore, in sensor network communication, for different size of packets, Equation 4.14 holds true $\forall t_s = K$, where K is data sending interval.

$$t_d(i_{n2}) > t_d(i_{n1}) \tag{4.14}$$

Using Equation 4.14, Equation 4.12 can be rewritten as

$$t_{jre} = t_i - \left[(t_p)-(t_c)+t_{jre}\right]-t_d \tag{4.15}$$

where t_d represents time delay introduced due to sending time, access time, propagation time, and reception time (we consider propagation time to be negligible for WSNs wireless range compared with radio waves' speed). More information can be found in Reference [40].

Time Prediction and Transformation

At any point in time, if parent node has the information of the slope of the child clock, then parent can compute its child's predicted offset (P_o) by using Equation 4.16.

$$P_o = \left[m_I - m_C\right] \times t \tag{4.16}$$

$$t_i = \left(t_j - t_{j-1}\right) + t_{jre} \tag{4.17}$$

where:

t_j is the current time
t_{j-1} is the time when t_{jre} was computed for node j

In general, Equation 4.17 can be used to send the current time of a parent node to child node. If the child node is a new node, then $t_i = t_j$, else it has already been known to parent j and can make use of Equation 4.17. In case of multihop scenario, to determine the time of node at level L distance is given by Equation 4.18:

$$P_o(n) = \left[1 - \sum_{n=1}^{L} m_n\right] \times t \tag{4.18}$$

where:

$m_n = m_1 + m_2 + \cdots + m_{n-1} + m_n$ up to nth node
m_1 is the skew between parent and its child at level 1
m_2 is the skew between child at level 1 and child at level 2 and so on, up to L levels

If the slopes of all the nodes are available (considering nodes maintained their slopes ever after their power cycle), then time transformation can be applied on the data available from all the nodes at the base station. If δ_{max} and δ_{min} represent maximum and minimum offset limits, then, base station node can determine the exact occurrence of an event with respect to its time by estimating the occurrence (P_{ce}) of an event using Equation 4.19.

$$P_{ce} = t \times \left(m_b - m_n \right) \tag{4.19}$$

where n represents node number (excluding base station slope m_b).

COMPUTING RESYNCHRONIZATION INTERVALS

To lower the relative error between children and respective parents, a dynamic inter-synchronization interval scheme is essential. A predetermined, fixed intersynchronization interval may often satisfy the error requirements; however, as the system scales into bigger entity, it is less likely that this static intersynchronization interval to efficiently confine to the error limits.

At any instance, the time predicted for a child j by parent node i is given by Equation 4.17. The resynchronization interval is estimated using upper limit of time differences (δ_{max}). Let δ be a threshold limit, a measure of time differences between the parent and the child, then $\delta = |t_i - t_{j_{te}}|$, gives the threshold value. If $\delta > \delta_{max}$, then parent unicasts a message containing the child time. Therefore, using δ_{max}, synchronization interval can be set. The reason for having t_p and t_c is to measure the drifts in short terms, whereas $t_{j_{te}}$ is a measure of long-term (resynchronization) interval. This helps to detect any large deviations in the child's clock. In turn, child's clock can be corrected using Equation 4.17. Using this method, it adds dynamism to calculate resynchronization intervals.

DETERMINING CONSUMED ENERGY

Energy consumed by a sensor node, in general, is a combination of two cardinal factors: (1) energy consumed linearly when the sensor node is functioning (processor—executing operations during awake and sleep times, with other member components) and (2) energy expended in transmitting and receiving a packet.

Let the energy consumed by a node for processing operations be E_P J and E_C J due to communication activities. Let the data transmission rate of a node (i.e., radio) be d_{tx}, then the time it takes for a byte to be transmitted is $t_{tx} = 1/d_{tx}$. Let p such bytes form a transmitting packet size and q bytes form a receiving packet size. Additionally, for generality, let the data reception rate be $t_{rx} = t_{tx}$, same as data transmission rate. Let the current required for transmitting a packet be C_{tx}, for receiving a packet C_{rx} and the battery voltage be V_{bat}. Energy expended by a node is given by Equations 4.20 through 4.22.

$$E_T = E_P + E_C \tag{4.20}$$

$$= \left[E_{pw} + E_{ps} \right] + \left[E_{tx} + E_{rx} \right] \tag{4.21}$$

$$= \left[E_{pw} + E_{ps} \right] + \left[\left(C_{tx} \times V_{bat} \times t_{tx} \times p \right) + \left(C_{rx} \times V_{bat} \times t_{rx} \times q \right) \right] \tag{4.22}$$

where:

E_T is the total energy consumption of the node

E_{pw} is the energy expended due to node's awake status and E_{ps} is due to node's sleep status

$\left[E_{tx} + E_{rx}\right]$ due to transceiving operations

SIMULATION RESULTS

Our results were verified using Castalia-3.2 simulator. OMNeT++ is a discrete event simulation environment. It is an extensible, modular, component-based C++ simulation library and framework essentially build simulators [53]. At the core of the OMNeT++, it provides option for component architecture model that can be programmed using C++. At higher level, these components can be linked together (using .NED) [53]. Castalia is a discrete event simulator that runs on OMNeT++ simulation framework. It can be used to simulate WSNs, body area networks and also low-power embedded devices. It is a not specific to any sensor and provides a range of realistic node behaviors. It provides flexibility in its modularity, reliability, and celerity of execution that are partly attributed to OMNeT++ support [37].

Simulations were conducted for a total of 30,000 s. A total number of 12 nodes were included to form a cluster with node 0 being the cluster head as shown in Figure 4.8. All the nodes were defaulted to receive state upon completion of transmission. CC2420 radio parameters, "IDEAL" conditions as specified in Castalia

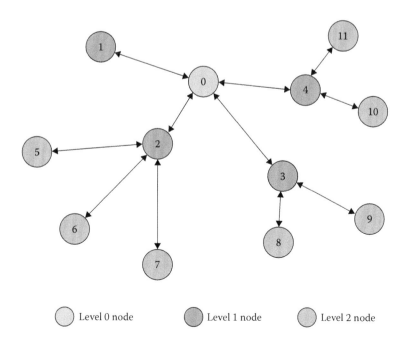

FIGURE 4.8 A common hierarchical structure for TPSN, LTS, and VCTS.

simulator were used. We implemented radio disk model with a radius of 22 m. Neither medium access control nor any routing protocol was used. Application layer time stamps were used instead of medium access control layer. Synchronization error obtained for three algorithms (TPSN, LTS, and VCTS) are shown in Figure 4.9a and b, for a resynchronization interval of 300 s and 1,000 s, respectively. Total energy

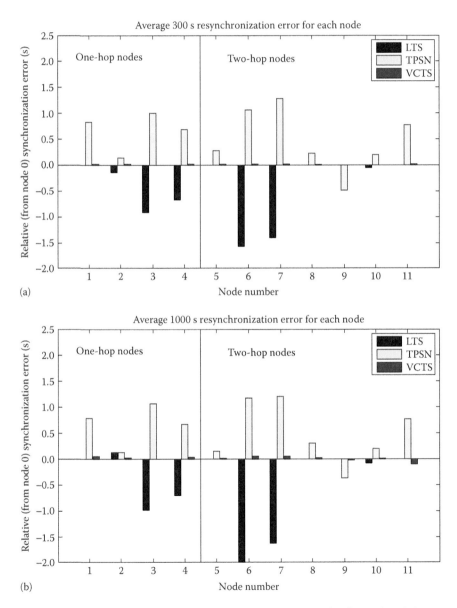

FIGURE 4.9 (a) Synchronization error for 300 s resynchronization interval and (b) synchronization error for 1,000 s resynchronization interval.

consumed and energy consumption for communication only, by these methods are shown in Figure 4.10a and b. Energy for the communications was calculated offline using E_C of Equation 4.20 and combined with E_P, obtained from simulator for 30,000 s. For calculating E_C, all data messages were set to 50 bytes in length, $d_{tx} = 250$ kbps, $V_{bat} = 3.6$ V, $C_{tx} = 17.4$ mA, and $C_{rx} = 18.8$ mA.

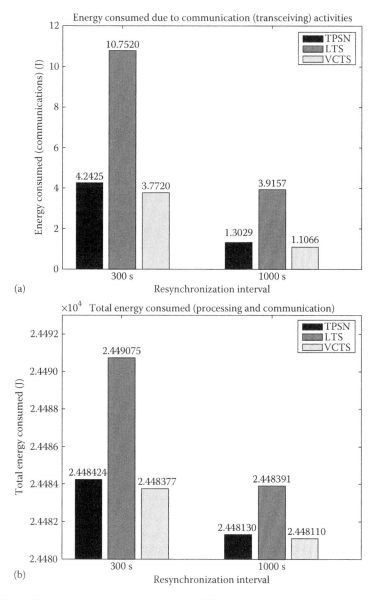

FIGURE 4.10 (a) Energy expended by three different protocols (for transcceiving operations) and (b) total energy expended by three different protocols: TPSN, LTS, and VCTS.

Discussion

Initially, synchronization error due to different resynchronization intervals will be discussed followed by their energy consumption.

First, for 300 s resynchronization interval, TPSN maintains a moderately increased error from one-hop (0.526633 s) to two-hop (0.613173 s) distanced node. The one-hop error (0.573529 s), for LTS, increased steadily to two-hop distance error (1.011589 s). On the other hand, virtual clock-based time synchronization (VCTS) maintained a relatively unvaried error from one-hop (0.009964 s) to two-hop (0.009074 s). Next, for 1,000 s resynchronization interval, TPSN performed better by declining its error to 0.590545 s from 0.660078 s. However, LTS followed its former position to reach a maximum error of 1.223768 s, during two-hop, from almost half the one-hop error. Nevertheless, VCTS had a moderate increase in error from 0.032712 s (one-hop) to 0.038614 s (two-hop). From Figure 4.9a and b, it can be inferred that, while TPSN and LTS maintained a steady trend in error values, VCTS quadrupled its error from 300 s to 1,000 s resynchronization interval. However, these errors are much below the values of both TPSN and LTS.

The energy consumed due to combined activities—nodes' awake state, sleep state, and communication—is shown in Figure 4.10a and b, and that the energy spent in communication only is depicted in Figure 4.10. The energy consumed by the entire network for different protocols, for 30,000 s is approximately 24,480 J, and the remaining is contributed by message exchanges. For 300 s resynchronization interval, energy expended by TPSN is 4.2425 J, LTSN is 10.7520 J, and VCTS is 3.7720 J. Likewise, energy consumed for 1,000 s interval are 1.3029 J, 3.9157 J, and 1.1066 J, respectively.

The proposed method performs better than TPSN and LTS, both in terms of synchronization error and energy efficiency. It is evident from the results that LTS relaxes accuracy constraints while reducing accuracy, confining to authors claim. It is important to note that energy can be saved at the cost of sacrificing synchronization error in all the three methods. Low-energy adaptive clustering hierarchy (LEACH) is a clustering-based energy distribution protocol that distributes the load of local cluster head for communication between cluster head and the base station. It uses randomized rotation of cluster head to balance the local cluster head energy [22]. Numerous LEACH-based clustering algorithms are also implemented [1,14,21,30,32,51,56,60]. The combination of LEACH (for cluster-energy distribution) and VCTS (for time synchronization between nodes in a given cluster) can assist to design better energy-efficient time synchronization approaches as shown in Figure 4.11.

An appropriate time synchronization algorithm is instrumental in devising an efficient sensor network. Maintaining the reliability of the network data, mainly the occurrence of the event, helps to detect an event of interest and act upon rightly. A datum that has anomalies in its time-stamp due to synchronization error is of less conducive. From an SHM perspective, this is crucial for civil and infrastructure engineers. Hence, it is of great importance that we minimize the synchronization error introduced by the low-cost solution. The use of precision clocks, certainly, accounts for expensive solution. Instead, a low-cost solution, at the expense of computation would be a remarkable progress. Therefore, time synchronization while adhering to

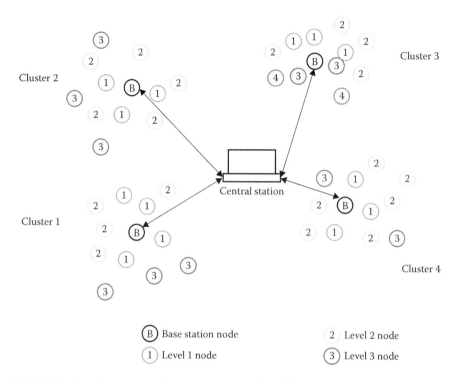

FIGURE 4.11 Schematic of cluster formation using LEACH. Numbers within the node represent hierarchical levels. B is the chosen cluster-head in any given round by LEACH.

stringent energy awareness is what is required. The accuracy of synchronized time among nodes, coupled closely (in time), ameliorates the system's ability to determine an event. Eliminating relative drifts and nondeterministic delays (send time, access time, propagation time, receive time), considered simultaneously, would provide promising results in synchronizing time. In the proposed approach, we are maintaining a relative offset errors of child nodes, for a parent node to use the datum from the child and accommodate it to its local clock timings. Energy can be reduced by means of reducing communication overhead for synchronizing tasks. Also, time of a child can be predicted and error can be maintained at minimum level. The performance of the proposed algorithm is significantly better than TPSN and LTS.

SHM USING EMERGING IoT

IoT is an emerging technology that unifies sensors, sensor nodes, and the Internet backbone into a single platform. The embedded devices with sensors can communicate with each other using the Internet. Users can also control these sensors in real time over the Internet. Cloud computing technology has also immensely fostered the growth of IoT. The state-of-the-art IoT infrastructure includes cloud platform for storage and retrieval of sensor data. Many useful application programming interfaces

(APIs) have been made available open to users for ease of use and open source. For instance, Libelium [65] provides IoT-based solutions, where the embedded devices (Waspmotes) are programmable using integrated development environment that significantly reduce the development time. They also have a platform called Meshlium, which is Linux machine that has the Internet connection (3G, 4G), database server, memory storage, and wireless transceiver (ZigBee and DigiMesh to receive data from Waspmotes). The users can further install the required packages as the platform is a Linux operating system. One can store the data in local database or send the data to external database using different protocols. Cloud platforms like Xively [66] provide modern data transfer services with representational state transfer APIs, can be used to upload data to cloud server for visualization of the sensor data.

FLEXIBILITY WITH NETWORKING SIMILAR TO THE INTERNET

The embedded devices with sensors have the ability to communicate mutual and form a network among themselves. The network can be formed dynamically on the availability of the resources (such as power, node failure). This feature provides a flexibility in networking. The same versatility is extended when connecting to the Internet. Each of the nodes can be made to connect to the Internet and communicate. This is highly advantageous because the failure of gateway node in otherwise static network, the critical sensor information will be lost. The ability of ad hoc network formation overcomes the problem of node failure and hence the data loss. Depending on the energy availability, any node can become a cluster head in controlling the network as observed in LEACH [22].

IPV6 ADDRESSING

The Internet ("network of networks") uses unique address to identify the devices connected to the network. Usually we use names, for instance, http://www.google. com to access services from Google. However, devices on the Internet do not understand this language. The machines use Internet Protocol version 4 (IPv4) to resolve the names. A domain name server matches the names with IPv4 address and helps the devices on the Internet to communicate with each other. IPv4 address (32 bit addressing scheme) can handle $2^{32} \simeq 4.3$ billion unique devices. Since the inception of the Internet, there are an estimated 3 billion users (refer to Figure 4.12) covering 40% of population, with at least a million users being added per year. Additionally, there are many devices ("objects" or "things") connected to the Internet A staggering 8.7 billion devices were reported to be connected to the Internet [63], already exceeding the IPv4 limit of 4.3 billion. Because of networking strategies (subnetwork mask), many devices can be connected to the same IP address using Network Address Translator (NAT). However, with IoT devices expected at 50 billion devices by 2020 [63] and unique identifiers to connect to the devices, new addressing schemes are required. ToF handles more number of unique devices, IPv6 standards [62] was proposed, which has the addressing (128-bit) capability of handling $2^{128} \simeq 3.4 \times 10^{38}$ addresses. Many devices also have the mechanism to understand both IPv4 and IPv6 addressing, making them

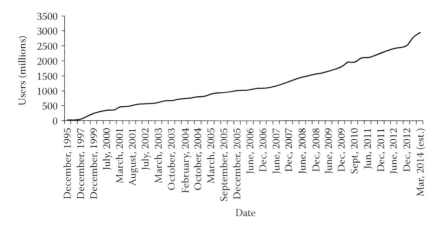

FIGURE 4.12 Growth of the Internet (in terms of users) since 1995 [64].

to communicate during transition from IPv4 to IPv6. The IoT devices possess low power and ad hoc networking, hence having NAT strategy is not ideal. The future IoT devices are being envisioned to have IPv6 address so that the devices can be uniquely identified.

New Sensors with Internet Connectivity

With IPv6 addressing, the new devices and sensor themselves can be assigned unique address. This provides a utility for the users a high resolution and finer-scale knowledge of the sensors and controls. The devices can connect to the Internet in ad hoc mechanism and eliminates many networking issues when there is a failure with individual devices. The device themselves have radio frequency communications, which is used for ad hoc networking among the nodes. The dynamic nature of nodes having the communication and networking strategies almost provides a vision of machine-to-machine communicating architecture. This adds increased flexibility in networking among the nodes and also to and from the nodes. The flexibility will be similar to the Internet. The bridges and the buildings can be monitored in an unprecedented way. The data resolution for scientific and engineering study will be unparalleled compared with the traditional monitoring.

Big Data Analytics

As number of sensors and sensor locations increase, we run into the problem of "Big Data." Deluge of data from different sensors and different modalities will be problematic for conclusive analysis. Big data analytics deals with handling multimodal and heterogeneous data for meaningful analysis. The existing data analysis algorithms prove to be inefficient in terms of processing time and processing techniques. A new line of analysis is the Big Data analysis. The algorithms are handle sophisticated homogenous, heterogeneous, and multimodal data effectively to provide a better analytics. There have been considerable efforts by the research communities in addressing Big Data problem.

A CASE STUDY OF SHM

In this section, a case study of continuous monitoring of SHM using WSN have been presented. The aim of the study [17] was to address the following challenges in SHM: (1) a three-dimensional 14-bay truss structure with existing damage was verified, (2) continuous online SHM to measure ambient vibrations, and (3) SHM through distributed computing using densely deployed sensors.

In a traditional SHM system, a central base station would acquire the data from sensors deployed at different locations of the structure. Those sensors that are close to damage were likely to be more pronounced in their outputs. Additionally, the long wires running would complicate the measurement process. The data acquisition becomes further burdened with large amounts of data to be transported and processed at the central station. With the advent of smart sensors [47,48]—sensors with programmable microprocessors, memory, and processing—has reduced load of central data processing, communication overhead while proving continuous SHM status.

The damage to the structure is detected by using flexibility-based damage localization method. The damage load vectors (DLVs) help determine the structures with damages. The sensors for undamaged structure with static forces produce zero stress at the sensor locations. This helps to localize the local structural damage. To verify this, a 5.6 m long three-dimensional truss structure was used. Steel tubes with 1.09 cm inner diameter and 1.71 cm external diameter formed the member of truss. The truss was excited vertically using Ling Dynamic Systems shaker. The response of the systems were measured using accelerometers (sensitivity of 100 mv/g, 1–4000 Hz and range of ±50). Spectrum analyzers were used to measure the responses.

In traditional SHM, the excitation requires external input and also the installation of machinery becomes expensive. Continuous online SHM would be an ideal solution. However, the DLVs for online method are not easily obtained. To overcome this, modal normalization constants were obtained for the construction of flexibility matrix. However, the construction of the flexibility matrix for a damaged structure is not feasible. To this end, extension of DLV was proposed to handle online damage [18].

To monitor large structures, the traditional way of using long cables is cumbersome and expensive. Also, from a safety and reliability perspective, continuous monitoring is necessary. WSN approach demands more bandwidth as more sensor nodes are deployed. To overcome this problem, distributed computing system was employed. Small clusters of sensor nodes work in tandem to process the sensor data and the "manager" sends the essential and limited data about the structure to the central station. Neighboring managers need to communicate with each other before sending the data. If the structure is not undamaged, "Ok" message about the location of the structure is sent to the central station by the manager. Extended DLV was employed by the nodes in the clusters to determine the damage to the structure in the locality.

A DEMONSTRABLE APPLICATION OF IoT

IoT can further ameliorate the above-mentioned approach by providing cloud-based services such as software as a service, platform as a service, and infrastructure as a service. The IPv6 address to individual sensor nodes makes the sensors authentic before providing such services. Figure 4.13 shows the SHM scheme using IoT. Devices (or "things") will be able to directly interact with the cloud services for data uploading and analytic for real-time processing at the node level. A wide range of analytics in real time can be provided to the engineers, scientists and the users simultaneously. The central station can also be eliminated.

Further, carbon nanotubes as structure will be enable different sensing modalities based on their structural properties. The "smartness" will be embedded into the structure itself. This would help reduce the number of sensors required and consequently the cost. Additionally, with sufficient advancements, carbon nanotubes would be help to detect the local damages to the structure through global measuring techniques because of their intrinsic properties.

Video cameras can also be used for SHM. With low-cost cameras and advanced video processing algorithms, some of the visible signs of the structural damage can be easily detected using computer vision algorithms. Figure 4.14 shows the four video cameras monitoring the bridges. Additionally, usage of bridges by pedestrians and traffic can be monitored in real-time. Pedestrian monitoring aids in detecting the loitering behavior, occupancy of the bridges by pedestrians at different times of the day and their general behavior. Computer vision algorithms can also assists in monitoring traffic by: (1) analyzing the vehicles for lane departure, (2) categorizing the vehicles based on size, load, and passengers, (3) detection of vehicle hazards such as smoke and reckless driving, (4) as early warning systems for incoming traffic about status of the traffic flow on the bridges and underground tunnels, and (5) also in automatic

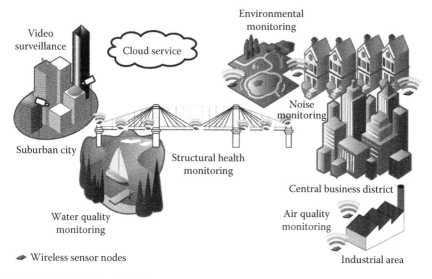

FIGURE 4.13 SHM using IoT.

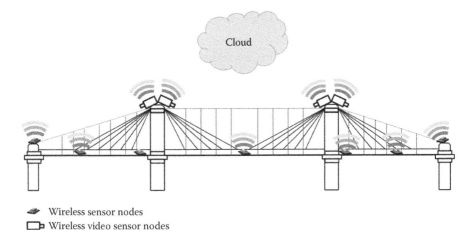

◆ Wireless sensor nodes
▢► Wireless video sensor nodes

FIGURE 4.14 SHM using IoT. All devices are addressable using IPv6 standards. Devices will make use of cloud services to store data and use analytics for real-time processing of the data at the node level.

collision avoidance systems between vehicles and structures. Video camera networks will add rich source of information to the SHM to the existing SHM sensor data.

CONCLUSION

IoT is an emerging technology that has promising future in continuous monitoring and controlling the connected devices. SHM can be achieved in real-time and rich analytics. However, sensor nodes face the problem of time synchronization error. An appropriate time synchronization algorithm is instrumental in devising an efficient sensor network. Maintaining the reliability of the network data, mainly the occurrence of the event, helps to detect an event of interest and act upon rightly. A data that has anomalies in its time-stamp due to synchronization error is of less conducive. From an SHM perspective, this is crucial for civil and infrastructure engineers. Hence, it is of great importance that, we minimize the synchronization error introduced by the low-cost solution. The use of precision clocks, certainly, accounts for expensive solution. Instead, a low-cost solution, at the expense of computation would be a remarkable progress. Therefore, time synchronization while adhering to stringent energy awareness is what is required. The accuracy of synchronized time among nodes, coupled closely (in time), ameliorates the system's ability to determine an event. Eliminating relative drifts and nondeterministic delays (send time, access time, propagation time, and receive time), considered simultaneously, would provide promising results in synchronizing time. In the proposed approach, we are maintaining a relative offset errors of child nodes, for a parent node to use the data from the child and accommodate it to its local clock timings. Energy can be reduced by means of reducing communication overhead for synchronizing tasks. Also, time of a child can be predicted and error can be maintained at minimum level. The performance of the proposed algorithm is significantly better than TPSN and LTS.

REFERENCES

1. Abdulsalam, H. M., and Kamel, L. K. W-LEACH: Weighted low energy adaptive clustering hierarchy aggregation algorithm for data streams in wireless sensor networks. In *IEEE International Conference on Data Mining Workshops*, pp. 1–8. IEEE, 2010.
2. Akyildiz, I. F., Su, W., Sankarasubramaniam, Y., and Cayirci, E. Wireless sensor networks: A survey. *Computer Networks*, 38(4):393–422, 2002.
3. Allan, D. W., and Weiss, M. A. Accurate time and frequency transfer during common-view of a GPS satellite. *Proceedings of 34th Annual Frequency Control Symposium, USAERADCOM*, Electronic Industries Association, Fort Monmouth, WJ, May, 1980.
4. Bengio, Y. Learning deep architectures for AI. *Foundations and Trends in Machine Learning*, 2(1):1–127, 2009.
5. Bently, D. E., and Hatch C. T. *Fundamentals of Rotating Machinery Diagnostics*. The American Society of Mechanical Engineers, 2002.
6. Buyya, R., Yeo, C. S., Venugopal, S., Broberg, J., and Brandic, I. Cloud computing and emerging it platforms: Vision, hype, and reality for delivering computing as the 5th utility. *Future Generation Computer Systems*, 25(6):599–616, 2009.
7. Chang, P. C., Flatau, A., and Liu, S. C. Review paper: Health monitoring of civil infrastructure. *Structural Health Monitoring*, 2(3):257–267, 2003.
8. Damolini, S. *Carbon Nanotubes and Their Application to Very Long Bridges*. PhD thesis, Massachusetts Institute of Technology, Cambridge, MA, 2009.
9. Elson J., Girod, L., and Estrin, D. Fine-grained network time synchronization using reference broadcasts. In *Proceedings of the 5th Symposium on Operating Systems Design and Implementation*, volume 36, pp. 11–19. ACM, 2002.
10. Elson, J., *Time Synchronization in Wireless Sensor Networks*. PhD thesis, UCLA, 2003.
11. Estrin, D., Govindan, R., Heidemann., J., and Kumar, S. *Next Century Challenges: Scalable Coordination in Sensor Networks*, pp. 263–270. ACM, 1999.
12. European Commission's Directorate General for Communications Networks, Content and Technology (DG Connect). *A Digital Agenda for Europe*. 2010. https://ec.europa.eu/digital-agenda/en (last accessed January 22, 2016).
13. Falvo, M. R., Clary, G. J., Taylor, R. M., Chi, V., Brooks, F. P., Washburn, S., and Superfine, R. Bending and buckling of carbon nanotubes under large strain. *Nature*, 389(6651):582–584, 1997.
14. Farooq, M. O., Dogar, A. B., and Shah, G. A. MR-LEACH: Multi-hop routing with low energy adaptive clustering hierarchy. In *4th International Conference on Sensor Technologies and Applications*, pp. 262–268. IEEE, 2010.
15. Farrar, C. R., and Lieven, N. A. J. Damage prognosis: The future of structural health monitoring. *Philosophical Transactions of the Royal Society A: Mathematical, Physical and Engineering Sciences*, 365(1851):623–632, 2007.
16. Ganeriwal, S., Kumar, R., and Srivastava, M. B. Timing-sync protocol for sensor networks. In *Proceedings of the 1st International Conference on Embedded Networked Sensor Systems*, pp. 138–149. ACM, 2003.
17. Gao, Y., and Spencer, B. F., Jr. Structural health monitoring strategies for smart sensor networks. NSEL Report Series NSEL-011, Department of Civil and Environmental Engineering, University of Illinois at Urbana-Champaign, 2008.
18. Gao, Y., and Spencer, B. F., Jr. Online damage diagnosis for civil infrastructure employing a flexibility-based approach. *Smart Materials and Structures*, 15(1):9, 2006.
19. Greunen, J. V., and Rabaey, J. Lightweight time synchronization for sensor networks. In *Proceedings of the 2nd ACM International Conference on Wireless Sensor Networks and Applications*, pp. 11–19. ACM, 2003.

20. Gubbi, J., Buyya, R., Marusic, S., and Palaniswami, M. Internet of things (IoT): A vision, architectural elements, and future directions. *Future Generation Computer Systems*, 29(7):1645–1660, 2013.
21. Heinzelman, W. B., Chandrakasan, A. P., and Balakrishnan, H. An application-specific protocol architecture for wireless microsensor networks. *IEEE Transactions on Wireless Communications*, 1:660–670, 2002.
22. Heinzelman W. R., Chandrakasan, A., and Balakrishnan, H. Energy-efficient communication protocol for wireless microsensor networks. In *Proceedings of the 33rd Annual Hawaii International Conference on System Sciences*, volume 2, pp. 1–10. IEEE, 2000.
23. Iijima, S. Helical microtubules of graphitic carbon. *Nature*, 354(6348):56–58, 1991.
24. Jorio, A., Dresselhaus, G., and Dresselhaus, M. S. *Carbon Nanotubes: Advanced Topics in the Synthesis, Structure, Properties and Applications*, volume 111. Springer, Berlin, Germany, 2008.
25. Kahn, J. M., Katz, R. H., and Pister, K. S. J. *Next Century Challenges: Mobile Networking for "Smart Dust,"* pp. 271–278. ACM, New York, 1999.
26. Kim, B.-K., Hong, S.-H., Hur, K., and Eom, D.-S. Energy-efficient and rapid time synchronization for wireless sensor networks. *IEEE Transactions on Consumer Electronics,* 56(4):2258–2266, 2010.
27. Kim, S., Pakzad, S., Culler, D., Demmel, J., Fenves, G., Glaser, S., and Turon, M. Health monitoring of civil infrastructures using wireless sensor networks. In *6th International Symposium on Information Processing in Sensor Networks*, pp. 254–263. IEEE, 2007.
28. Kim, S. *Wireless Sensor Networks for Structural Health Monitoring.* Master's thesis, UC-Berkeley, CA, 2005.
29. Krishnamurthy, V., Fowler, F., and Sazonov, E. The effect of time synchronisation of wireless sensors on the modal analysis of structures. *Smart Materials and Structures*, 17:1–13, 2008.
30. Lindsey, S., and Raghavendra, C. S. Pegasis: Power-efficient gathering in sensor information systems. In *IEEE Aerospace Conference Proceedings*, volume 2, pp. 1125–1130. IEEE, 2002.
31. Li, F., Cheng, H. M., Bai, S., Su, G., and Dresselhaus, M. S. Tensile strength of single-walled carbon nanotubes directly measured from their macroscopic ropes. *Applied Physics Letters*, 77(20):3161–3163, 2000.
32. Loscrì, V., Morabito, G., and Marano, S. A two-level hierarchy for low-energy adaptive clustering hierarchy (TL-LEACH). In *IEEE 62nd Vehicular Technology Conference*, pp. 1809–1813. IEEE, 2005.
33. Maróti, M., Kusy, B., Simon, G., and Lédeczi, Á. The flooding time synchronization protocol. In *Proceedings of the 2nd International Conference on Embedded Networked Sensor Systems*, pp. 138–149. ACM, 2004.
34. Mills, D. L. Internet time synchronization: The network time protocol. *IEEE Transactions on Communications*, 39:1482–1493, 1991.
35. Montgomery, D. C. *Introduction to Statistical Quality Control*. John Wiley & Sons, New York, 2007.
36. The Hon. Wayne Swan MP. Intergenerational report: Australia to 2050: Future challenges, Australian government treasury report. Technical report, 2010. http://archive.treasury.gov.au/igr/igr2010/Overview/pdf/IGR_2010_Overview.pdf (last accessed January 24, 2016).
37. NICTA. *Castalia?* 2011 (verified on July 10, 2011). https://castalia.forge.nicta.com.au/index.php/en/index.html (last accessed January 20, 2016).
38. Oberlin, A., Endo, M., and Koyama, T. Filamentous growth of carbon through benzene decomposition. *Journal of Crystal Growth*, 32(3):335–349, 1976.
39. Pister, K. S. J., Kahn, J, K., Boser, B. E. Smart dust: Wireless networks of millimeter-scale sensor nodes. Highlight Article in *1999 Electronics Research Laboratory Research Summary*, p. 2, 1999.

40. Rao, A., Gubbi, J., Ngo, N., Nguyen, J., and Palaniswami, M. Energy efficient time synchronization in WSN for critical infrastructure monitoring. In *Trends in Networks and Communications*, volume 197 of *Communications in Computer and Information Sciences*, pp. 314–323, 2011.

41. Römer, K., Blum, P., and Meier, L. Time Synchronization and Calibration in Wireless Sensor Networks, in *Handbook of Sensor Networks: Algorithms and Architectures*, pp. 199–237. John Wiley & Sons, New York, 2005.

42. Salvetat, J.-P., Briggs, G. A. D., Bonard, J.-M., Bacsa, R. R., Kulik, A. J., Stöckli, T., Burnham, N. A., and Forró, L. Elastic and shear moduli of single-walled carbon nanotube ropes. *Physical Review Letters*, 82(5):944, 1999.

43. Shahzad, K., Ali, A., and Gohar, N. D. ETSP: An energy-efficient time synchronization protocol for wireless sensor networks. In *22nd International Conference on Advanced Information Networking and Applications—Workshops*, pp. 971–976, 2008.

44. Shull, P. J. *Nondestructive Evaluation: Theory, Techniques, and Applications*. CRC Press, New York, 2002.

45. Sichitiu, M. L., and Veerarittiphan, C. Simple, accurate time synchronization for wireless sensor networks. *IEEE Wireless Communications and Networks*, 2:1266–1273, 2003.

46. Sohn, H., Farrar, C. R., Hemez, F. M., Shunk, D. D., Stinemates, D. W., Nadler, B. R., and Czarnecki, J. J. *A Review of Structural Health Monitoring Literature: 1996–2001*. Los Alamos National Laboratory, Los Alamos, NM, 2004.

47. Spencer, B. F., Ruiz-Sandoval, M., and Gao, Y. Frontiers in structural health monitoring. In *Proceedings of the China-Japan Workshop on Vibration Control and Health Monitoring of Structures and Third Chinese Symposium on Structural Vibration Control,* Shanghai, China, 2002.

48. Spencer, B. F., Ruiz-Sandoval, M. E., and Kurata, N. Smart sensing technology: Opportunities and challenges. *Structural Control and Health Monitoring*, 11(4):349–368, 2004.

49. Sundararaman, B., Buy, U., and Kshemkayani, A. D. Clock synchronization for wireless sensor networks: A survey. *Ad Hoc Networks*, 3:281–323, 2005.

50. Sundmaeker, H., Guillemin, P., Friess, P., and Woelfflé, S. *Vision and Challenges for Realising the Internet of Things*. Publications Office of the European Union, Luxembourg, EUR-OP, 2010.

51. Thein, M. C. M., and Thein, T. An energy efficient cluster-head selection for wireless sensor networks. In *2010 International Conference on Intelligent Systems, Modelling and Simulation*, pp. 287–291. IEEE, 2010.

52. Treacy, M. M. J., Ebbesen, T. W., and Gibson, J. M. Exceptionally high young's modulus observed for individual carbon nanotubes. *Nature*, 381(6584):678–680, 1996.

53. András Varga. *What is omnet++?* 2011 (verified on July 10, 2011). https://omnetpp.org/intro/what-is-omnet (last accessed January 20, 2016).

54. Wang, Y., Lynch, J. P., and Kae, K. H. A wireless structural health monitoring system with multithreaded sensing devices. *Structure and Infrastructure Engineering*, 3:103–120, 2007.

55. Worden, K., and Dulieu-Barton, J. M. An overview of intelligent fault detection in systems and structures. *Structural Health Monitoring*, 3(1):85–98, 2004.

56. Xiangning, F., and Yulin, S. Improvement on LEACH protocol of wireless sensor network. In *2007 Fourth International Conference on Sensor Technologies and Applications*, pp. 260–264. IEEE, 2007.

57. Yakobson, B. I., Brabec, C. J., and Bernholc, J. Nanomechanics of carbon tubes: Instabilities beyond linear response. *Physical Review Letters*, 76(14):2511, 1996.

58. Yang, S. *Researchers Create Wireless Sensor Chip The Size Of Glitter*. Available at http://berkeley.edu/news/media/releases/2003/06/04_sensor.shtml, 2003.

59. Yan, L., Zhang, Y., Yang, L. T., and Ning, H. *The Internet of Things: From RFID to the Next-Generation Pervasive Networked Systems*. CRC Press, Boca Raton, FL, 2008.

60. Yan, L.-S., Pan, W., Luo, B., Liu, J.-T., and Xu, M.-F. Communication protocol based on optical low-energy-adaptive-clustering-hierarchy (o-leach) for hybrid optical wireless sensor networks. In *Asia Communications and Photonics Conference and Exhibition*, volume 7633, pp. 1–6. IEEE, 2009.

61. Yu, M.-F., Lourie, O., Dyer, M. J., Moloni, K., Kelly, T. F., and Ruoff, R. S. Strength and breaking mechanism of multiwalled carbon nanotubes under tensile load. *Science*, 287(5453):637–640, 2000.

62. Ip version 6 addressing architecture. Available at https://tools.ietf.org/html/rfc4291, 2006.

63. Connections counter: The internet of everything in motion. Available at http://newsroom .cisco.com/feature-content?type=webcontent articleId=1208342.

64. Internet world stats: Usage and population statistics. Available at http://www .internetworldstats.com/emarketing.htm.

65. Libelium. Available at http://www.libelium.com/.

66. Xively. Available at https://xively.com/.

5 Laser Ultrasonic Imaging for Damage Visualization and Damage Accumulation Evaluation

Jinhao Qiu, Chao Zhang, Jinling Zhao, and Hongli Ji

CONTENTS

INTRODUCTION

Developing advanced nondestructive evaluation (NDE) to detect possible damages is of great importance for various engineering structures to keep its safety and reliability. Several detecting techniques including ultrasonic C-scan,[1] ultrasonic Lamb wave technique,[2] X-ray inspection,[3] thermography,[4] eddy-current detection,[5] and so on have been proposed and developed. Among them, the ultrasonic-wave-based techniques show its great potential in NDE industry due to their effectiveness of estimating the location, type, and size of damages.[6-10]

The ultrasonic C-scan approach was proposed in early stage to detect the internal damages.[11] One or more probes are usually used to generate and sense bulk ultrasonic wave, which propagates in the structure along the thickness direction as shown in Figure 5.1a. The collected data of transmitted and reflected waves are processed in time domain or frequency domain to detect the internal damages in structures. However, the detection result, which depends on the experience and skill of inspectors, suffers from the complex interference of multiple reflected and diffracted waves in thin plate-like structure. In addition, direct-contact scan and indirect-contact scan through the media (such as water and oil) are costly and labor intensive especially in large-scale inspection applications.

Apart from ultrasonic C-scan approach, the Lamb wave-based NDE techniques that have been widely used in plate-like structures also show appealing features in detecting damages because Lamb wave has the advantages of high sensitivity to various damage types and the capability of long transmission distance.[12,13] And in general, Lamb wave can be generated and received by piezoelectric sensor pairs[14] that are distributed over the inspection region as shown in Figure 5.1b. By analyzing

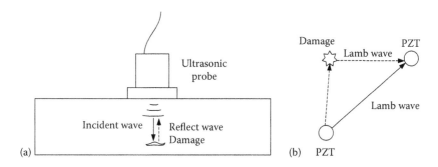

FIGURE 5.1   Different inspection principles using: (a) bulk wave and (b) Lamb wave.

the changes of Lamb wave propagation characteristics or using the time-of-flight information of the scattered waves, the damage position can be detected. However, contact-type transducers that are usually surface mounted or embedded on the test structures lead to some limitation of this method in wide applications that: (1) the excitation and sensing positions are fixed at several discrete points. To improve the inspection quality and extend the inspection region, a large amount of piezoelectric sensors are required; (2) transducer installation and cabling can be costly and labor intensive especially in the large-scale inspection applications; (3) many contact transducers are not applicable under harsh environments such as high temperature and radioactive conditions; and (4) the added weight of the transducers and the cables is not cost-effective in practical application.

To address these shortcomings in the ultrasonic C-scan and Lamb wave-based NDE technique, ultrasonic propagation visualization technique based on a laser ultrasonic system provides new paradigms, which have been extensively studied and will be widely used in practical applications.[15–17] According to the different methods of exciting and sensing ultrasonic wave, ultrasonic propagation visualization can be achieved in four different schemes.[18] The ultrasonic excitation can be performed using pulse laser or fixed piezoelectric transducer and the ultrasonic sensing can be achieved by the laser interferometry or fixed piezoelectric sensor. When the pulse laser and the laser interferometry are combined to construct the inspection system, noncontact laser ultrasonic technique[18,19] can be achieved as shown in Figure 5.2. The advantages of this technique can be concluded as following: (1) ultrasonic wavefield can be constructed through a series of two-dimensional images that can provide more information about the damage; (2) the damage diagnosis is a base-line free

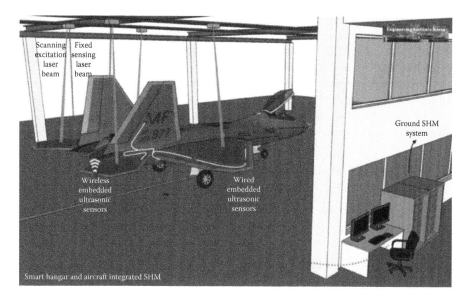

FIGURE 5.2 Future damage detection using laser ultrasonic technique for the next-generation composite aircraft. (Data from J. R. Lee et al., Repeat scanning technology for laser ultrasonic propagation imaging. *Measurement Science and Technology*, 2013, 24:085201.)

technique that is less venerable to the changing environmental and operational conditions; and (3) the complete noncontact inspection strategy is more competitive for onsite use and more practical for large-scale inspection.

The basic idea of the ultrasonic propagation visualization is to scan one of the excitation and sensing points and fixed the other. By processing the collected data, the ultrasonic propagation, which displays the ultrasonic wave travels from the fixed position to the inspection region, can be constructed. Then, the damage can be directly identified by observing the anomalous wave in the snapshots of the wave propagation at different time points.[20] To highlight the damage-induced anomalous waves and evaluate the damage by a digital intensity image, many signal processing methods and imaging algorithms have been proposed to evaluate the damage.[21–23] In this chapter, the damage detection approach based on the anomalous incident wave (AIW) energy map using laser ultrasonic technique is introduced. By eliminating the reflected wave from the original wavefield, the incident wave has been extracted to estimate the damage. To quantify the energy changes caused by the damage, the traveling wave energy has been removed through adjacent incident waves subtraction. Finally, a digital image of the AIW energy is constructed to show the size and shape of the damage. The experiments in both metal and composite structure demonstrate the proposed approach.

Furthermore, fatigue damage caused by stress cycling can compromise the mechanical integrity and safety of a composite structure, and therefore it is essential to detect the damage accumulation via NDE methods. Ultrasonic techniques commonly using ultrasound propagation, reflection, and transmission characteristics are directly related to the properties of materials, and thus are widely utilized for NDE of structure damages. Conventional ultrasonic transducers (e.g., piezoelectric disks) tend to have interfacial debonding with the structures during the fatigue process, while the above-mentioned laser ultrasonic technique enables noncontact and in-situ NDE of the accumulative damage of composites. In addition, the laser-generation-based method provides high spatial resolution and thus can be used to visualize the wave propagation and to measure phase velocity precisely. The latter half of this chapter will focus on measuring some characteristic parameters (including phase velocity) of Lamb waves in composites during the fatigue process with the introduced laser ultrasonic techniques and correlating these parameters to fatigue damages.

LASER ULTRASONIC IMAGING SYSTEM

Laser ultrasonic imaging system can be divided into three parts: (1) ultrasonic generation; (2) ultrasonic reception; and (3) the damage evaluation and visualization process. For ultrasonic generation, both the pulse laser and the piezoelectric transducer can be used to generate Lamb waves in the plate-like structure. The ultrasonic waves can be generated using the Nd-YAG laser through thermo-elastic effect, which is little affected by the surface irregularity and the incident angle of the excitation laser beam. However, this pulse ultrasonic source is a fixed broadband wave source, which posed certain limitations. The surface-mounted or embedded piezoelectric transducer for ultrasonic generation through a contact way could be an alternative. The major advantage is that any arbitrary waveform such as a narrowband toneburst signal can be generated, which is less affected by the dispersive effect and has a higher energy level compared with the pulse laser.

The ultrasonic reception can be realized through both contact and noncontact ways. By using the surface-mounted or embedded piezoelectric sensor, a high SNR (signal-to-noise ratio) ultrasonic signal can be obtained experimentally. However, it is limited under harsh environments and curved structures. The noncontact ultrasonic reception can be achieved by the laser interferometry. For example, the one-dimensional laser Doppler vibrometer (LDV, OFV-505/5000) measures the out-of-plane displacement up to 25 MHz based on the Doppler frequency-shift effect of laser. Even though the SNR of the ultrasonic signal measured by the LDV is lower than that of the piezoelectric sensor, the ultrasonic wave can still be constructed by averaging the results of the multitimes measurement.

The damage evaluation and ultrasonic propagation visualization can be constructed using four different inspection strategies as shown in Figure 5.3.

1. *Fixed-point PZT sensor and scanning pulse laser*—The system of the first scheme in Figure 5.3 consists of a Q-switched diode-pumped solid state laser, a galvanometric laser mirror scanner, a focused acoustic emission (AE) sensor, a programmable filter, a digitizer, and a computer for hardware control and image processing. A schematic diagram of the entire experimental system configuration is shown in Figure 5.4. The AE sensor providing ultrasonic reception was fixed and focused at a particular sensing region on the specimen, whereas the laser beam acting as ultrasonic generation

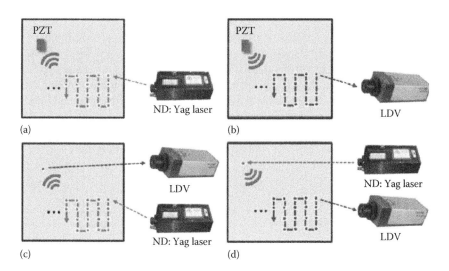

FIGURE 5.3   Four different laser scanning schemes for wavefield visualization: (a) Fixed-point PZT sensor and scanning pulse laser; (b) Fixed-point PZT excitation and scanning LDV sensing; (c) Fixed-point pulse laser and scanning LDV sensing; and (d) Fixed-point LDV sensing and scanning pulse laser. (Data from Y. K. An et al., Complete noncontact laser ultrasonic imaging for automated crack visualization in a plate. *Smart Materials and Structures*, 2013, 22:025022.)

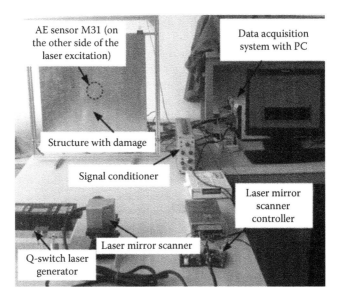

FIGURE 5.4   Experimental setup.

was directed to the targeted specimen with the help of a laser mirror scanner for remote and controllable scanning of the targeted area. The filters, digitizer, and computer hardware are used for data acquisition and image processing for damage evaluation and visualization. All devices are packed in a controller.

During the data acquisition process, the different ultrasonic responses between movable ultrasonic source point A and fixed sensor point B can be obtained experimentally as Case 1 in Figure 5.5. Based on the linear reciprocal theorem,[24,25] the wave propagation from point A to B can be directly converted into the propagation from point B to A. Thus, the obtained data in Case 1 can represent the wave propagation from the fixed laser point B to an inspection region, which is distributed with a series of AE sensors as shown in Case 2. Then, the responses of each scanning point at the

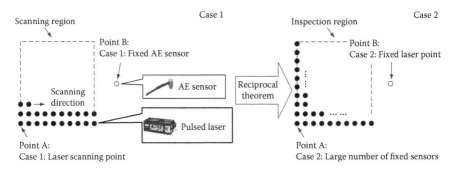

FIGURE 5.5   Schematic diagram of laser ultrasonic scanning technique.

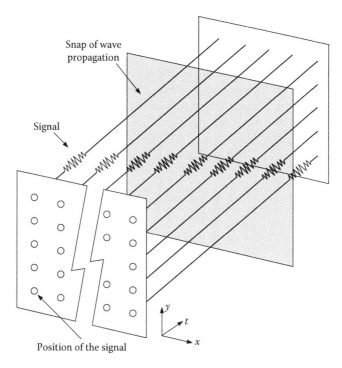

FIGURE 5.6   Data structure in wave propagation visualization.

same time t are plotted on an intensity snapshot to represent the wavefield at time point t. Finally, the snapshots are continually displayed in a time series and show the wave propagation in the inspection region as shown in Figure 5.6.[26]

2. *Fixed-point PZT excitation and scanning LDV sensing*—Different from strategy 1, the data acquisition from this inspection strategy can be directly regarded as the Case 2 in Figure 5.5. The waveform generated by the piezo-electric transducer can be controlled by the input signal, which is more flexible in inspection application. When the sensing laser beam is perpendicular to the polished target surface, the SNR of the ultrasonic signals is high due to the majority of the incident laser beam that reflected straight back to the receiver. However, the SNR of the ultrasonic decreases as the incident angle of the laser beam increases. Thus, multiple ultrasonic signals often need to be measured at a single sensing point and averaged to improve the SNR.[19]

3. *LDV sensing and pulse laser excitation*—When fixing the LDV sensing point and scanning the pulse laser in the inspection region, the ultrasonic propagation can be constructed sharing the same principle with strategy 1. On the contrary, the situation that is to scan the LDV sensing point and fix the pulse laser is the same with strategy 2. Both of them can achieve a complete noncontact inspection.

LASER ULTRASONIC IMAGING FOR DAMAGE VISUALIZATION

UNDERSTANDING OF THE WAVE PROPAGATION IN THE DAMAGE REGION

Assumed that the ultrasonic wave passes through a damage as shown in Figure 5.7. The incident wave $w_i(x, t)$ propagates along the positive x-direction and the reflected wave $w_r(x, t)$ propagates along the negative x-direction. Generally, the harmonic wave can be given by the following equations:[18]

$$\begin{cases} w_i(x,t) = A\cos(\omega t - kx) & x \leq x_0 - \dfrac{l_0}{2} \\[3mm] w_i(x,t) = B\cos(\omega t - kx + \alpha) & x > x_0 + \dfrac{l_0}{2} \end{cases} \tag{5.1}$$

$$\begin{cases} w_r(x,t) = C\cos(\omega t + kx + \beta) & x \leq x_0 - \dfrac{l_0}{2} \\[3mm] w_r(x,t) = 0 & x > x_0 + \dfrac{l_0}{2} \end{cases} \tag{5.2}$$

where:
    ω is the angular frequency
    k is the wavenumber
    x_0 and l_0 represent the position and width of the slit
    α and β are the phase shifts caused by the damage
    A represents the incident wave amplitude while B and C represent the amplitudes
        of residual incident wave by transmission and reflected wave

It should be mentioned that Equations 5.1 and 5.2 have already ignored the amplitude attenuation caused by the wave propagating and the actual ultrasonic waves would be more complex. However, the hypothesis in pure harmonic waveform equations can be used for simplicity without the loss of generality. Because of the complex damage boundary, the ultrasonic waves in the damage area cannot be expressed simply. The amplitude and waveform in the damage area are different from the waves in the health area.

From Equation 5.1, the incident wavefield $w_i(x, t)$ is not continuous in the region $[x_0 - l_0/2, x_0 + l_0/2]$. Considering regions 1 and 3 in Figure 5.7, the difference of the

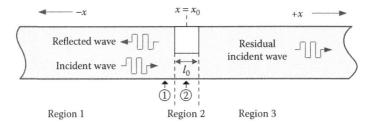

FIGURE 5.7   Schematic diagram of the ultrasonic wave interact with a damage.

incident wave between two adjacent points is the phrase lag caused by the different traveling distance. This phrase difference can be compensated by adding a time-lag Δt to adjacent incident wave

$$\Delta t = \Delta x / v \tag{5.3}$$

where Δx is the spacing between two adjacent point and v is the wave velocity. However, the incident wave in region 2 has large energy attenuation and complex phrase variation.

DAMAGE DETECTION ALGORITHM USING AIW ENERGY MAP

To represent the discontinuous characteristic of the incident wavefield in the damage area, the AIW is defined as the difference between two incident wave signals obtained at the adjacent points. Taking two incident wave signals $w_i(x, i)$ and $w_i(x + \Delta x, i)$ as an example, the AIW at the position x can be calculated by

$$\Delta w_i\left(x,i\right) = w_i\left(x,i\right) - w_i\left(x+\Delta x, i + d_{r\,\max}\right) \tag{5.4}$$

where Δx is the minimum spatial resolution of the laser scanning process and the time-lag $d_{r\,\max}$ can be calculated through the cross-correlation $r(d)$.

$$r\left(d\right) = \sum_{i=0}^{N} w_i\left(x,i\right) w_i\left(x+\Delta x, i + d\right) \tag{5.5}$$

$d_{r\,\max}$ represents the time that the incident wave travels from x to $x + \Delta x$ that makes $r(d_{r\,\max})$ reach the maximum. Considering that the incident wave propagates in the continuous medium, two incident waves at the adjacent points are similar with each other. By using Equation 5.4, the traveling incident wave in regions 1 and 3 as shown in Figure 5.7 can be eliminated largely. As a result, the energy changes in region 2 caused by the damage will be highlighted and the discontinuous characteristic of Equation 5.1 is extracted. To evaluate the size and shape of the damage by a digital image, the AIW energy E_{AIW} can be calculated by the sum of squared AIW.[27]

$$E_{AIW}\left(x\right) = \sum_{i=1}^{N} \Delta w_i^2\left(x,i\right) \tag{5.6}$$

Compared with calculating anomalous wave directly from the original wavefield,[26,28] the AIW extracts the wave propagation changes caused by the damage as well. However, due to the incident wave is much stronger than reflected wave, $d_{r\,\max}$ only denotes the time delay of the incident waves. Because the reflected wave and incident wave have the opposite time delays between the adjacent points, the reflected wave is also enlarged by this process. As the reflected wave propagating toward the health area in region 1, the energy of the anomalous wave is also distributed near the damage area, which reduces the resolution of damage image. After filtering the reflected wave, the AIW energy eliminates the reflected wave energy in health area that makes the shape of the damage much clearer.

As shown in Figure 5.7, a portion of incident wave is reflected with the propagation direction reversed. Because the damage is nonpenetrated, the other portion of incident wave passes through the damage and propagates in the original direction. Thus, the one-dimensional wavefield $w(x, t)$ near the damage area can be given by adding incident wavefield $w_i(x, t)$ and reflected wavefield $w_r(x, t)$.

$$w(x,t) = w_i(x,t) + w_r(x,t) \tag{5.7}$$

To construct the AIW map, the incident wavefield w_i should be extracted from the original wavefield w. From Equations 5.1 and 5.2, the major difference between incident and reflected waves is the sign in front of the wavenumber k that indicates the wave propagation direction is along positive or negative x-direction. To obtain the incident wave, two-dimensional Fourier transform (2D-FT) converts the wavefield from space–time (x–t) domain into wavenumber–frequency (k–ω) domain to separate the waves with different propagation directions as[29,30]

$$W(k,\omega) = \int_{-\infty}^{+\infty} \int_{-\infty}^{+\infty} w(x,t)e^{-j(kx+\omega t)}dxdt \tag{5.8}$$

where $W(k, \omega)$ is the wavefield in wavenumber–frequency domain and j is the imaginary unit. The wavefield is divided into two parts: (1) the incident wave portion that is in the quadrant with $k\omega < 0$ and (2) the reflected wave portion that is in the quadrant with $k\omega > 0$. Two-dimensional inverse Fourier transform can then be applied to recover the time domain signals of these two components as follows:

$$w_{i(r)}(x,t) = \frac{1}{2\pi} \int_{-\infty}^{+\infty} \int_{-\infty}^{+\infty} W(k,\omega)\Phi_{I(R)}e^{j(kx+\omega t)}dkd\omega \tag{5.9}$$

where $\Phi_{I(R)}$ is a window function defined as follows, used to separate the incident and reflected waves.

$$\Phi_I = \begin{cases} 0 & k\omega \geq 0 \\ 1 & k\omega < 0 \end{cases} \tag{5.10}$$

$$\Phi_R = \begin{cases} 0 & k\omega \geq 0 \\ 1 & k\omega < 0 \end{cases} \tag{5.11}$$

The principle is explained using 1D wave-field based 2D-FT as above. Similarly, the above 1D wave decomposition method can be extended to 2D case by means of Three-dimensional Fourier transform (3D-FT) and Three-dimensional inverse Fourier transform (3D-IFT), for obtaining the wavefield by a laser ultrasonic scanning technique. Because the sensor response is collected by digital acquisition, the wave-field takes the form of discrete data. Thus, the Fourier transform can be realized by fast Fourier transform (FFT).

Experimental Validation

To examine the feasibility of the proposed laser ultrasonic scanning system and image processing technique, the scheme using fixed-point AE sensor and scanning pulse laser as shown in Figure 5.4 is employed for ultrasonic imaging and damage evaluation. Experiments of both aluminum structure and composite structure were carried out.

Impact Damage Visualization in Aluminum Structure

An aluminum plate with a nonpenetrating slit is used to experimentally validate the proposed method of damage imaging as shown in Figure 5.8. The slit with the length of 20 mm and width of 2 mm is located 100 mm away from the AE sensor. The depth of the artificial slit is 2 mm, while the thickness of the plate is 3 mm. The dimensions of inspection region are 100×100 mm². To imitate the situation of internal damage, the laser scanning points are placed on the opposite side of the slit. According to that the spatial interval is 1 mm, 10,201 points are scanned that costs 8.5 minutes at the max repeat frequency 20 Hz of the laser generator. The sampling period is set as 60 μs to ensure all the waves can pass through the inspection region.

The wavefields at different times are shown in Figure 5.9. Both symmetrical (S_0) and asymmetrical (A_0) modes of ultrasonic waves can be distinguished by the different propagation speeds in Figure 5.9a. The amplitude of the A_0 mode wave, which is much larger than that of the S_0 mode wave, plays a decisive role in damage evaluation. After 40 μs as shown in Figure 5.9b, the A_0 mode wave passes through the slit. By observing the wave scattering caused by the damage, the position of the damage can be faintly identified. However, the shape and size of damage cannot be identified due to the complex wavefield near the damage. According to the different components of the wavefield, four regions can be distinguished in Figure 5.9b. Region I is far away from the damage and the wave propagates from the source. Region II includes both incident and reflected waves. Because of the interference phenomenon caused by the waves with different directions, there are anomalous fringes in region II. Region III

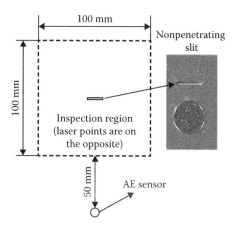

FIGURE 5.8 Experimental schematic.

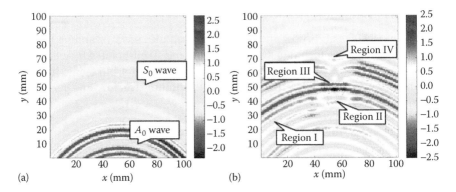

FIGURE 5.9 Experimental wavefields at different times: (a) $t = 25\ \mu s$ and (b) $t = 40\ \mu s$.

represents the actual damage area where the waves change largely. In region IV, the incident wave passes through the nonpenetrating slit partly and the diffraction wave occurs at the edge of the slit. Because of the interference effect, a few anomalous fringes can be found in region IV.

3D-FFT is used to analyze the wavefield as shown in Figure 5.10. The wavefield in the space–time domain is converted into the wavenumber–frequency domain. Figure 5.10a is the cross section of the wavefield at $f = 150$ kHz. Because the amplitude of the incident wave is much larger than reflected wave, the main energy locates in the region $k_y < 0$. It should be mentioned that the inspection region does not include the whole wavefield and the main propagation direction is along the axis y. Thus, the wave energy only focuses on part of the energy circle. Because of the direction of the incident wave is along positive y-direction, the rectangular window Φ_I can be written as

$$\Phi_I = \begin{cases} 0 & k_y \omega \geq 0 \\ 1 & k_y \omega < 0 \end{cases} \tag{5.12}$$

The cross section of the incident wavefield and the reflected wavefield are shown in Figure 5.10b and c.

The separated wavefields in space–time domain with different propagation directions are obtained by 3D-inverse fast Fourier transform. Both incident wavefield and reflected wavefield at the same time with the wavefield in Figure 5.9b are shown in Figure 5.11a and b. Compared with Figure 5.9b, the incident wavefield is much simpler than the original wavefield. The changes in wave propagation exist only after the waves pass through the damage area. Because of the elimination of the interference caused by the reflected wave, Figure 5.11a shows the damage position much clearer. However, the diffracted wave and transmitted wave also propagate behind the damage area.

Figure 5.12 shows the results obtained by wave energy map calculation. As the wave propagation snap in Figure 5.9b, the energy distribution of the original wavefield can also be divided into four regions in Figure 5.12a. Region I, which is less affected by the damage, represents the energy attenuation while the waves propagate. The energy in region II is oscillating obviously due to the interference caused by incident and reflected waves. Region IV indicates the area of energy reduction

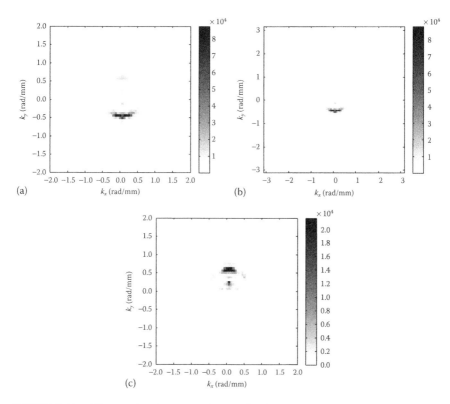

FIGURE 5.10 Wavenumber-frequency domain filtering using 3D-FFT: (a) original wavefield; (b) incident wavefield; and (c) reflected wavefield at $f = 150$ kHz.

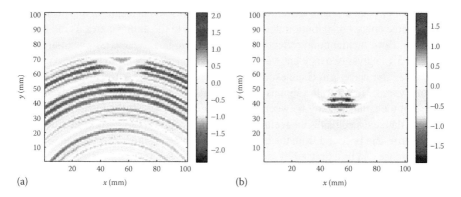

FIGURE 5.11 Experimental wave images at $t = 40$ μs: (a) incident wave and (b) reflected wave.

after the waves pass through the damage. Only the energy changes in region III show the shape of the damage that can hardly be distinguished from other regions. Figure 5.12b shows the incident wave energy map, while the reflected wave energy map is illustrated in Figure 5.12c. The damage position can be identified where the energy decreases in Figure 5.12b or the energy generates in Figure 5.12c. However,

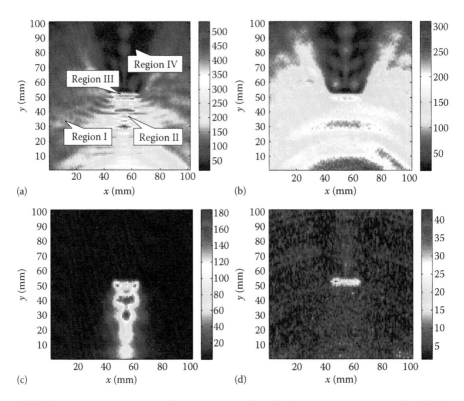

FIGURE 5.12   Damage images for a nonpenetrating slit using wave energy map: (a) original wavefield; (b) incident wavefield; (c) reflected wavefield; and (d) anomalous incident wavefield.

neither of them can evaluate the damage shape. The reason that results in this phenomenon is that the energy distribution does not be limited in the damage area as the waves propagate. Both incident and reflected waves travel in healthy and damaged areas.

Figure 5.13a gives two incident waves at the adjacent points with an interval Δy in the health area. The waveform changes as the wave propagates due to the dispersive effect in Lamb waves. However, the signals in position (90,62) and (90,63) are highly similar because of the small spacing between them. Figure 5.13b shows the incident waves in the actual damage area that is the rectangle [44, 64] × [52, 54]. The significant difference in waveforms can be found from the incident waves at the positions (46,52) and (46,53). This difference can be extracted by extending Equation 5.4 to three dimensional.

$$\Delta w_i\left(x, y, i\right) = w_i\left(x, y, i\right) - w_i\left(x, y + \Delta y, i + d_{r\max}\right) \tag{5.13}$$

To obtain a suitable time-lag $d_{r\max}$, all the cross-correlation functions in the inspection region are calculated. The time-lags at the different points that make the cross-correlation functions reach the maximum are obtained. However, the time-lag varies with the position changes. Take one hundred positions in the health area (with $x = 90$) and the damage area (with $x = 46$), respectively, as examples. It can be found that the time-lags in the health area trend to a constant 0.4 μs, while the time-lags are

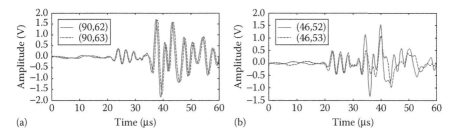

FIGURE 5.13   Two incident waves at the adjacent points in: (a) health area and (b) damage area.

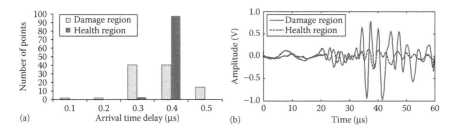

FIGURE 5.14   Anomalous incident waves according to Equation 5.13: (a) $d_{r\,max}$ at the different points in both health area and damage area and (b) AIW in both health area and damage area.

distributed from 0.1 to 0.5 μs in the damage area. In Figure 5.14a, the resolution of the time-lags is 0.1 μs that is the sampling interval. To eliminate the influence on the time-lag calculation caused by the damage, $d_{r\,max}$ is chosen by a voting mechanism. Each position votes for the time-lag according to its cross-correlation function and the time-lag that captures a majority of the votes will be used to match the arrival time. Even though the uniform time-lag cannot represent the time-lag at any point, the error is less than one sampling interval when the inspection region is far away from the AE sensor. Figure 5.14b shows the AIW results in damage position (46,52) and health position (90,62). After matching the arrival time by a time-lag 0.4 μs, the AIW is much larger in damage position. As shown in Figure 5.12d, the AIW energy map shows the damage shape clearly.

Impact Damage Visualization in Composite Structure

A symmetric carbon-fiber-reinforced polymer (CFRP) laminated plate with a stacking configuration of $[45°/–45°/0°/90°]_s$ is used to validate the damage evaluation method.[27] The dimensions of the CFRP plate are 480×480 mm^2 and the thickness is 1 mm. To make an artificial damage, a drop-weight impact test is performed according to Test Method D7136.[31] Damage is imparted through out-of-plane, concentrated impact (perpendicular to the plane of the CFRP plate) using a drop weight with a hemispherical striker tip as shown in Figure 5.15. The impact energy E_{impact} can be calculated by

$$E_{impact} = mgh \tag{5.14}$$

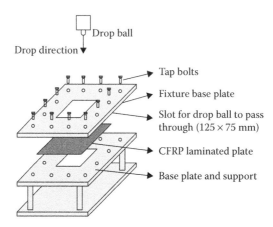

FIGURE 5.15   Schematic diagram of the drop-weight impact test for a CFRP plate.

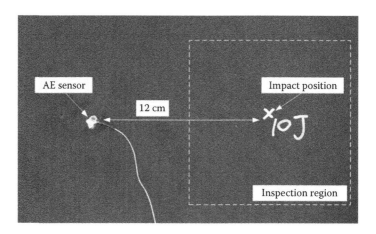

FIGURE 5.16   Picture of the CFRP plate with a damage caused by a 10 J impact.

where m is the mass of the drop weight, g is the acceleration due to gravity and h is the drop-height. In this chapter, the impact energy is 10 J and the impact damage is difficult to be detected visually. However, the delamination, split, and crack exist inside the structure.

As shown in Figure 5.16, the impacted side of the CFRP laminated plate is scanned by the movable laser point and the inspection region is 100×100 mm^2. The AE sensor is placed on the opposite side of the CFRP plate with a distance 12 cm away from the center of inspection region. The spatial interval of the laser scan is 1 mm and the signal is filtered with the band-pass frequency from 100 to 200 kHz.

Two wavefields at the different time points are shown in Figure 5.17. Compared with the ultrasonic wave in the metal structure, the ultrasonic wave in the CFRP plate has low SNR due to the large material damp in the composite structure and the wave velocity is also lower. To ensure both of the S_0 and A_0 waves can pass

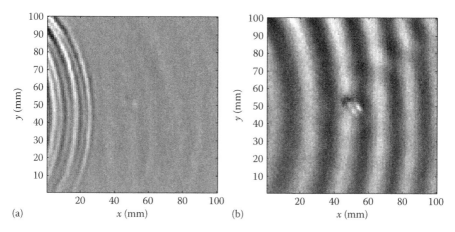

(a)   (b)

FIGURE 5.17   Experimental wavefields at different times: (a) $t = 60$ μs and (b) $t = 160$ μs.

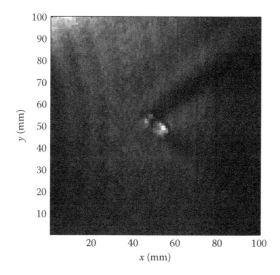

FIGURE 5.18   Damage evaluation using the original wave energy map.

through the inspection region, the sampling period is set to 200 μs. Same with the phenomenon in the metal plate with a slit, the amplitude changes around the damage region that can detect the position of the damage.

The original wave energy map is constructed as shown in Figure 5.18. The wave energy decreases as the wave propagating. The anomalous wave energy region shows the impact damage faintly. To evaluate the inspection result, Figure 5.19 shows the damage image obtained by a conventional immersion ultrasonic C-scan method. Compared with Figure 5.19, the large amount of the wave energy in the 'health' area reduces the resolution of damage image in Figure 5.18.

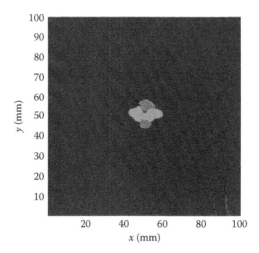

FIGURE 5.19   Damage evaluation using immersion ultrasonic C-scan method.

After filtering the reflected wave, the AIW energy map can be obtained according to Equation 5.6. Compared with calculating anomalous wave directly from the original wavefield, AIW not only extracts the wave propagation changes caused by the impact damage, but also removes the negative influence from the reflected waves in matching the arrival times of two adjacent waves. Figure 5.20 shows the damage evaluation result using AIW energy map.

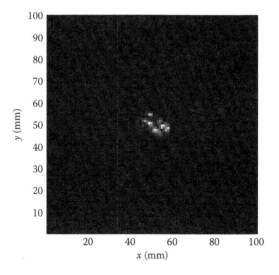

FIGURE 5.20   Damage evaluation using the AIW energy map.

LASER ULTRASONIC METHOD FOR RECONSTRUCTING
MATERIAL PROPERTIES OF COMPOSITES

Generally, the key point of NDE for fatigue damages in composites is to find a specific parameter that is sensitive to the damage accumulation during the fatigue process. Many researchers have investigated that the stiffness of composites will experience a decrease during the fatigue process.[32–34] Seale et al. found that Lamb wave speed and Young's modulus of the testing composite specimen show reduction during the fatigue process, as shown in Figure 5.21, and correlate the reduction to the crack density.[35]

Although Figure 5.21 shows decent trend of the wave velocity and Young's modulus during the fatigue process, the experimental implementation is quite questionable.

1. Usually, strain gauges can be used to measure the elastic modulus and piezoelectric disks to estimate wave speed of Lamb waves in the specimens. The received Lamb waves travels from the actuator to the receiver, while the strain gauges can only measure the strain in a local area. This indicates that the measured velocities and elastic modulus in Figure 5.18 are not of the same position in the specimens.
2. Another issue is that the size of the transducer is 1.27 cm, which is almost of the same level as the width of the specimen (3.8 cm). The spatial resolution is not high enough to measure phase velocity of S_0 mode (whose wavelength is about 10 mm in this reference).
3. Meanwhile, the authors neglect the boundary reflection and A_0 mode in the received time domain signals.

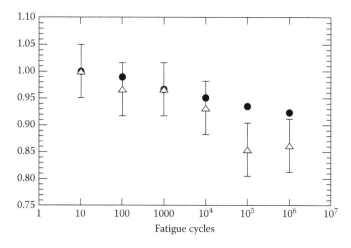

FIGURE 5.21 Plot of normalized modulus (●) and normalized velocity squared (△) versus fatigue cycles for a fatigued sample.

Introduction of Laser-Generation-Based Methodologies for Damage Accumulation Evaluation

Speed velocity of lower symmetric Lamb wave mode and the elastic modulus are closely related to the fatigue damage of composite specimens.[35] A laser ultrasonic technique based method to evaluate fatigue damage accumulation for composites and overcame the mentioned shortcomings[35] has been introduced.[36] The laser ultrasonic system takes advantages of the noncontact laser beam and can generate ultrasonic guided waves in structures at different positions over a large area.

The authors[36] focused on measuring three parameters (amplitudes of Lamb waves, phase velocity, and elastic modulus) in composites during the fatigue process and correlating these variables to fatigue damages. The three parameters were obtained by the following methods. An AE transducer of flat frequency response and high sensitivity was fixed on the structure to detect the wave signals. The amplitude of Lamb waves when the laser pulse pointed at a specific position was recorded during the fatigue process. The phase velocities of a certain Lamb wave mode can be measured via the position–time diagram.[37] Because the phase velocity depends on the elastic modulus of the material, the stiffness coefficients were reconstructed with genetic algorithms (GAs) by minimizing the difference between the measured and calculated phase velocities.

Consider the questions in the study,[35] the laser ultrasonic technique is more suitable for fatigue accumulated damage evaluation in composites due to the following conditions.

1. The phase velocities are measured based on the position–time diagram along the laser beam scanning line. The reconstruction of the material properties is by inverting the measured phase velocities, which implies that the measure elastic modulus and the phase velocities are of the same region.
2. The laser beam is of a diameter of 2 mm, which enables the high spatial resolution of the position–time diagram for phase velocity measurement.
3. Rather than neglect the trailing signal, the authors provided several hints to extract the phase velocities precisely in the small sized specimens.

This section will explain the details of reconstructing the nine stiffness coefficients in composites, while the fatigue damage accumulation evaluation based on this section will be presented in section "Reconstruction Method."

The reconstruction for the stiffness coefficients follows the procedure shown in Figure 5.22. The GA-based reconstruction for the stiffness coefficients in composites requires the knowledge of measured phase velocities and the calculated velocities. The calculation of the phase velocity dispersion curves when given stiffness coefficients will be illustrated in section "Calculation of Dispersion Curves," while the details of measuring phase velocities using the laser ultrasonic system are shown in section "Measurement of Phase Velocities." Because of the anisotropy characteristics of composite materials, the phase velocities vary with the wave propagation directions. Sensitivity analysis about the region where the phase velocities are most sensitive to the change of the stiffness coefficients is essential to simplify the reconstruction procedure and will be conducted in section "Sensitivity Analysis." Then the GA-based

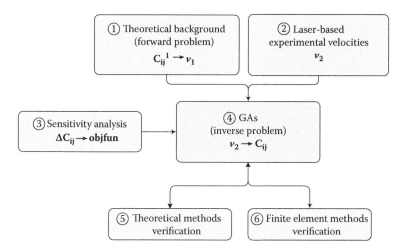

FIGURE 5.22 Flowchart of the process for reconstructing stiffness coefficients in composites.

reconstruction method is detailed. Because the experimental composite plates are of unknown materials, the reconstructed material properties cannot be verified directly (section "Reconstruction Method"). Two indirect verification methods will be discussed in section "Results and Discussion" by comparing the phase velocities and the wavefront between the experimental ones and the reconstructed ones.

CALCULATION OF DISPERSION CURVES

Dispersion characteristic means that wave velocities of a specific mode in a waveguide depend on the frequency-thickness product and that new modes will emerge when frequency rises. Wave velocity (including phase velocities and group velocities) dispersion curves are the basic requirement for ultrasound-based NDE for damages in waveguides.

Composite laminates are a complex waveguide due to their anisotropy and multilayer properties. Most used methodologies for modeling Lamb waves in composite laminates contain 3D elasticity theory with transfer/global matrix technique,[38–44] equivalent single layer theory (ESLT),[45–50] and semi-analytical finite element,[51–54] A 3-node (each with three displacement degree of freedom [DoF]) one-dimensional element was developed[54] to calculate dispersion curves and wave structures in composites efficiently. Those who are interest in semi-analytical finite element method (FEM) could refer to the study[54] for details. This section will briefly introduce 3D elasticity theory and ESLTs to model Lamb waves in composites.

3D Elasticity Theory with Global Matrix Techniques

Lamb Waves in a Composite Lamina

The displacement fields of a local orthogonal Cartesian coordinate system are expressed in Equation 5.15. In this coordinate system shown in Figure 5.23, x_1-axis is the propagation direction of c_p, and x_3-axis is the laminate stacking direction.

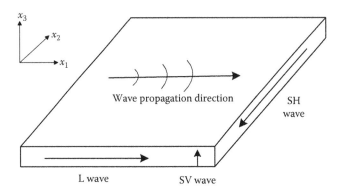

FIGURE 5.23 Longitudinal and shear wave modes in a local coordinate system (c_p direction).

First, the properties of each layer are transformed from the material orientation system to this local coordinate one, where k is the wavenumber along axis x_1, c is the magnitude of c_p, α is still an unknown parameter, and V, W are the relative displacement amplitudes.[41]

$$\left(u_1, u_2, u_3\right) = \left(1, V, W\right) U e^{ik\left(x_1 + \alpha x_3 - ct\right)} \tag{5.15}$$

Combining constitutive equations, strain–displacement equations with the equilibrium equations yields

$$K_{11}\left(\alpha\right)u_1 + K_{12}\left(\alpha\right)u_2 + K_{13}\left(\alpha\right)u_3 = 0$$

$$K_{12}\left(\alpha\right)u_1 + K_{22}\left(\alpha\right)u_2 + K_{23}\left(\alpha\right)u_3 = 0 \tag{5.16}$$

$$K_{13}\left(\alpha\right)u_1 + K_{23}\left(\alpha\right)u_2 + K_{33}\left(\alpha\right)u_3 = 0$$

where

$$K_{11}\left(\alpha\right) = C_{11} - \rho c^2 + C_{55}\alpha^2$$

$$K_{12}\left(\alpha\right) = C_{16} + C_{45}\alpha^2$$

$$K_{13}\left(\alpha\right) = \left(C_{13} + C_{55}\right)\alpha$$

$$K_{22}\left(\alpha\right) = C_{66} - \rho c^2 + C_{44}\alpha^2 \tag{5.17}$$

$$K_{23}\left(\alpha\right) = \left(C_{36} + C_{45}\right)\alpha$$

$$K_{33}\left(\alpha\right) = C_{55} - \rho c^2 + C_{33}\alpha^2$$

Nontrivial solutions of u demand $|K_{mn}(\alpha)| = 0$, which yields the sixth-order polynomial equation $\alpha^6 + L_1\alpha^4 + L_2\alpha^2 + L_3 = 0$. This equation has six roots labeled as α_q ($q = 1, 2, ..., 6$) where $\alpha_2 = -\alpha_1$, $\alpha_4 = -\alpha_3$, $\alpha_6 = -\alpha_5$. These three pairs of α stand for the x_3-direction wavenumber coefficient of longitudinal wave, vertical shear wave, and horizontal shear wave, respectively, as illustrated in Figure 5.23. Hence, amplitudes of the relative displacements are

$$V_q = \frac{K_{11}(\alpha_q)K_{23}(\alpha_q) - K_{13}(\alpha_q)K_{12}(\alpha_q)}{K_{13}(\alpha_q)K_{22}(\alpha_q) - K_{12}(\alpha_q)K_{23}(\alpha_q)}$$

$$W_q = \frac{K_{11}(\alpha_q)K_{23}(\alpha_q) - K_{13}(\alpha_q)K_{12}(\alpha_q)}{K_{12}(\alpha_q)K_{33}(\alpha_q) - K_{23}(\alpha_q)K_{13}(\alpha_q)}$$

(5.18)

The displacement and stress components in an anisotropic plate are represented as a liner summation of the six waves:

$$(u_1, u_2, u_3) = \sum_{q=1}^{6} (1, V_q, W_q) U_{1q} e^{ik(x_1 + \alpha_q x_3 - ct)}$$

(5.19)

$$(\sigma_{33}, \sigma_{13}, \sigma_{23}) = ik \sum_{q=1}^{6} (D_{1q}, D_{2q}, D_{3q}) U_{1q} e^{ik(x_1 + \alpha_q x_3 - ct)}$$

When the common factor $e^{ik(x_1 - ct)}$ is suppressed, the characteristic equation of lamina can be written in expanded matrix form:

$$\begin{Bmatrix} u_1 \\ u_2 \\ u_3 \\ \sigma_{33}/ik \\ \sigma_{13}/ik \\ \sigma_{23}/ik \end{Bmatrix} = \begin{bmatrix} E_1 & E_2 & E_3 & E_4 & E_5 & E_6 \\ V_1 E_1 & V_2 E_2 & V_3 E_3 & V_4 E_4 & V_5 E_5 & V_6 E_6 \\ W_1 E_1 & W_2 E_2 & W_3 E_3 & W_4 E_4 & W_5 E_5 & W_6 E_6 \\ D_{11} E_1 & D_{12} E_2 & D_{13} E_3 & D_{14} E_4 & D_{15} E_5 & D_{16} E_6 \\ D_{21} E_1 & D_{22} E_2 & D_{23} E_3 & D_{24} E_4 & D_{25} E_5 & D_{26} E_6 \\ D_{31} E_1 & D_{32} E_2 & D_{33} E_3 & D_{34} E_4 & D_{35} E_5 & D_{36} E_6 \end{bmatrix} \begin{Bmatrix} U_{11} \\ U_{12} \\ U_{13} \\ U_{14} \\ U_{15} \\ U_{16} \end{Bmatrix}$$

(5.20)

where

$$D_{1q} = C_{13} + C_{36} V_q + C_{33} \alpha_q W_q$$

$$D_{2q} = C_{55}(\alpha_q + W_q) + C_{45} \alpha_q V_q$$

$$D_{3q} = C_{45}(\alpha_q + W_q) + C_{44} \alpha_q V_q$$

$$E_q = e^{ik\alpha_q x_3}, q = 1, 2, \ldots, 6$$

(5.21)

Global Matrix Method

To obtain the global matrix of a laminate, the 6 × 6 matrix and the displacement vector on the right of the characteristic Equation 5.20 are simply expressed as [S] and {U}, respectively. As an example, the displacement and stress components are continuous at the interface between the top surface of layer 2 and the bottom surface of layer 3.

$$[S_{2t}]\{U_2\} = [S_{3b}]\{U_3\} \tag{5.22}$$

where the subscript '2t' means the top surface of layer 2, and '3b' means the bottom surface of layer 3. Then, taking a five-layer laminate for example, one can obtain the global function from Equation 5.22.

$$\begin{bmatrix} [S_{1t}] & [-S_{2b}] & & & \\ & [S_{2t}] & [-S_{3b}] & & \\ & & [S_{3t}] & [-S_{4b}] & \\ & & & [S_{4t}] & [-S_{5b}] \end{bmatrix} \begin{Bmatrix} \{U_1\} \\ \{U_2\} \\ \{U_3\} \\ \{U_4\} \\ \{U_5\} \end{Bmatrix} = \{0\} \tag{5.23}$$

Phase Velocity Curves

In the global matrix Equation 5.23, nontrivial solutions of vector demand the determinant of the global matrix, which is a function of the phase velocity (c_p), the frequency (f), and the direction of c_p (θ), to be zero. That leads to a characteristic equation, solution of which gives the c_p-f curves (dispersion curves) in a rectangular coordinate system for a given angle θ or the c_p-θ curves in a polar coordinate system for a given frequency f.

As a matter of fact, phase velocity is seldom used in experiments. However, it is the foundation to calculate other properties of Lamb waves theoretically.

Group Velocity Curves

c_g is an important vector in source location based on Lamb waves. The magnitude of c_g is required in experiments adopting time-of-flight-based techniques. On the other hand, the directions of c_g are necessary to determine the sources in complicated structures using the directivity of sensors.

In isotropic materials, propagation characteristic is independent on propagation direction. The magnitude of c_g can be computed easily using $c_g = d\omega/dk$. The direction of c_g is the same as that of c_p. However, c_p in anisotropic material varies with the direction of wave propagation. According to the study[55], the group velocity in composites can be written as

$$\begin{Bmatrix} c_{gx} \\ c_{gy} \end{Bmatrix} = \begin{bmatrix} \cos\theta & -\sin\theta \\ \sin\theta & \cos\theta \end{bmatrix} \begin{Bmatrix} \dfrac{\partial\omega}{\partial k} \\ \dfrac{\partial\omega}{k\partial\theta} \end{Bmatrix} \tag{5.24}$$

where θ is the direction of c_p, $k_x = k\cos\theta$, and $k_y = k\sin\theta$.

TABLE 5.1

Stiffness Coefficient of Composite Lamina Made of Graphite (65%) and Epoxy (35%)

C_{11} (GPa)	C_{12} (GPa)	C_{13} (GPa)	C_{22} (GPa)	C_{23} (GPa)	C_{33} (GPa)	C_{44} (GPa)	C_{55} (GPa)	C_{66} (GPa)
155.43	3.72	3.72	16.34	4.96	16.34	3.37	7.48	7.48

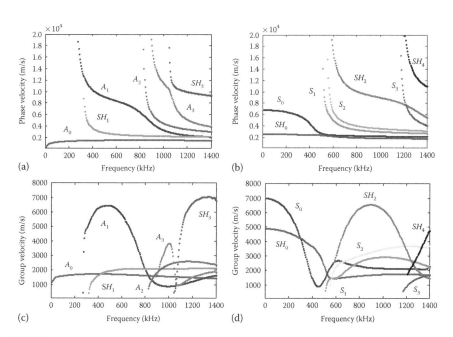

FIGURE 5.24 Dispersion curves of Lamb waves when c_p propagates along 30° in laminate $[+456/-456]_S$: (a) c_p curves of antisymmetric modes; (b) c_p curves of symmetric modes; (c) c_g curves of antisymmetric modes; and (d) c_g curves of symmetric modes.

Dispersion Curves Calculated by Global Matrix Method

The equations described in the previous sections were implemented using MATLAB®, because it can seamlessly combine symbolic and numeric computation. The material in this study is epoxy (35%)/graphite (65%) composite shown in Table 5.1, with a density of 1,700 kg/m³. A laminate $[+45_6/-45_6]_S$ is used and the thickness of each lamina is 0.125 mm. Dispersion curves of Lamb waves propagating along 30° in laminate $[+45_6/-45_6]_S$ are drawn in Figure 5.24. All Lamb waves have cut-off frequencies except the fundamental modes (A_0, S_0, and SH_0).

Equivalent Single Layer Theory

Although the above solutions can provide accurate results, the complexity of 3D elasticity theory depends heavily on the stacking sequence of the plies (along the plane of symmetry or not) and the number of plies. The ESLTs show great efficiency over the 3D elasticity theory especially for laminates of complicated stacking sequence and large number of plies. Also, the approximate plate theories are more efficient to solve large-scale problems, such as the reconstruction of the unknown stiffness coefficients in composites based on Lamb waves phase velocities. Such optimization problems based on GAs, which have large number of population and generations, can be very time-consuming using 3D elasticity theory.

To improve computational efficiency, many researchers have applied various ESLTs to estimate Lamb waves properties in composites. In ESLTs, the material properties of the constituent layers are combined to form a hypothetical single layer whose properties are equivalent to through-the-thickness integrated sum of its constituents. The classical plate theory based on the Kirchhoff hypothesis has been generally recognized to be accurate only if the wavelength is about ten times of the laminate thickness. Composite laminates often exhibit significant transverse shear deformation due to their low transverse shear stiffness. The classical plate theory that neglects the inter-laminar shear deformation is not valid over the high frequency range where wavelength is near the laminate thickness. The Mindlin plate theory[45] that includes transverse shear and rotary inertia effects provides accurate prediction of the lowest antisymmetric wave mode. Because constant transverse shear deformation is assumed in the first-order shear deformation theories (FSDTs)[46,47] (including the Mindlin plate theory), the higher modes of wave propagation cannot be described accurately using the FSDTs. More accurate theories such as higher-order shear deformation theories (HSDTs) assume quadratic,[48] cubic,[49] and higher variations of displacement through the entire thickness of the laminate.

Because both the FSDTs and some HSDTs neglect stresses free boundary condition on the top and bottom surfaces of the panel, a complicated scheme was developed to calculate the shear correction factors.[48] The correction factors are not unique for different lamination. They vary when the laminate properties (stacking sequence of the plies, the number of plies, and the ply properties) change. The usual procedure for determining the correction factors is to match specific cut-off frequencies from the approximate theories to the ones obtained from the exact theory. So the real CPU time of the conventional ESLTs should include the time one needs to calculate the exact solutions. To avoid computing the complex shear correction factors, Reddy et al.[56] adjusted the displacement field considering the vanishing of transverse shear stresses on the top and bottom of a general laminate. But when Lamb waves propagate in panels, the stresses free condition on the panel surfaces refers not only to the vanishing of transverse shear stresses but also to the vanishing of normal stress.

Inspired by Reddy, Zhao et al.[50] deduced two new third-order plate theories considering the transverse shear deformation and stresses free boundary condition to model Lamb waves efficiently in composite laminates. These two new third-order plate theories are explained in this section.

1. *Displacement field for Lamb waves*—Consider a laminate of constant thickness h composed of anisotropic laminas perfectly boned together. The origin of a global Cartesian coordinate system is located at the middle x–y plane with z-axis being normal to the mid-plane, so two outer surfaces of the laminate are at $z = \pm h/2$. A packet of the transient Lamb waves propagates in the composite laminate in an arbitrary direction θ, which is defined

relative to x-axis. Each lamina with an arbitrary orientation in the global coordinate system can be considered as a monoclinic material having x–y as a plane of symmetry, thereby the stress–strain relations of a single lamina can be expressed as the following matrix form.

$$\begin{Bmatrix} \sigma_1 \\ \sigma_2 \\ \sigma_3 \\ \sigma_4 \\ \sigma_5 \\ \sigma_6 \end{Bmatrix} = \begin{bmatrix} C_{11} & C_{12} & C_{13} & 0 & 0 & C_{16} \\ C_{12} & C_{22} & C_{23} & 0 & 0 & C_{26} \\ C_{13} & C_{23} & C_{33} & 0 & 0 & C_{36} \\ 0 & 0 & 0 & C_{44} & C_{45} & 0 \\ 0 & 0 & 0 & C_{45} & C_{55} & 0 \\ C_{16} & C_{26} & C_{36} & 0 & 0 & C_{66} \end{bmatrix} \begin{Bmatrix} \varepsilon_1 \\ \varepsilon_2 \\ \varepsilon_3 \\ \varepsilon_4 \\ \varepsilon_5 \\ \varepsilon_6 \end{Bmatrix} \tag{5.25}$$

where subscript '1' donates 'x', '2' donates 'y', '3' donates 'z', '4' donates 'yz', '5' donates 'xz', and '6' donates 'xy'. When the global coordinate system (x, y, z) does not coincide with the principal material coordinate system (x', y', z), the 6×6 stiffness matrix **C** in (x, y, z) system can be obtained from the lamina stiffness matrix **C'** in (x', y', z) system by multiplying the transforming matrix.

We begin with the displacement field in the study.[49] The variables ϕ_x, χ_x, ϕ_y, χ_y, ψ_z, and ϕ_z will be determined with the boundary condition that the stresses, $\sigma_{zz} = \sigma_3$, $\sigma_{yz} = \sigma_4$, and $\sigma_{xz} = \sigma_5$ vanish on the top and bottom surfaces of laminate panels. For plates laminated of orthotropic layers, $\sigma_4 = 0$ and $\sigma_5 = 0$ are equivalent to that the corresponding strains (ε_4 and ε_5) are zero on the surfaces. But for the condition $\sigma_3 = 0$ on the surfaces, the strain components (ε_1, ε_2, ε_3, and ε_6) are coupled with the stiffness coefficients (C_{13}, C_{23}, C_{33}, and C_{36}) for the two laminas on the top and bottom surfaces according to the constitutive equation in Equation 5.25. For symmetric laid-up laminates, the boundary condition can be expressed in the form of displacement parameters as

$$\begin{cases} \left(v_{,z} + w_{,y} \right)_{z=\pm h/2} = 0 \\ \left(u_{,z} + w_{,x} \right)_{z=\pm h/2} = 0 \\ \left[Q_{13} u_{,x} + Q_{23} v_{,y} + Q_{33} w_{,z} + Q_{36} \left(u_{,y} + v_{,x} \right) \right]_{z=\pm h/2} = 0 \end{cases} \tag{5.26}$$

where Q_{13}, Q_{23}, Q_{33}, and Q_{36} respectively donate the stiffness coefficients C_{13}, C_{23}, C_{33}, and C_{36} of the two laminas on the top and the bottom surfaces. Assume the solution forms of Lamb waves as

$$\begin{aligned} &\{ u_0, \psi_x, \phi_x, \chi_x, v_0, \psi_y, \phi_y, \chi_y, w_0, \psi_z, \phi_z \} \\ &= \{ U_0, \Psi_x, \Phi_x, X_x, V_0, \Psi_y, \Phi_y, X_y, W_0, \Psi_z, \Phi_x \} \exp i \left(k_x x + k_y y - \omega t \right) \end{aligned} \tag{5.27}$$

where ω is the angular frequency, wave vector $\mathbf{k} = [k_x, k_y]^T$ points to the direction of wave propagation (θ) in the x–y plane. The Symbolic Math Toolbox of MATLAB is utilized to solve the six equations in Equation 5.26. The six variables ϕ_x, χ_x, ϕ_y, χ_y, ψ_z, and ϕ_z are expressed by the left five independent variables, then displacement field is modified as

$$u = u_0 + z\psi_x + z^2\left(4Q_{13}u_{0,xx} + 4Q_{36}u_{0,xy} + 4Q_{36}v_{0,xx} + 4Q_{23}v_{0,xy}\right)/a_1$$

$$+ z^3\left(-96Q_{33}\psi_x + 12h^2Q_{13}\psi_{x,xx} + 16h^2Q_{36}\psi_{x,xy} + 4h^2Q_{23}\psi_{x,yy}\right.$$

$$\left. + 8h^2Q_{36}\psi_{y,xx} + 8h^2Q_{23}\psi_{y,xy} - 96Q_{33}w_{0,x}\right)/a_2$$

$$v = v_0 + z\psi_y + z^2\left(4Q_{13}u_{0,xy} + 4Q_{36}u_{0,yy} + 4Q_{36}v_{0,xy} + 4Q_{23}v_{0,yy}\right)/a_2$$

$$+ z^3\left(-96Q_{33}\psi_y + 12h^2Q_{23}\psi_{y,yy} + 16h^2Q_{36}\psi_{y,xy} + 4h^2Q_{23}\psi_{x,xx}\right. \qquad (5.28)$$

$$\left. + 8h^2Q_{36}\psi_{x,yy} + 8h^2Q_{13}\psi_{x,xy} - 96Q_{33}w_{0,y}\right)/a_2$$

$$w = w_0 - z\left(8Q_{13}u_{0,x} + 8Q_{36}u_{0,y} + 8Q_{36}v_{0,x} + 8Q_{23}v_{0,y}\right)/a_1$$

$$+ z^2\left(4Q_{13}w_{0,xx} + 8Q_{36}w_{0,xy} + 4Q_{23}w_{0,yy} - 8Q_{13}\psi_{x,x} - 8Q_{36}\psi_{x,y}\right.$$

$$\left. - 8Q_{23}\psi_{y,y} - 8Q_{36}\psi_{y,x}\right)/a_3$$

where

$$a_1 = Q_{13}h^2k_x^2 + 2Q_{36}h^2k_xk_y + Q_{23}h^2k_y^2 + 8Q_{33}$$

$$a_2 = 3Q_{13}h^4k_x^2 + 6Q_{36}h^4k_xk_y + 3Q_{23}h^4k_y^2 + 72Q_{33}h^2 \qquad (5.29)$$

$$a_3 = Q_{13}h^2k_x^2 + 2Q_{36}h^2k_xk_y + Q_{23}h^2k_y^2 + 24Q_{33}$$

The displacement field described in Equation 5.28 is in connection with the stiffness coefficients of the two laminas on the top and the bottom surfaces and with the wave vector. The five independent variables $(\psi_x, \psi_y, w_0, u_0, v_0)$ donate three antisymmetric modes and two symmetric modes, respectively. Computation of stress correction factors is avoided because the displacement function is deduced with consideration of the stresses free boundary condition.

2. *Equations of motion*—With the linear strain–displacement relations, the equations of motion of the higher order theory can be derived using the principle of virtual displacement of Hamilton's principle.

$$\int_0^T (\delta U - \delta T + \delta W)\, dt = 0 \qquad (5.30)$$

where U, T, and W are virtual strain energy, virtual kinetic energy, and virtual work done by applied forces, respectively. For Lamb waves modeling, the condition that the stresses are free on the surfaces is considered, thus the virtual work W is zero. With substitution of stress and strain components into Equation 5.30, the final integral equation for plate elasticity is given as

$$
\int_0^T \int_{\Omega_0} \int_{-h/2}^{h/2} \left(\sigma_1 \delta\varepsilon_1 + \sigma_2 \delta\varepsilon_2 + \sigma_3 \delta\varepsilon_3 + \sigma_4 \delta\varepsilon_4 + \sigma_5 \delta\varepsilon_5 + \sigma_6 \delta\varepsilon_6 \right) dz\, dA\, dt
$$

$$
- \int_0^T \int_{\Omega_0} \int_{-h/2}^{h/2} \rho \left(\dot{u}\delta\dot{u} + \dot{v}\delta\dot{v} + \dot{w}\delta\dot{w} \right) dz\, dA\, dt = 0
$$

(5.31)

Plate inertias and stress resultants per unit length are defined in the following:

$$
I_{1,\dots,7} = \sum_{n=1}^{N} \int_{z_n}^{z_{n+1}} \rho_n \left(1, z, z^2, z^3, z^4, z^5, z^6 \right) dz
$$

$$
\left(N_i, M_i, P_i, S_i \right) = \sum_{n=1}^{N} \int_{z_n}^{z_{n+1}} \sigma_i \left(1, z, z^2, z^3 \right) dz \quad (i = 1,2,3,4,5,6)
$$

(5.32)

where index n refers to the layer number in a laminate. Using fundamental lemma of calculus of variation, the equations of motion can be derived.

To express the equations of motion with displacement field parameters, the constitutive equation of laminate with arbitrary lay-up can be utilized:

$$
\begin{Bmatrix} \mathbf{N} \\ \mathbf{M} \\ \mathbf{P} \\ \mathbf{S} \end{Bmatrix} = \begin{bmatrix} [A] & [B] & [D] & [F] \\ & [D] & [E] & [F] \\ & & [F] & [H] \\ \text{Sym} & & & [J] \end{bmatrix} \begin{Bmatrix} \varepsilon^0 \\ \varepsilon^1 \\ \varepsilon^2 \\ \varepsilon^3 \end{Bmatrix}
$$

(5.33)

In Equation 5.31, the stress and moment resultant vectors are defined as

$$
\mathbf{N} = \{ N_1 \ N_2 \ N_3 \ N_4 \ N_5 \ N_6 \}^T
$$

$$
\mathbf{M} = \{ M_1 \ M_2 \ M_3 \ M_4 \ M_5 \ M_6 \}^T
$$

$$
\mathbf{P} = \{ P_1 \ P_2 \ P_3 \ P_4 \ P_5 \ P_6 \}^T
$$

$$
\mathbf{S} = \{ S_1 \ S_2 \ S_3 \ S_4 \ S_5 \ S_6 \}^T
$$

(5.34)

The elements of stiffness matrixes [A], [B], [D], [E], [F], [H], and [J] are

$$\left(A_{ij}, B_{ij}, D_{ij}, E_{ij}, F_{ij}, H_{ij}, J_{ij}\right) = \sum_{n=1}^{N} \int_{z_n}^{z_{n+1}} C_{ij}\left(1, z, z^2, z^3, z^4, z^5, z^6\right) dz \tag{5.35}$$

$$(i, j = 1, 2, 3, 4, 5, 6)$$

The strain vectors represent

$$\varepsilon^i = \left\{\varepsilon_1^i \ \varepsilon_2^i \ \varepsilon_3^i \ \varepsilon_4^i \ \varepsilon_5^i \ \varepsilon_6^i\right\}^T \quad (i = 0, 1, 2, 3) \tag{5.36}$$

where ε_j^i ($i = 0,1,2,3$) are defined as the strain coefficients:

$$\varepsilon_j = \varepsilon_j^0 + z\varepsilon_j^1 + z^2\varepsilon_j^2 + z^3\varepsilon_j^3 \quad (j = 1, 2, 3, 4, 5, 6) \tag{5.37}$$

and the strain components ε_j in Equation 5.37 can be obtained according to the displacement filed of Equation 5.28.

Substituting Equations 5.27 and 5.33 into the equations of motion yields a generalized eigen-value problem. Five real positive eigen-values related to three antisymmetric and two symmetric Lamb modes can be obtained from the 5×5 order characteristic matrix. The determinant of the matrix is a function of the phase velocity (c_p), the frequency (f), and the propagation direction (θ). That leads to a characteristic equation, solution of which gives the c_p-f curves (dispersion curves) in a rectangular coordinate system for a given angle θ or the c_p-θ curves in a polar coordinate system for a given frequency f.

3. *Accuracy comparison with the accurate solutions*—As the propagation characteristics of A_0 mode in low frequency range are more essential to experiment researchers, the difference between the ESLTs results and the exact solutions in low frequency range are drawn in Figure 5.25. Note that the material properties and the stacking sequence are the same in section "3D Elasticity Theory with Global Matrix Techniques." It is obvious that the new HSDT gives the best estimation of A_0 mode in low frequency range compared with other existing plate theories. The new HSDT proposed in Equation 5.28 can depict two symmetric modes. Although there is one term in w describing symmetric modes in the new HSDT, this term has no independent variable. As shown in Figure 5.26, the new HSDT is not appropriate for modeling symmetric Lamb modes.

4. *A new HSDT of six-DoFs*—As mentioned above, the new HSDT of five-DoFs is not appropriate for modeling symmetric Lamb modes due to the absence of independent variables describing symmetric modes in displacement component w. Thus, to depict symmetric modes more precisely, an independent variable ψ_z is introduced in w of a new six-DoFs HSDT.

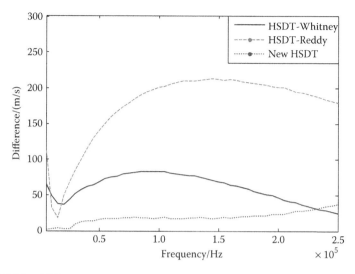

FIGURE 5.25 Difference with the 3D elasticity results for A_0 mode in low frequency range among the HSDT by Whitney, the HSDT by Reddy, and the new HSDT.

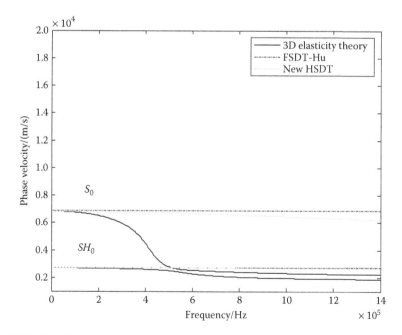

FIGURE 5.26 Comparison of dispersion curves for symmetric modes among 3D elasticity theory, the FSDT by Hu and the new HSDT.

Considering the stresses free boundary condition, the authors deduced the displacement functions of the new six-DoFs HSDT as

$$u = u_0 + z\psi_x + z^2 \left(4Q_{13}u_{0,xx} + 4Q_{36}v_{0,xx} - 8Q_{33}\psi_{z,x}\right.$$

$$\left. + 4Q_{36}u_{0,xy} + 4Q_{23}v_{0,xy}\right)/a_3$$

$$+ z^3 \left(-96Q_{33}\psi_x + 12h^2Q_{13}\psi_{x,xx} + 16h^2Q_{36}\psi_{x,xy} + 4h^2Q_{23}\psi_{x,yy}\right.$$

$$\left. + 8h^2Q_{36}\psi_{y,xx} + 8h^2Q_{23}\psi_{y,xy} - 96Q_{33}w_{0,x}\right)/a_2$$

$$v = v_0 + z\psi_y + z^2 \left(4Q_{13}u_{0,xy} + 4Q_{36}u_{0,yy} - 8Q_{33}\psi_{z,y}\right.$$

$$\left. + 4Q_{36}v_{0,xy} + 4Q_{23}v_{0,yy}\right)/a_3 \tag{5.38}$$

$$+ z^3 \left(-96Q_{33}\psi_y + 12h^2Q_{23}\psi_{y,yy} + 16h^2Q_{36}\psi_{y,xy} + 4h^2Q_{13}\psi_{y,xx}\right.$$

$$\left. + 8h^2Q_{36}\psi_{x,yy} + 8h^2Q_{13}\psi_{x,xy} - 96Q_{33}w_{0,y}\right)/a_2$$

$$w = w_0 + z\psi_z + z^2 \left(4Q_{13}w_{0,xx} + 8Q_{36}w_{0,xy} + 4Q_{23}w_{0,yy}\right.$$

$$\left. - 8Q_{13}\psi_{x,x} - 8Q_{36}\psi_{x,y} - 8Q_{23}\psi_{y,y} - 8Q_{36}\psi_{y,x}\right)/a_3$$

$$+ z^3 \left(-32Q_{33}\psi_z + 4Q_{13}h^2\psi_{z,xx} + 4Q_{23}h^2\psi_{z,yy} - 32Q_{13}u_{0,x}\right.$$

$$\left. - 32Q_{36}u_{0,y} - 32Q_{23}v_{0,y} - 32Q_{36}v_{0,x} + 8Q_{36}h^2\psi_{z,xy}\right)/\left(a_3h^2\right)$$

where the independent variables ψ_x, ψ_y, w_0 describe antisymmetric Lamb modes, the other three independent variables u_0, v_0, ψ_z describe symmetric Lamb modes, and a_1, a_2, a_3 are defined as in Equation 5.29. Phase velocities of antisymmetric Lamb modes obtained by the new five-DoFs HSDT are the same as those calculated by the new six-DoFs HSDT, because the terms depicting antisymmetric Lamb modes in Equations 5.28 and 5.38 are all the same. Emphasis is laid on the estimation accuracy for lower symmetric modes, and the dispersion curves of the first three symmetric modes from different theories are drawn in Figure 5.27.

Unlike the HSDT of five-DoFs, the new six-DoFs HSDT can predict better steep descent of S_0 mode, which proves that the increasing number of independent variables is essential to improve accuracy. Reddy's HSDT is half-satisfying the stresses free boundary condition. In comparison to Reddy's theory, the new six-DoFs HSDT takes the full boundary condition into consideration and gives better estimation. The FSDT by Zak has a linear displacement field with respect to thickness coordinate z. Each of the three displacement components u, v, w in Equation 5.38 has two more terms than those of Zak's displacement field. Although the two terms do not include any independent variables, the dispersion curves of the new

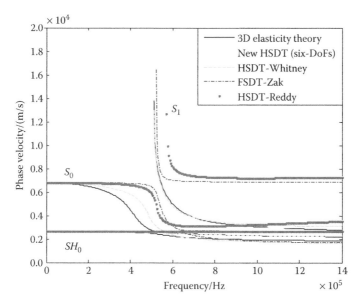

FIGURE 5.27   Comparison of symmetric Lamb modes dispersion curves from different ESLTs and the 3D elasticity theory.

six-DoFs HSDT are still much closer to the exact solutions compared with those of Zak's theory.

5. *Efficiency comparison between difference theories*—To compare the computation efficiency between the ESLTs and 3D elasticity theory, the relative computing time (with respect to FSDT by Mindlin) of the existing plate theories and the 3D elasticity theory are listed in Table 5.2. The CPU time of the conventional ESLTs includes the time one needs to calculate the correction factors. All the results are solved by the ergodic searching method with the same searching range and the same step interval. The computing time of plate theories depends on the order of characteristic matrix and on the complexity of each matrix element. The order of the characteristic matrix is determined by the independent variables in the displacement field. Statistics in Table 5.2 indicate that, for the C^0 continuity displacement functions by Mindlin, Hu, Zak, and Whitney, the computing time is linear to the matrix order.

TABLE 5.2
Relative Computing Time of ESLTs and 3D Elasticity Theory

Theory	Mindlin	Hu	Zak	Whitney	Reddy	New HSDT (5-DoFs)	New HSDT (6-DoFs)	3D Elasticity Theory
Matrix order	3	5	6	8	7	5	6	/
Relative CPU time	1.0 + 1,659.5	1.7 + 1,659.5	2.0 + 1,659.5	2.7 + 1,659.5	7.0	139.7	214.4	1,659.5

The terms of C^1 continuity make the matrix element more complex, which leads to the computing time by Reddy's theory seven times of that by Mindlin's. The two new HSDTs satisfying the vanishing of stresses on the laminate surfaces have displacement fields containing several C^2 continuity components. The complicated C^2 continuity matrix elements make the two new HSDTs most time consuming among all the existing ESLTs. Even so, the new HSDTs still run much faster than the 3D elasticity theory.

Measurement of Phase Velocities

Figure 5.28 shows the experimental apparatus for the laser ultrasonic system. A pulsed Q switch Nd:YAG laser ($\lambda = 532$ nm, energy ≈ 50 mJ) with a pulse duration of 8 ns was used as the generation source. In anisotropic plates, Lamb wave velocities vary with the wave propagation directions. To obtain the propagation information along a specific direction, the laser source scans the surface of the target plate along a series of circles (sharing a same center point) with a radial interval $\Delta r = 1$ mm and a circumferential interval $\Delta\theta = 2°$. Each of the laser irradiations generates a Lamb wave without any ablation damage. The AE sensor, pasted on the center of the circles (on the other surface of the plate), detects the Lamb waves generated at each grid point irradiated by the laser. The AE sensor used for the out of plane deformation measurements has a flat frequency response in the frequency band of approximately 25 kHz–15 MHz and has a diameter of about 3 mm. The amplification system of the AE sensor can be regarded as a linear one and the sampling rate is 10 MS/s. All of the signals detected are stored on a PC for the later visualization of the ultrasonic propagations. A snapshot of the scanned area at any given propagation time can be obtained by collecting all of the amplitudes at the corresponding time from the waveform data stored on the PC. According to the reciprocity theorem, the snapshot is identical to the waveform propagating from the AE sensor, while it is actually constructed by the waves propagating in the reverse directions.

The specimens are two CFRP plates (a unidirectional one and a cross-ply one) whose thickness is 1 mm and density is 1,600 kg/m³ but material properties are to be identified.

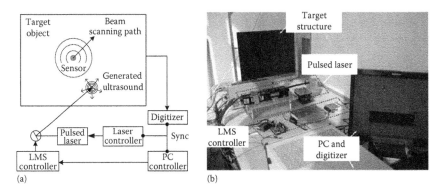

FIGURE 5.28 Laser ultrasonic system: (a) Schematic of the experimental setup and (b) Photo of the experimental setup. (LMS denotes laser mirror scanner.)

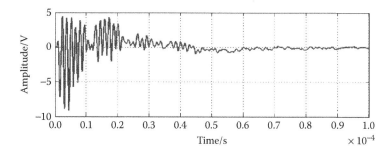

FIGURE 5.29   Experimental signal sensed by the AE transducer when the laser source points at the position of the sensor.

The experimental signal sensed by the AE transducer when the laser source points at the position of the sensor, which can be regarded as the generation signal in the later FEM model, is shown in Figure 5.29. Figure 5.30 shows the experimental examples of the two-dimensional snapshots of the Lamb wave propagation in the unidirectional plate at 12 μs based on the laser ultrasonic system. The first snapshot is without any signal process and the propagation of A_0, SH_0, and S_0 modes, whose wavefront curves are different, can be clearly identified. Lamb waves propagate fastest along the 0° principal direction and the anisotropic properties can be observed directly based on this snapshot.

To obtain velocities at a given frequency, a wavelet transform method was utilized to obtain the narrow band signal. Figure 5.30b shows the snapshot of Lamb wave propagation in the unidirectional plate at 500 kHz center frequency. A position–time diagram is created by combining the time-domain signals from a series of points along a line across the transducer. The velocity can then be estimated as the slope of a chosen phase. The measurement of the slope is based on finding the series of the phase peaks along the propagation direction, as the open squares and circles in Figure 5.31 indicate. By applying least-squares fitting (LSF) to the series of phase peaks, the phase velocities of S_0 and A_0 modes (500 kHz) along 0° direction were calculated as 9,402.4 m/s and 1,698.7 m/s, respectively. The slopes along 90° direction (210 kHz) are 2,502.0 m/s for S_0 mode and 832.2 m/s for A_0 mode, according to Figure 5.31b.

Sensitivity Analysis

GAs for the Inversion of Phase Velocities

The GA-based inverse method of reconstruction starts with a population of randomly guessed candidate values of the stiffness coefficients set, each of that are then evaluated using the error function. A few of them with the lowest error function are preserved, while the others are placed with new trials. GAs parameters are listed in Table 5.3. The error function to be minimized, during the inversion of the ultrasonic phase velocities, can be represented as follows:

$$\text{err}(C) = \frac{1}{n}\sum_{i-1}^{n}\left|\frac{v_i^e - v_i^t}{v_i^e}\right| \tag{5.39}$$

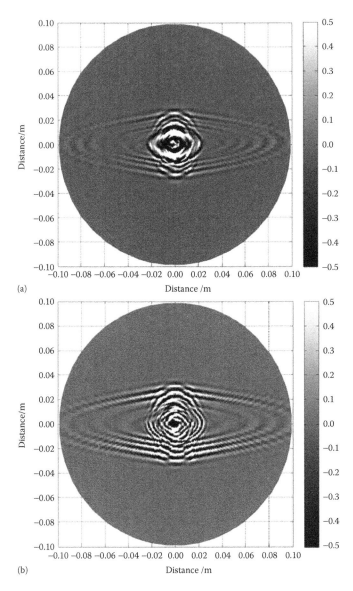

FIGURE 5.30 Visualized images of the unidirectional plate based on the laser ultrasonic system: (a) snapshot of Lamb waves propagation (12 μs) and (b) snapshot of Lamb waves propagation at 500 kHz (12 μs).

The objective function above describes the relative error between the experimental velocity v^e and the theoretical ones v^t. When the propagation direction is specified, n denotes the number of the chosen frequencies.

A basic knowledge that the propagation direction of phase velocity is not same as that of group velocity in anisotropic plate,[57] as shown in Figure 5.32, should be taken

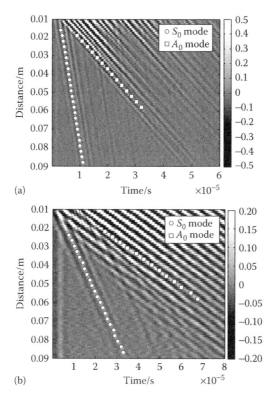

FIGURE 5.31   The position–time diagram of the unidirectional plate based on the laser ultrasonic system: (a) along the 0° direction at 500 kHz and (b) along the 90° direction at 210 kHz.

TABLE 5.3

GAs Parameters Used in the Inversion Procedure

Number of generation	1,000
Population size	20
Number of elites	2
Crossover rate	0.8
Mutation rate	0.2
Crossover type	Single point

into account in GAs. The snapshot in Figure 5.32 is based on an FEM simulation of Lamb wave propagation at a given frequency in a unidirectional plate whose properties are listed in Table 5.4. The direction of group velocity is along the white dot line that implies the sensing line in experiments, while the relevant phase velocity travels perpendicular to the wave front. Because the measured velocity v^e is obtained from

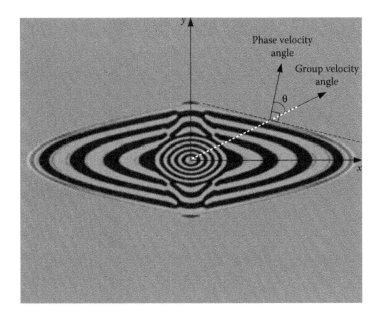

FIGURE 5.32   Schematic of the different propagation angle between phase velocity and group velocity in a unidirectional plate (based on an FEM snapshot).

TABLE 5.4

Typical Stiffness Coefficients (units: GPa) for Two CFRP Plates with Thickness of 1 mm and Density of 1,600 kg/m³

Parameter	C_{11}	C_{12}	C_{13}	C_{22}	C_{23}	C_{33}	C_{44}	C_{55}	C_{66}
Unidirectional	155.43	3.72	3.72	16.34	4.96	16.34	3.37	7.48	7.48
Cross-ply	77.40	6.24	5.81	73.40	5.81	12.43	4.26	4.26	5.00

the group angle, the relevant theoretical phase velocity v'_t, which is perpendicular to the wave front, should be transformed to the group velocity coordinate by

$$v_t = v'_t/\cos(\theta) \tag{5.40}$$

where θ is the angle between the phase velocity and the group velocity.

Sensitivity Analysis of Phase Velocities to the Change of Stiffness Coefficients

In the study,[58] the authors reconstructed the material properties using a series of angles from 0° to 90° (considering the symmetric characteristic of the chosen plate) with an increment of 0.2°, which increases the complexity of the inversion problem a lot. To simplify the reconstruction process and improve the efficiency, a sensitivity analysis should be performed to identify regions where the change of the material properties significantly affects the measured Lamb wave velocities. In addition, the relative sensitivity of unknown parameters (elastic moduli) becomes important in a multiparameter optimization problem as described in this paper. If the sensitivities of the nine parameters are markedly different from each other, then the performance of the GAs may not be reliable, that is, the error of the reconstructed elastic moduli will be larger for some of the elastic moduli, to which the velocities are less sensitive.

To simplify the inversion procedure and improve the reliability of the reconstructed multiple parameters, the sensitivity analysis of the phase velocity to the change of the nine stiffness coefficients along 0° to 90° is conducted at a specified frequency, as illustrated in Figure 5.33. The horizontal coordinate denotes the group velocity direction and the phase velocities have been transformed to the group velocity direction. The vertical coordinate denotes the relative difference of the phase velocities when each stiffness coefficient is increased by 20%.

During the sensitivity analysis, the material properties of the two plates are as in Table 5.4 that shows the typical stiffness coefficients of a unidirectional and a cross-ply CFRP plates. However, this does not mean that the experimental specimens are of the same material properties as in Table 5.4.

According to Figure 5.33a and Figure 5.24c, for both plates, phase velocities of A_0 mode in 0° direction are most sensitive to C_{11} and C_{55}. Phase velocities of A_0 mode in 90° direction are most sensitive to C_{22} and C_{44}, while phase velocities of A_0 mode in the 20°–60° region are sensitive to C_{12} and C_{66}. However, A_0 mode velocities are much less sensitive to the left three parameters (C_{13}, C_{23}, and C_{33}) and thus A_0 mode velocities cannot be used to reconstruct the left three stiffness coefficients efficiently.

As to S_0 mode, as shown in Figures 5.33b and 5.24d, velocities in 0° direction are more sensitive to C_{11} and C_{13}, velocities in 90° direction are more sensitive to C_{22}, C_{23}, and C_{33}. Velocities in 20° direction are most sensitive to C_{12} and C_{66} for the unidirectional plate, while velocities in 45° direction are most sensitive to C_{12} and C_{66} for the cross-ply one.

Although both of A_0 and S_0 modes velocities in 0° direction are sensitive to C_{11}, S_0 mode is chosen to reconstruct C_{11} due to its higher objective function value (higher sensitivity) compared with that of A_0 mode. Similarly, S_0 mode velocities are more sensitive to C_{22}, C_{12}, and C_{66} and will be inversed to reconstruct these three parameters. By comparison and combination of A_0 mode and S_0 mode, the reconstructing method about which Lamb wave mode and which propagation direction should be chosen is illustrated in Tables 5.5 and 5.6. The C_{11} and C_{22} using S_0 mode velocities are reconstructed first and regarded as known when inverting C_{44} and C_{55} using A_0 mode velocities. To avoid nonuniqueness of solution for the multiparameter problem, phase velocities at a serious of frequencies (dispersion curves) will be utilized in GAs whose objective function is described in Equation 5.39.

RECONSTRUCTION METHOD

Section "Genetic Algorithms for the Inversion of the Phase Velocities" explains the GAs to reconstruct the stiffness coefficients. The reconstruction of the nine stiffness coefficient was carried out on two different CFRP plates. With MATLAB Optimization Tool, the material properties were identified by inverting the

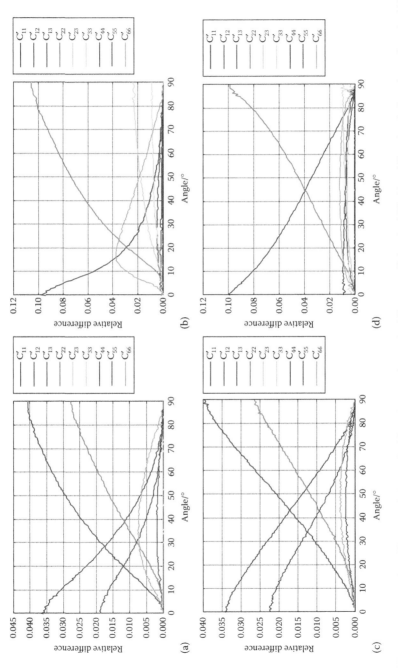

FIGURE 5.33   The relative difference of phase velocities when varying each stiffness coefficient by 20%: (a) unidirectional plate: A_0 mode at $f = 100$ kHz; (b) unidirectional plate: S_0 mode at $f = 300$ kHz; (c) cross-ply plate: A_0 mode at $f = 100$ kHz; and (d) cross-ply plate: S_0 mode at $f = 300$ kHz.

TABLE 5.5
Reconstruction Method of Stiffness Coefficients in a Unidirectional Plate

Parameter to Be Reconstructed	C_{11}	C_{12}	C_{13}	C_{22}	C_{23}	C_{33}	C_{44}	C_{55}	C_{66}
Using Lamb wave mode	S_0	S_0	S_0	S_0	S_0	S_0	A_0	A_0	S_0
Using direction	0°	20°	0°	90°	90°	90°	90°	0°	20°

TABLE 5.6
Reconstruction Method of Stiffness Coefficients in a Cross-Ply Plate

Parameter to Be Reconstructed	C_{11}	C_{12}	C_{13}	C_{22}	C_{23}	C_{33}	C_{44}	C_{55}	C_{66}
Using Lamb wave mode	S_0	S_0	S_0	S_0	S_0	S_0	A_0	A_0	S_0
Using direction	0°	45°	0°	90°	90°	90°	90°	0°	45°

experimental velocities shown in Figure 5.34. According to the reconstruction method in Tables 5.5 and 5.6, C_{55} and C_{44} are taken from the reconstruction results along 0° and 90° direction of A_0 mode, respectively. C_{11} and C_{13} are identified using the S_0 mode velocities along 0° direction. Based on the S_0 mode velocities along 90° direction, C_{22}, C_{23}, and C_{33} are calculated. As shown in Figure 5.34b, phase velocities in the unidirectional plate along 20° direction are sensitive not only to C_{12} and C_{66} but also to others. To assure the accuracy of the reconstructed value of C_{12} and C_{66}, the other seven parameters should be identified first and substituted into GAs to continue the reconstruction of C_{12} and C_{66}.

RESULTS AND DISCUSSION

Reconstructing Results

The reconstruction results are listed in Tables 5.7 and 5.8. The relative phase velocities error along 0° and 90° of S_0 mode is less than 1% and thus the reconstructed results (C_{11}, C_{13}, C_{22}, C_{23}, and C_{33}) are of high accuracy according to the sensitivity analysis in Figure 5.33. The objective function value along 90° direction of A_0 mode is always the highest, which might result from the inaccuracy and the unsmooth dispersion curve of the experimental velocities along 90° direction of A_0 mode.

Verification of the Reconstructed Results

To verify the accuracy of the reconstructed results in Tables 5.7 and 5.8, the nine parameters are substituted into a commercial software ABAQUS, as the material properties, to simulate the propagation characteristics of Lamb waves. A normal-to-plane nodal force, whose time distribution is taken as the experimental signal sensed by the AE transducer when the laser source points at the position of the sensor, as shown in Figure 5.28, is set as the load condition in the FEM simulation. The model is pinned at the boundaries. The element is of a liner solid type and the element size is set as 0.3 mm. The maximum time step of the explicit scheme is set as 0.01 μs.

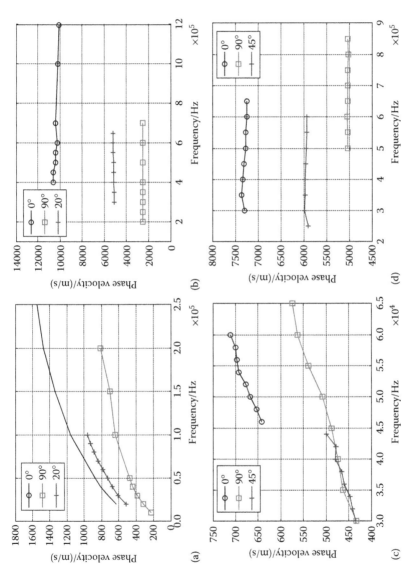

FIGURE 5.34 The experimental phase dispersion curves along specific directions: (a) unidirectional plate: A_0 mode; (b) unidirectional plate: S_0 mode; (c) cross-ply plate: A_0 mode; and (d) cross-ply plate: S_0 mode.

TABLE 5.7

Reconstructed Parameters and the Objective Function of the Unidirectional Plate

Lamb Wave Mode and Direction		Parameter	Reconstructed Value (GPa)	Objective Function Value (% Error)
A_0	0°	C_{55}	5.99	1.74
	90°	C_{44}	2.10	4.06
S_0	0°	C_{11}	175.72	0.74
		C_{13}	14.53	
	90°	C_{22}	11.76	0.34
		C_{23}	8.54	
		C_{33}	13.81	
	20°	C_{12}	10.08	2.71
		C_{66}	6.33	

TABLE 5.8

Reconstructed Parameters and the Objective Function Value of the Cross-Ply Plate

Lamb Wave Mode and Direction		Parameter	Reconstructed Value (GPa)	Objective Function Value (% Error)
A_0	0°	C_{55}	1.51	0.79
	90°	C_{44}	0.58	4.88
S_0	0°	C_{11}	93.60	0.24
		C_{13}	10.10	
	90°	C_{22}	65.70	0.20
		C_{23}	11.70	
		C_{33}	16.40	
	45°	C_{12}	11.01	1.82
		C_{66}	3.99	

The simulation results are compared with the experimental images, as shown in Figure 5.35. The wave curves of the simulated three basic Lamb wave modes coincide well with the experimental ones.

Meanwhile, the calculated and measured velocities curves of S_0 mode are compared in Figure 5.36. As illustrated in Figure 5.36a and c, the theoretical and experimental dispersion curves along 0° and 90° directions of S_0 mode coincide well. There exists difference between the theoretical and experimental dispersion curves along other directions (20° for the unidirectional plate and 45° for the cross-ply one), which is mainly because that the reconstruction process of these directions is based on the 0° and 90° directions results and the previous error is brought in.

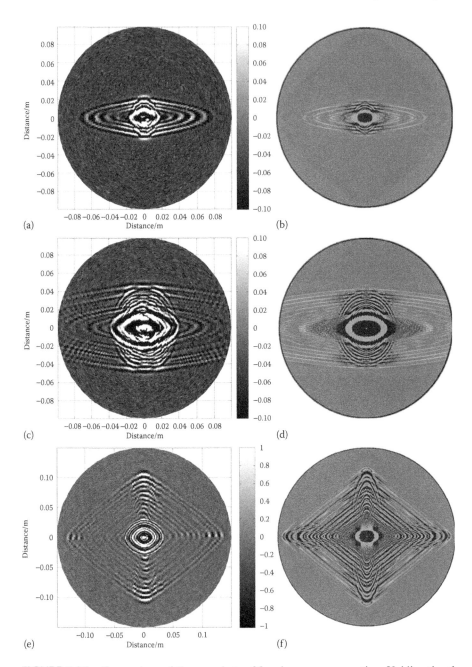

FIGURE 5.35 Comparison of the snapshots of Lamb waves propagation. Unidirectional plate: (a)–(d); cross-ply plate: (e)–(f). (a) Experimental image (9 μs); (b) FEM results based on the reconstructed material properties (9 μs); (c) experimental image (20 μs); (d) FEM results based on the reconstructed material properties (20 μs); (e) experimental image (20 μs); and (f) FEM results based on the reconstructed material properties (20 μs).

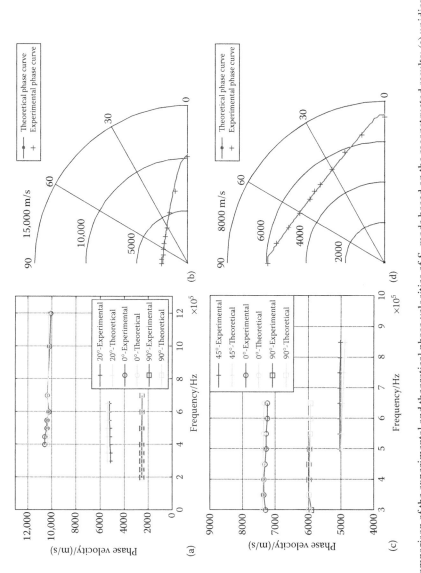

FIGURE 5.36 Comparison of the experimental and theoretical phase velocities of S_0 mode based on the reconstructed results: (a) unidirectional plate: phase dispersion curves; (b) unidirectional plate: phase curve at $f = 500$ kHz; (c) cross-ply plate: phase dispersion curves; and (d) cross-ply plate: phase curve at $f = 500$ kHz.

Information along only three directions is utilized in the inversion procedure and thus velocities along other directions need to be verified. For the unidirectional plate, the experimental velocities of S_0 mode ($f = 500$ kHz) along other seven directions (10°, 30°, 40°, 50°, 60°, 70°, and 80°) were extracted from the laser ultrasonic results and agree well with the theoretical ones, as shown in Figure 5.36b. In Figure 5.36d, the measured velocities of S_0 mode ($f = 500$ kHz) in the cross-ply plate along other eight directions (10°, 20°, 30°, 40°, 50°, 60°, 70°, and 80°) are taken into account and match the theoretical ones.

LASER ULTRASONIC METHOD FOR DAMAGE ACCUMULATION EVALUATION

Section "Laser Ultrasonic Method for Reconstructing Material Properties of Composites" details the theoretical model for Lamb waves, the measurement of phase velocities, and the reconstruction process for the stiffness coefficients in composites. As illustrated in Figure 5.21, composite structures will experience material degradation during the fatigue process. Zhao et al.[36] studied the relationships between three characteristic parameters (amplitude of Lamb waves, phase velocities, and the modulus of the specimen) and the fatigue loading cycle for composite laminates. This section will introduce the laser-generation-based method as a new way for nondestructive damage accumulation evaluation for composites. Note that the different damage types (matrix cracks, delaminations or fiber fracture) are not distinguished in [36], because microscopy or other NDE techniques might be needed.

METHODOLOGY

Experimental Setup

The whole experimental setup to analyze the relationship between Lamb wave characteristics and the fatigue loading cycle for composites is shown in Figure 5.37. The system contains two parts: (1) the fatigue test system (fatigue test machine) to conduct the fatigue test and (2) the NDE system (laser ultrasonic system) to evaluate the fatigue state of the specimen.

The composite sample studied was T700/3001 graphite epoxy with a stacking sequence of $[0/90]_{4S}$ whose density is 1,600 kg/m³. The specimen is of a length of 120 mm (not including the length of the reinforced sheets), a width of 25 mm, and an average thickness of 2.25 mm. The specimen was subjected to tension–tension fatigue test at a frequency of 5 Hz, and the upper stress was taken to be 65% of the ultimate strength (708.5 MPa) while the lower stress was 0 MPa.

Before the NDE process, other two same specimens were fatigued to failure under the above fatigue load condition to obtain the average fatigue life (18,000 cycle), which is the required information to determine the load cycle increments. To study the influence of the initial fatigue damage and the final sudden failure to the three parameters (amplitude of the Lamb waves, phase velocity, and the modulus), the initial and the ultimate load cycle increments were designed to be of high resolution. The specific fatigue load cycle is arranged in Table 5.9. The NDE test was conducted once before any fatigue load imposed on the specimen, after that the tension–tension

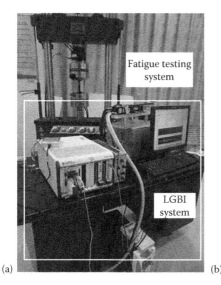

FIGURE 5.37 Photo of experimental setup: (a) Fatigue test system and (b) NDE test system.

TABLE 5.9

Fatigue Load Cycle Increments for the [0/90]₄ₛ Composite Sample (Units: Cycle)

Load Cycle Period	Increment	Load Cycle Period	Increment
0–100	20	5,000–10,000	1,000
100–400	50	10,000–16,000	2,000
400–1,000	100	16,000–17,000	1,000
1,000–2,000	200	17,000–17,400	200
2,000–5,000	500	17,400–18,000	50

load was applied with a specific cycle increment according to Table 5.9. During each pulse, the stress of the whole specimen will be released back to 0 MPa, and the laser will scan the specimen along a line across the AE sensor to conduct the NDE test.

Measurement of the Three Parameters

Based on the introduction in section "Measurement of Phase Velocities," all of the three parameters were obtained via the position–time diagram. The amplitude of Lamb waves can be acquired simply as a peak value at a fixed position (the value at a specific coordinate (p, t) from the position–time diagram). The phase velocity will be calculated as the slope of a chosen phase, while the modulus of composites can be estimated by inverting the measured phase velocities. The details are as follows.

During the NDE test process, the laser scanned a line starting at the position of the AE sensor and parallel to the fatigue tension direction. The scanning space increment is as small as 0.5 mm to assure the spatial resolution while the sampling frequency of the receiving signals is as high as 10 MHz to assure the time resolution. Figure 5.38 shows the position–time diagram of the composite specimen without being fatigue tested. In Figure 5.38a and b, the discontinuity at the position 22 mm away from the sensor is because that the laser pulse cannot propagate to the specimen well through the rubber band (used to fix the extensometer).

The position–time diagram containing A_0 and S_0 modes in Figure 5.38a is much more complicated than that in large plate (e.g., Figure 5.31) due to the continuous reflection from the specimen boundaries. The reason is that Lamb wave of low frequency, as 300 kHz in Figure 5.38a, always has large wavelength. When the wavelength of Lamb wave is larger than the width of the specimen, the reflection wave from the boundary will interfere with the propagation wave before it can propagate for a single-complete-wavelength distance. To assure the first phase peak (which is to be used to evaluate the amplitude and the phase velocity) not interfering

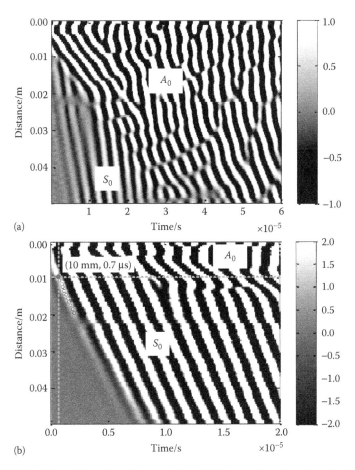

(a)

(b)

FIGURE 5.38   The position–time diagram of the $[0/90]_{4S}$ specimen based on the laser ultrasonic system: (a) at $f = 300$ kHz and (b) at $f = 700$ kHz.

with the reflected waves, the component filtered at high frequency (as 700 kHz in Figure 5.38b) should be utilized to make sure that the measured wavelength is less than the width of the specimen. In addition, the further the scanning point is away from the AE sensor, the sooner propagation waves received by the sensor will interfere with the reflection waves. In this way, signals from the distance near to the AE sensor should be preferentially selected.

As the solid circle in Figure 5.38b, the amplitudes can be evaluated as the value at coordinate (10 mm, 0.7 μs) in the position–time diagram (at the position 10 mm away from the sensor and at a propagation time = 0.7 μs), which is the first phase peak of S_0 mode filtered at 700 kHz (whose value is 2.55 V). The first peaks from a distance of 10–20 mm were searched (as the open circles in Figure 5.38b) and used to calculate the phase velocity (7,074.4 m/s) by LSF method. The wavelength of S_0 mode at $f = 700$ kHz can then be calculated as 10.11 mm, which is much smaller than the width of the specimen. The modulus of the laminate along the tension direction is then estimated to be 80.08 GPa by inverting phase velocities of S_0 mode.

DAMAGE ACCUMULATION EVALUATION

The specimen was then subjected to fatigue and NDE test with loading cycle intervals according to Table 5.9. The three parameters (amplitude, phase velocity, and the elastic modulus) were estimated periodically using method mentioned above. The fatigue life is 16,886 loading cycles (while the predicted fatigue life is 18,000 cycles), which made it impossible to observe the parameters of the final failure (by method shown in Table 5.9). Being divided by the original values at 0 load cycle, the normalized three parameters varying to the fatigue cycle were shown in Figure 5.39.

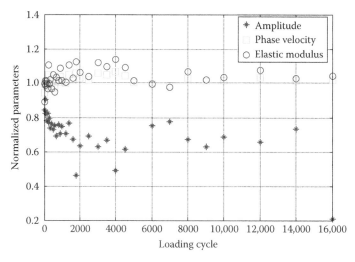

FIGURE 5.39   The relationships between the three normalized parameters and the fatigue loading cycle.

Unexpectedly, the phase velocities and the elastic modulus do not change obviously during the fatigue test. Amplitudes, on the contrary, experience a steep reduction during the early stage of the fatigue test. The normalized amplitude value at 1,600 cycle is as low as 0.67, which means that the amplitude (at 10 mm away from the AE sensor) dropped by 33% during the early 10% fatigue life.

The sharp and continuous drop of the amplitude value during the early fatigue life illustrates that certain types of fatigue damages appear and accumulate. However, the method cannot reliably distinguish between different types of damage, and therefore it is necessary to use microscopy or other NDE techniques to characterize the damage later.

The GA-based reconstruction method could be used to study how the three different stiffness coefficients (C_{11}, C_{13}, and C_{33}) vary to the fatigue load cycle. Moreover, the fatigue damage types and conditions are expected to be identified based on the measured amplitude, phase velocity or the stiffness coefficients.

CONCLUSIONS

The laser ultrasonic technique for NDE is exposed in this chapter. Based on the four schemes of scanning the laser, the wavefield that represents the waves propagate in the structure can be visualized through a series of the images with the different time points. Compared with the ultrasonic C-scan and the Lamb wave-based NDE approach, this technique can provide more information about the ultrasonic propagation in the damage region and it is a base-line free damage detection technique that is less venerable to the changing environmental and operational conditions. Furthermore, the complete noncontact inspection strategy is more competitive for onsite use and more practical for large-scale inspection.

The laser ultrasonic system is introduced in Section "Laser Ultrasonic Imaging System." Four different inspection strategies are usually used to achieve the wavefield visualization in recent research. The scheme of the fixed-point AE sensor and scanning the pulse laser has been described in detail and it has been implemented to validate the NDE approach. In section "Laser Ultrasonic Imaging for Damage Visualization," a damage detection approach based on calculating the AIW energy map has been proposed. Both the frequency–wavenumber analysis and the damage characteristic extraction have been used to evaluate the size and shape of the damage. The experiments in metal structure and composite material have been carried to validate the proposed method. Sections "Laser Ultrasonic Method for Reconstructing Material Properties of Composites" and "Laser Ultrasonic Method for Damage Accumulation Evaluation" review a laser-generation-based ultrasound method for NDE of damage accumulation in composites. The variation of three parameters (amplitude of Lamb waves, phase velocity of S_0 mode, and the elastic modulus of the specimen) were measured during the tension–tension fatigue process. The high spatial resolution of the laser system contributes to the accurate measurement of the amplitudes and the phase velocity in spite of the complex reflection from the specimen boundaries. The measurement of the elastic modulus is based on inverting the experimental phase velocities via GAs. The experimental results of a $[0/90]_{4S}$ laminate indicate that the phase velocity and the material property do not change remarkably during the fatigue test process. However, the amplitude of Lamb waves dropped sharply by 33% during the early 10% fatigue life, which means that the amplitude is a promising characteristic parameter that can be used to predict early fatigue damage for composites.

REFERENCES

1. B. W. Drinkwater, P. D. Wilcox. Ultrasonic arrays for non-destructive evaluation: A review. *NDT&E International*, 2006, 39(7):525–541.
2. X. Zhao, R. L Royer, S. E Owens, J. L Rose. Ultrasonic lamb wave tomography in structural health monitoring. *Smart Materials and Structures*, 2011, 20(10):105002.
3. D. J. Bull, L. Helfen, I. Sinclair, S. M. Spearing, T. Baumbach. A comparison of multi-scale 3D x-ray tomographic inspection techniques for assessing carbon fibre composite impact damage. *Composites Science and Technology*, 2013, 75:55–61.
4. A. Maier, R. Schmidt, B. Oswald-Tranta, R. Schledjewski. Non-destructive thermography analysis of impact damage on large-scale CFRP automotive parts. *Materials*, 2014, 7(1):413–429.
5. J. Cheng, H. Ji, J. Qiu, T. Takagi, T. Uchimoto, N. Hu. Novel electromagnetic modeling approach of carbon fiber-reinforced polymer laminate for calculation of eddy currents and eddy current testing signals. *Journal of Composite Materials*, 2014:0021998314521475.
6. S. Chatillon, G. Cattiaux, M. Serre, O. Roy. Ultrasonic non-destructive testing of pieces of complex geometry with a flexible phased array transducer. *Ultrasonics*, 2000, 38(1):131–134.
7. S. J. Song, H. J. Shin, Y. H. Jang. Development of an ultra sonic phased array system for nondestructive tests of nuclear power plant components. *Nuclear Engineering and Design*, 2002, 214(1):151–161.
8. J. E. Michaels. Detection, localization and characterization of damage in plates with an in situ array of spatially distributed ultrasonic sensors. *Smart Materials and Structures*, 2008, 17(3):035035.
9. V. Giurgiutiu. Embedded NDT with piezoelectric wafer active sensors, nondestructive testing of materials and structures. Springer, the Netherlands, 2013, pp. 987–992.
10. R. M. Levine, J. E. Michaels. Model-based imaging of damage with lamb waves via sparse reconstruction. *The Journal of the Acoustical Society of America*, 2013, 133(3):1525–1534.
11. J. Blitz, G. Simpson. *Ultrasonic methods of non-destructive testing.* Chapman & Hall, London, 1996.
12. V. Giurgiutiu, A. Zagrai, J. Bao. Damage identification in aging aircraft structures with piezoelectric wafer active sensors. *Journal of Intelligent Material Systems and Structures*, 2004, 15(9–10):673–687.
13. J. B. Ihn, F. K. Chang. Pitch-catch active sensing methods in structural health monitoring for aircraft structures. *Structural Health Monitoring*, 2008, 7(1):5–19.
14. V. Giurgiutiu. Tuned lamb wave excitation and detection with piezoelectric wafer active sensors for structural health monitoring. *Journal of Intelligent Material Systems and Structures*, 2005, 16(4):291–305.
15. J. Takatsubo, B. Wang, H. Tsuda, N. Toyama. Generation laser scanning method for the visualization of ultrasounds propagating on a 3-D object with an arbitrary shape. *Journal of Solid Mechanics and Materials Engineering*, 2007, 1:1405–1411.
16. S. Yashiro, J. Takatsubo, N. Toyama. An NDT technique for composite structures using visualized lamb-wave propagation. *Composites Science and Technology*, 2007, 67(15):3202–3208.

17. J. R. Lee, H. Jeong, C. C. Ciang, D. J. Yoon, S. S. Lee. Application of ultrasonic wave propagation imaging method to automatic damage visualization of nuclear power plant pipeline. *Nuclear Engineering and Design*, 2010, 240(10):3513–3520.

18. Y. K. An, B. Park, H. Sohn. Complete noncontact laser ultrasonic imaging for automated crack visualization in a plate. *Smart Materials and Structures*, 2013, 22(2):025022.

19. J. R. Lee, S. Y. Chong, N. Sunuwar, C. Y. Park. Repeat scanning technology for laser ultrasonic propagation imaging. *Measurement Science and Technology*, 2013, 24(8):085201.

20. S. Yashiro, J. Takatsubo, H. Miyauchi, N. Toyama. A novel technique for visualizing ultrasonic waves in general solid media by pulsed laser scan. *NDT&E International*, 2008, 41(2):137–144.

21. H. Sohn, D. Dutta, J. Y. Yang, H. J. Park, M. DeSimio, S. Olson, E. Swenson. Delamination detection in composites through guided wave field image processing. *Composites Science and Technology*, 2011, 71(9):1250–1256.

22. D. Dhital, J. R. Lee. A fully non-contact ultrasonic propagation imaging system for closed surface crack evaluation. *Experimental Mechanics*, 2012, 52(8):1111–1122.

23. M. Morii, N. Hu, H. Fukunaga, J. H. Li, Y. L. Liu, S. A. Alamusi, J. H. Qiu. A new inverse algorithm for tomographic reconstruction of damage images using lamb waves. *Computers Materials and Continua*, 2011, 26(1):37.

24. M. Fink, C. Prada. Acoustic time-reversal mirrors. *Inverse Problems*, 2001, 17(1):R1.

25. J. D. Achenbach. *Reciprocity in Elastodynamics*. Cambridge University Press, Cambridge, 2003.

26. J. R. Lee, C. C. Chia, C. Y. Park, H. Jeong. Laser ultrasonic anomalous wave propagation imaging method with adjacent wave subtraction: Algorithm. *Optics and Laser Technology*, 2012, 44(5):1507–1515.

27. C. Zhang, J. Qiu, H. Ji. Laser ultrasonic imaging for impact damage visualization in composite structure. In *Proceedings of the 7th European Workshop on Structural Health Monitoring*, Nantes, France, 2014.

28. C. C. Chia, J. R. Lee, C. Y. Park, H. M. Jeong. Laser ultrasonic anomalous wave propagation imaging method with adjacent wave subtraction: Application to actual damages in composite wing. *Optics and Laser Technology*, 2012, 44(2):428–440.

29. M. Ruzzene. Frequency–Wavenumber domain filtering for improved damage visualization. *Smart Materials and Structures*, 2007, 16(6):2116.

30. T. E. Michaels, J. E. Michaels, M. Ruzzene. Frequency–Wavenumber domain analysis of guided wavefields. *Ultrasonics*, 2011, 51(4):452–466.

31. ASTM Standard D7136/d7136m–05. Standard test method for measuring the damage resistance of a fiber-reinforced polymer matrix composite to a drop-weight impact event. West Conshohocken, PA: ASTM International, 2005.

32. N. Laws, G. J. Dvorak, M. Hejazi. Stiffness changes in unidirectional composites caused by crack systems. *Mechanics of Materials*, 1983, 2(2):123–137.

33. S. L. Ogin, P. A. Smith, P. W. R. Beaumont. Matrix cracking and stiffness reduction during the fatigue of a (0/90) s GFRP laminate. *Composites Science and Technology*, 1985, 22(1):23–31.

34. H. A. Whitworth. A stiffness degradation model for composite laminates under fatigue loading. *Composite Structures*, 1997, 40(2):95–101.

35. M. D. Seale, B. T. Smith, W. H. Prosser. Lamb wave assessment of fatigue and thermal damage in composites. *The Journal of the Acoustical Society of America*, 1998, 103(5):2416–2424.

36. J. Zhao, J. Q., J. Zong, W. Yao, H. Ji. Non-destructive evaluation of damage accumulation for composites using a laser-generated lamb waves method. In *Proceedings of the 9th Asian-Australasian Conference on Composite Materials*, Suzhou, China, 2014.

37. H. Nishino, T. Tanaka, K. Yoshida, J. Takatsubo. Simultaneous measurement of the phase and group velocities of lamb waves in a laser-generation based imaging method. *Ultrasonics*, 2012, 52(4):530–535.

38. M. Castaings, B. Hosten. Delta operator technique to improve the Thomson–Haskell-method stability for propagation in multilayered anisotropic absorbing plates. *The Journal of the Acoustical Society of America*, 1994, 95(4):1931–1941.

39. B. Hosten, M. Castaings. Transfer matrix of multilayered absorbing and anisotropic media. Measurements and simulations of ultrasonic wave propagation through composite materials. *The Journal of the Acoustical Society of America*, 1993, 94(3):1488–1495.

40. M. J. S. Lowe. Matrix techniques for modeling ultrasonic waves in multilayered media. *IEEE Transactions on Ultrasonics, Ferroelectrics, and Frequency Control*, 1995, 42(4):525–542.

41. A. H. Nayfeh. *Wave propagation in layered anisotropic media: With application to composites*. Elsevier, Amsterdam, the Netherlands, 1995.

42. S. I. Rokhlin, L. Wang. Ultrasonic waves in layered anisotropic media: Characterization of multidirectional composites. *International Journal of Solids and Structures*, 2002, 39(16):4133–4149.

43. J. L. Rose. *Ultrasonic waves in solid media*. Cambridge University Press, New York, 2004.

44. J. Zhao, J. Qiu, H. Ji, N. Hu. Four vectors of lamb waves in composites: Semianalysis and numerical simulation. *Journal of Intelligent Material Systems and Structures*, 2013:1045389X13488250.

45. R. D. Mindlin. Influence of rotary inertia and shear on flexural motions of isotropic elastic plates. *ASME Journal of Applied Mechanics*, 1951, 18:31–38.

46. Y. Liu, N. Hu, C. Yan, X. Peng, B. Yan. Construction of a Mindlin pseudospectral plate element and evaluating efficiency of the element. *Finite Elements in Analysis and Design*, 2009, 45(8):538–546.

47. A. Żak. A novel formulation of a spectral plate element for wave propagation in isotropic structures. *Finite Elements in Analysis and Design*, 2009, 45(10):650–658.

48. J. M. Whitney, C. T. Sun. A higher order theory for extensional motion of laminated composites. *Journal of Sound and Vibration*, 1973, 30(1):85–97.

49. L. Wang, F. G. Yuan, Lamb wave propagation in composite laminates using a higher-order plate theory. In *Proceedings of the Society of Photo-Optical Instrumentation Engineers: The 14th International Symposium on Smart Structures and Materials and Nondestructive Evaluation and Health Monitoring*, San Diego, CA, 2007, pp. 65310I-65310I-12.

50. J. Zhao, H. Ji, J. Qiu. Modeling of lamb waves in composites using new third-order plate theories. *Smart Materials and Structures*, 2014, 23(4):045017.

51. P. E. Lagasse. Higher-order finite-element analysis of topographic guides supporting elastic surface waves. *The Journal of the Acoustical Society of America*, 1973, 53(4):1116–1122.

52. K. H. Huang, S. B. Dong. Propagating waves and edge vibrations in anisotropic composite cylinders. *Journal of Sound and Vibration*, 1984, 96(3):363–379.

53. H. Matt, I. Bartoli, F. L. D. Scalea. Ultrasonic guided wave monitoring of composite wing skin-to-spar bonded joints in aerospace structures. *Journal of the Acoustical Society of America*, 2005, 118(4):2240–2252.

54. H. Gao. Ultrasonic guided wave mechanics for composite material structural health monitoring. *Dissertation Abstracts International*, 2007, Vol. 68-05, Section: B, p. 3346.

55. L. Wang, F. G. Yuan. Group velocity and characteristic wave curves of lamb waves in composites: Modeling and experiments. *Composites Science and Technology*, 2007, 67:1370–1384.

56. D. H. Robbins, J. N. Reddy. Modelling of thick composites using a layerwise laminate theory. *International Journal for Numerical Methods in Engineering*, 1993, 36(4):655–677.

57. M. J. S. Lowe, G. Neau, M. Deschamps, Properties of guided waves in composite plates, and implications for NDE. *Review of Progress in Quantitative Nondestructive Evaluation*, 2004, Vol. 23, pp. 214–221.

58. J. Vishnuvardhan, C. V. Krishnamurthy, K. Balasubramaniam. Genetic algorithm based reconstruction of the elastic moduli of orthotropic plates using an ultrasonic guided wave single-transmitter-multiple-receiver shm array. *Smart Materials and Structures*, 2007, 16(5):1639–1650(12).

6 Sensor Networks for Structural Damage Monitoring
Sensors and Interrogation Techniques

Zhanjun Wu, Rahim Gorgin,
Dongyue Gao, and Yinan Shan

CONTENTS

To provide enough information for damage detection, a number of transducers are usually used to configure a sensor network. By acquiring Lamb waves at different locations of the monitoring area, sensors in the network can provide useful information for monitoring structural damages. Employing a sensor network instead of a single sensor will increase the identification confidence and improve tolerance to measurement noise.

TABLE 6.1
Comparison of Guided Wave Sensors with Other NDE Sensors

Sensor	Response Events	Effective Area	Objects	Pattern
Optical fiber	Load, impact, delamination	Local	Metal and composite	Online
Ultrasonic	Delamination, cracks	Global	Metal and composite	Offline
Piezoelectric	Load, impact, delamination, cracks	Global	Metal and composite	Online
AE sensor	Impact, cracks, delamination	Global	Metal and composite	Online
EMAT	Cracks, corrosion	Global	Metal and composite	Online
Microwave sensor	Water ingress	Local	Composite sandwich	Offline
Eddy sensor	Cracks, corrosion	Local	Metal	Offline

The efficiency of the sensor network depends on the density of sensors. Nevertheless, a very dense network is impractical for some applications and will increase the cost. Therefore, practicability and desired resolution must be taken into consideration to design a suitable sensor network for a special application.

In order to design a sensor network, different kinds of sensors can be used. The most commonly used transducers are piezoelectric sensors [1,2], optical fiber [3], lasers [4,5], and electromagnetic acoustic transducers (EMATs) [6], which are summarized in Table 6.1 and compared with other nondestructive evaluation (NDE) sensors. Although some types of sensors function well for maintenance checks when the equipment is offline for service, they are not compact enough to be permanently onboard the key equipment during its operation as required for structural health monitoring (SHM). This is particularly true in aerospace structures, in which the mass and space penalties associated with the additional sensors on the structure should be minimal.

In this chapter, some practical sensor networks using different kinds of sensor are presented.

FIBER BRAGG GRATING AND DISTRIBUTED NETWORKS

In the past decades, two major product revolutions have taken place due to the growth of the optoelectronics and fiber optic communications industries: the laser and the modern low-loss optical fiber [7]. Thus, researchers were able to do some of the first experiments on optical fibers, not for telecommunications—as had been the prime motivation for their development—but for sensor purposes. The application of optical fiber as a kind of sensor is a breakthrough of measurement field. So far, strain, temperature, and pressure are the most widely studied measurands, whereas the changes can be measured by the means of detecting the fluctuations of light intensity, phase, frequency, polarization state, and other parameters of the optical fiber. However, in various fields, new ideas are being continuously developed and tested not only on traditional measurements but also on new applications. The development and applications of optical fiber sensor technology can be found in References [8–11].

The end users simply desire sensor systems having good performances with reasonable price. Being small, noninvasive, immune to electromagnetic interference, erosion-resisting, and sensitive to many parameters of light, the fiber Bragg grating (FBG) can be used selectively to measure strain and temperature, as a relatively new sensor. In 1978, Hill and coworkers first discovered the ultraviolet photosensitivity of germanium-doped optical fiber and demonstrated the possibility of an intracore FBG [12,13]. They described a permanent grating written in the core of the fibers by the argon ion laser line at 488 nm launched into the fiber. This particular grating had a very weak index modulation, which was estimated to be on the order of 10^{-6} resulting in a narrowband reflection filter at the writing wavelength. However, by developing the side holographic exposure technique for FBG strain sensors production by Meltz [14–16] in 1989 and grating production online as the optical fiber is drawn [17], the FBG manufacturing technique have been refined and the devices are readily available and reproducible, which greatly stimulate the application of FBG sensors.

In the beginning of the twenty-first century, Mark Froggatt and his coworkers developed a method by which coherent optical frequency domain reflectometry (OFDR) can be used to "key" portions of fiber by measuring their complex Rayleigh backscatter signatures. They found that these complex keys can be used to locate specific fiber lengths embedded within a parallel optical network, and that they can further be used to interrogate any induced loss or temperature change in the identified portion of the network [18]. Also, they used OFDR to measure the group-index difference and the refractive-index difference between the fast and slow modes in polarization-maintaining optical fiber. This measurement enables a distributed measurement of the fiber's birefringence that is rapid and completely nondestructive [19–21]. Dawn Gifford, Brian Soller, and their coworkers made progress on high-resolution fiber reflectometry and distributed optical fiber sensing using Rayleigh backscatter. They introduced a tunable laser-based technique for submillimeter-resolution fiber reflectometry and showed how the technique is used to locate bends, breaks, bad splices, and poor connections in short-haul single- and multimode fiber links [22]. Their work also includes demonstrating the enhanced accuracy, precision, and dynamic range of the technique through measurements of several components [23]. In the beginning, application of this technique was temperature measurement, and the optical fiber was started to be treated as a kind of distributed optical fiber sensor. Later on in 2007, Dawn Gifford and his coworkers found that the Rayleigh scatter complex amplitude can be Fourier transformed to obtain the Rayleigh scatter optical spectrum, and shifts in the spectral pattern can be related to changes in strain or temperature.

This chapter addresses some key aspects concerning the FBG sensors, the distributed optical fiber sensors and sensor networks, including approaches to applying for SHM, and some representative applications.

SENSORS

In both scientific researches and industrial applications, the most commonly used optical fiber sensors are FBGs and distributed optical fiber sensors [24]. Although the FBGs have proven satisfactory performance on SHM by large amount of applications and long-term testings, the distributed optical fiber sensors are developing fast on maintenance process of key industrial equipment and other researches, as distributed optical fiber sensors own more measure points and are not affected by measurement range of measurands.

FBG Sensors

The basic principle of measurement commonly used in an FBG-based sensor system is to monitor the shift in wavelength of the returned signal, which is sensitive to the changes of measurands (e.g., strain, temperature) [25,26]. FBG sensing system is shown in Figure 6.1. Injecting spectrally broadband source of light into the fiber, a narrowband spectral component at the Bragg wavelength is reflected by the grating. In the transmitted light, this spectral component is missing.

The strongest interaction or mode coupling taking place at the Bragg wavelength λ is given by [14,16]

$$\lambda = 2n\Lambda \tag{6.1}$$

where:
   n is the modal index
   Λ is the grating period

Concluded by Equation 6.1, if any measurand affects the variation of the effective refractive index or of the grating period, it can result in the transformation of the wavelength of FBG. In the practiced application, the strain sensitivity factor

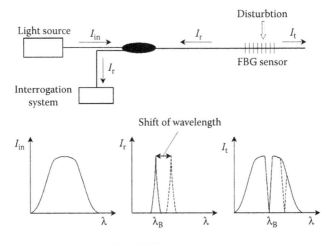

FIGURE 6.1   Sensing principle of the FBG sensors.

and temperature sensitivity factor of the FBG can be measured by experiments. All of the center wavelength shift of FBG due to the strain and temperature is expressed by

$$\Delta\lambda = K_e\Delta\varepsilon + K_T\Delta T \tag{6.2}$$

where:
K_e is the strain sensitivity factor
K_T is the temperature sensitivity factor

The separated FBG cannot distinguish the changes of the center wavelength of FBG due to strain and temperature, respectively. In order to eliminate the measurement error introduced by the temperature fluctuation of the environment, temperature compensation technology should be taken when FBG is used as a strain sensor.

The differential of Equation 6.1 is

$$d\lambda = 2\Lambda dn + 2n d\Lambda \tag{6.3}$$

Dividing Equation 6.3 by Equation 6.1, we get

$$\frac{d\lambda}{\lambda} = \frac{dn}{n} + \frac{d\Lambda}{\Lambda} \tag{6.4}$$

In linear elastic range, there is

$$\frac{d\Lambda}{\Lambda} = \varepsilon \tag{6.5}$$

where ε is strain.

Neglecting the influence of radial of deformation of optical fiber on its refractive index, the refractive index variation resulted from uniaxial deformation is given by [27]

$$\frac{dn}{n} = -\frac{n^2}{2}\left[p_{12} - \upsilon\left(p_{11} + p_{12}\right)\right]\varepsilon \tag{6.6}$$

where:
p_{11} and p_{12} are photoelastic constants
υ is the Poisson ratio

Let

$$P = \left[p_{12} - \upsilon\left(p_{11} + p_{12}\right)\right]\frac{n^2}{2} \tag{6.7}$$

Combining Equations 6.4, 6.5, and 6.6, we get

$$\frac{\Delta\lambda}{\lambda} = (1 - P)\varepsilon \tag{6.8}$$

which is the function of strain against wavelength shift assuming that temperature keeps constant. When the core of optical fiber is quartz, $n = 1.456$, $p_{11} = 0.121$, $p_{12} = 0.270$, $\upsilon = 0.17$, the value of P is about 0.22. Given that the Bragg wavelengths are 1545, 1550, and 1555 nm, respectively, it can be calculated with Equation 6.8 that the wavelength shifts caused by per microstrain almost remain constant, provided that the Bragg wavelength changes reasonably.

The strain-transducing features and temperature response of FBG have been tested [27,28], and the results are shown in Figures 6.2 and 6.3. They show good linear relationships between the Bragg wavelength and the strain, as well as the Bragg

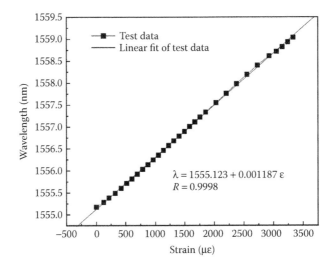

FIGURE 6.2   Strain response of FBG.

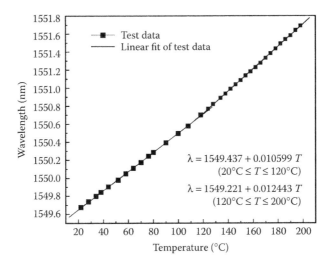

FIGURE 6.3   Temperature response of FBG.

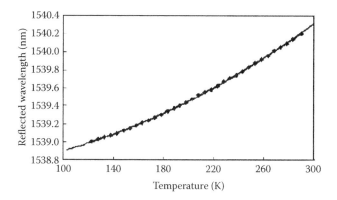

FIGURE 6.4 Bragg wavelength versus temperature for a FBG sensor with 1540 nm center wavelength.

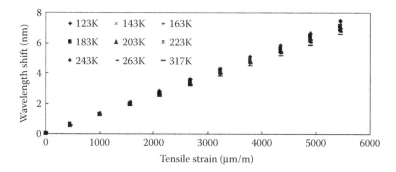

FIGURE 6.5 Bragg wavelength shift versus strain in the temperature range of 123–317K.

wavelength and the temperature, which prove FBG an ideal strain-sensing and a temperature-sensing element. Also, researchers have confirmed experimentally the nonlinear temperature dependence of the Bragg wavelength and strain responses of FBG sensors over a temperature range of 123–317 K [29]. It was found that the strain dependence was linear at constant temperature, and the strain sensitivity is temperature dependent, increasing with decreasing temperature. The results are shown in Figures 6.4 through 6.6.

Distributed Optical Fiber Sensors

They are different from discrete optical fiber sensors such as FBG sensors that distributed optical fiber sensors can help to overcome issues, for example, the strain and temperature profiles have a steep gradient, or spatially extended structures need to be precisely monitored through sensing solutions that enable a quasi-continuous distributed measurement along the length of sensing fibers. Raman and Brillouin scattering within the fibers have been established for long-distance applications, whereas the extremely low intensity of the scattered light used here makes it very difficult to increase the resolution [30]. Scanning the Rayleigh scattering along a length of optical

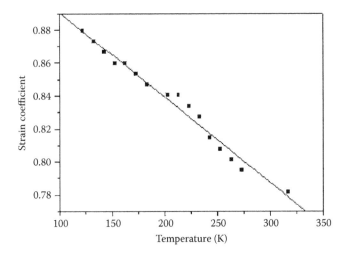

FIGURE 6.6   Strain coefficient versus temperature for an FBG sensor with 1535 nm center wavelength.

fiber enables a distributed sensor system in which every point along the fiber acts as a sensor, with a spatial resolution of millimeters when it is used as a strain sensor.

For a glass fiber, a fluctuating intensity profile of the Rayleigh backscattering along the glass fiber will be detected by OFDR technique [31–34]. The Rayleigh backscattering in optical fibers is caused by random fluctuations in the index profile along the fiber length, which is a static property for a given optical fiber and can be treated as a function of position in the optical fiber [19,35]. The baseline data is the Rayleigh backscattering signature of the fiber under test stored at an ambient state. A typical Rayleigh backscattering signature along an optical fiber segment is shown in Figure 6.7 [36].

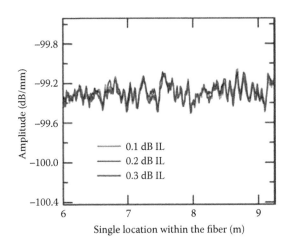

FIGURE 6.7   A typical Rayleigh backscattering signature along an optical fiber segment.

A change in strain or temperature results in a shift in the spectrum of light backscattered in the fiber. The locations where changes in strain and temperature happen can be ensured by comparing the Rayleigh backscattering signature with baseline. This shift in the spectrum in response to strain ε or temperature T is analogous to a shift in the resonance wavelength $\Delta\lambda$ or the spectral shift Δv, which is similar to FBG sensors [35]:

$$\frac{\Delta\lambda}{\lambda} = -\frac{\Delta v}{v} = K_T\Delta T + K_\varepsilon\varepsilon \tag{6.9}$$

where:

λ and v are the mean optical wavelength and frequency, respectively

K_T and K_ε are the temperature and strain calibration constants, respectively

SENSOR KITS AND NETWORK

Sensor Network Configuration

To provide enough information for basic measurement and other applications, a number of sensors are usually used to configure a sensor network. By acquiring information of measurands at different locations of the measuring area, sensors in the network can provide useful information for different applications.

The efficiency of the sensor network depends on the density of sensors. Nevertheless, a very dense network is impractical for some applications and will increase the cost. Therefore, practicability and desired resolution must be taken into consideration to design a suitable sensor network for a special application.

The sensor networks discussed in the sections that follow include the FBG sensor network and the distributed optical fiber sensor network, providing applications of identification and analysis.

FBG Sensor Network

Introduction

It is well known that transportation is the economic lifeline of a country, and bridges are the throat of communication. At present, there are thousands of main highways and bridges in China, among which over one-third are structurally defective, more or less damaged, and potentially threatened by functional degradation. Some of the key bridges need to be monitored, evaluated, and maintained, repaired, or reinforced. Large and important bridges are always designed to work for many decades or even longer. Accumulated damages inevitably result from fatigue, corrosion, or aging effects. Sometimes accidents could happen (e.g., Shenshui Bridge in Seoul, South Korea, in 1994; Zhaobaoshan Bridge in Ningbo, China, in 1998, and Qijiang Caihong Bridge, in Qijiang, China, in 1999; Gaoping Bridge in Taiwan, China, in 2000) [37,38]. Therefore, for bridges already in service, it is in urgent need to take effective measures to monitor and evaluate the health state, repair and control the damage, and warn regarding accidents beyond human control. For bridges to be built, it is necessary to take into consideration a long period of health monitoring and damage controlling while designing the bridges; thus, safety, operational suitability, and durability of the bridges can be ensured through the full life of bridges [12,38,39].

Traditional electricity-based sensors face difficulties when they are utilized in bridge structures, in which the environment of those sensors is very harsh. Long-period health monitoring systems consisting of those sensors always fail to provide reliable data for the purpose of safety evaluation, though the theory of evaluation is rather mature. FBGs have many advantages in this kind of application, such as immunity to electric/magnetic interference, absolute strain sensing capability, ease of multiplexing, and good compatibility with a range of structural materials. In addition, resistant to high temperature, moisture, and chemical corrosion also make it attractive under certain circumstances.

Hulan River Bridge is located on Hulan River, near Hulan county in Heilongjiang Province, which is the communication hinge of Harbin and Zhaoxing and was put to use in the summer of 2002. The overall length of the bridge is 420 m. The maximum span length is 42 m. It is rather a large bridge with prestressed steel-reinforced concrete box beam. The concrete box beam is prepared with post-tension method, and steel strands without bonding are used as main reinforcements. In the lower part of the bridge, the bored piles are adopted as the foundation, and column piers and ribbed abutments are used in order to meet the requirements of complex foundation and formidable construction situation. Due to construction complexity, short construction time, long span, heavy traffic, and large variations in temperature between winter and summer, it is representative to install *in situ* health monitoring system on this bridge.

System Description

In order to make the monitoring results representative, FBG strain sensors were adhered to costal fovea and embedded into two 42 m box beams, which are the second and fifth beams of the second span. The location of the FBG strain sensors is illustrated in Figure 6.8.

The outer diameter of the FBG is about 125 μm and made of quartz; hence, the FBG is rather fragile, especially when it is under shear stress. Thus, the protection technique here becomes very important to the success of monitoring. In order to avoid the damage of the FBG strain sensors during the casing, vibrating, and pressing of concrete and installation of the box beam, an installation method is developed. The installation procedure is described as follows: first, at an earlier selected location, the costal fovea of rebar was sanded and cleaned to make for a good bonding for FBG strain sensors and foil strain gauges. Then, two foil gauges and one FBG strain sensor were collocated and adhered into the costal fovea with 502# glue (acrylate series). Finally, those sensors were protected by silicon gel layer, gauze wrap, and J39 glue in turn. In order to avoid communication optical fiber breakdown

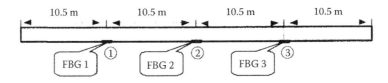

FIGURE 6.8   FBG installed along the longitudinal direction of box girder.

during concrete pouring, vibrating, and compressing, it was put into an aluminum/plastic pipe of 8 mm in diameter; then the pipe was fixed onto the structural rebar of rib plate of the box beam. The communication optical fiber with shield pipe was extended out of the box beam surface and then covered with a wooden box. With these measures, after the construction of the box beam, the survival of FBG and foil gauges was tested instantly. It was found that the five FBG strain sensors all worked well, whereas 3 of 10 foil gauges gave stable signals.

Monitoring Results

Prestress Process Monitoring of the Box Beam   The prestress process of the stranded wire affects the final quality of the box beam. During the process, pre-stressing forced relaxation, over stressing, and inverted arch may occur, which greatly influence the ultimate structural performance of the box beam. Therefore, the prestress process was monitored with those sensors embedded. The rebar rein-forced concrete box beam, embedded with FBG strain sensors and foil gauges, was prestressed after 10 days of curing. From the inspection made before prestress, it was found that the extended part of two FBG strain sensors embedded was unfortunately destroyed by mishandling of the site engineer, and only one foil gauge could give a stable signal. The prestress process was monitored by the remaining three FBG strain sensors and one foil gauge. The strand wires were tensioned at both ends with oil pressure jack and pinion rack. The prestress sequence of strand wires was crossed from bottom to top, left to right. The strain of every load measured by FBG strain sensors and foil gauge is recorded and is shown in Figures 6.9 and 6.10. Because the prestress duration is relatively short and the concrete has very low thermal conduc-tivity, the influence of the temperature on FBG strain sensors was neglected.

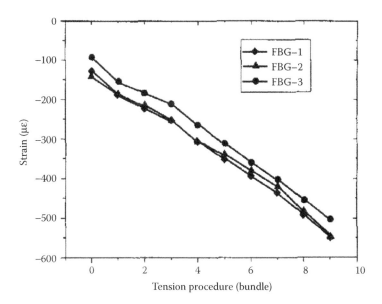

FIGURE 6.9   Results of the measurement of three FBG strain sensors.

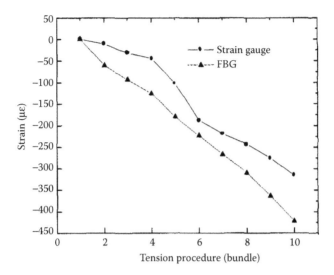

FIGURE 6.10   Comparison of electric strain gauge and FBG strain sensors.

From the test results, it can be seen that the signals monitored by FBG strain sensors are stable and reliable, and the post-tension process is well revealed. The results of the foil gauge are apparently unstable, which prove that FBG strain sensors have overwhelming advantages compared to traditional foil gauges, especially when they are embedded in concrete structures.

Quasi-Static Load Test of the Box Girder   After 28-day-long cure, the three FBG strain sensors remain intact, and only the foil gauge is out of use. In order to investigate the capability of the FBG strain sensors to monitor the state of the bridge after it is completed, the advantages of construction site are fully used. Two short box girders of 19.96 m length and 62 tons of weight are put on the surface of the box beam, where FBG strain sensors are embedded in turn as loading. The scheme loading and test results are shown in Figure 6.11 and Table 6.2, respectively.

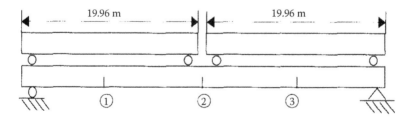

FIGURE 6.11   Sketch of dead load on the box girder.

TABLE 6.2
Results of FBG When a Box Girder Is under a Dead Load

Loading (tons)	FBG1 (µɛ)	FBG2 (µɛ)	FBG3 (µɛ)
62	69	108	45
132	112	213	100

Because the resolution of FBG can reach 1 pm, and hence 1 µɛ can be identified, and 1 ton weight can produce strain over 1 µɛ at the monitoring point, which can be deduced from the test results, any automobile with over 1 ton weight that passes on the bridge can be detected by the FBG strain sensors, which could be used to give statistic results of the traffic flow. In addition, fatigue damage could be deduced from the monitoring results, too, which could be used to assess the health condition of the bridge. This is the ultimate purpose of installation of FBG strain sensors on bridges.

Quasi-Static Load Test of the Bridge   Before the bridge was put into service, quasi-static load test of the bridge was carried out. During the test, local strain response monitoring was carried out with embedded FBG strain sensors. All sensors were tested prior to the monitoring process. Almost all the FBG strain sensors installed during construction survived, except one outputting unstable signal, which is probably caused by too much power optic loss because of excessive curvature of the transmission optical fiber. Meanwhile, only one of the foil strain gauges installed still works.

During the quasi-static load test process, two heavy trucks fully loaded with sand were moved to the midspan of the box beam in which FBG strain sensors were installed and remained there for a while. Strains monitored by FBG strain sensors are shown in Figures 6.12 and 6.13, and the temperature monitored by reference FBG is shown in Figure 6.14. Figure 6.15 shows that during the loading process, there are little changes of temperature, which should not be taken into account.

Service Monitoring of the Bridge with FBG Strain Sensors   The Hulan River was completed and put into service after 1 year of construction. With the FBG strain sensors' help, the deformation and temperature evolution of the bridge in service were monitored.

Local strain response of the bridge under random vehicle loads is shown in Figure 6.15. The strain history monitored can be utilized as load spectrum directly or indirectly to evaluate fatigue degradation of the bridge. Meanwhile, it can also be utilized to calculate the deflection of the bridge, which is an important information

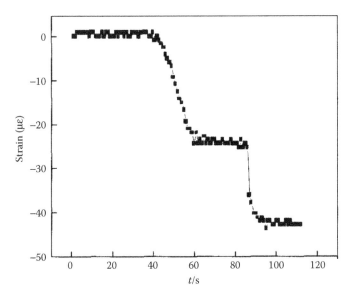

FIGURE 6.12 Evolution of strain of mid-span in the static test.

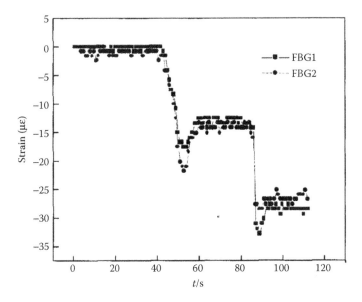

FIGURE 6.13 Evolution of strain of quarter-span in the static test.

for bridge health evaluation too. In addition, the strain history can reveal the traffic flow, which not only includes the number of vehicles passing and their passing speed but also loads distribution of the bridge in time domain, which are all key pieces of information for bridge management.

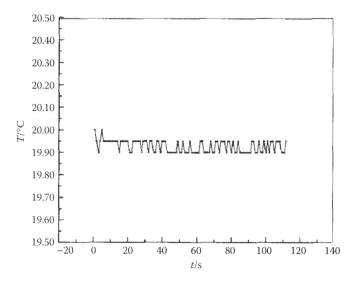

FIGURE 6.14   Temperature variance during the quasi-static test.

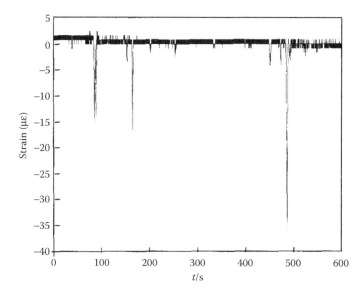

FIGURE 6.15   Evolution of strain of mid-span in 10 min.

The wavelength shift of the reference FBG temperature sensors, which are embedded in the box beam at the same depth as the FBG strain sensors, is shown in Figure 6.16. It can be seen that the variance frequency of the temperature is much lower than that of the strain. Therefore, if only vehicle loading is taken into consideration, it is obviously not necessary to make temperature compensation for bridge health and management monitoring employing FBG strain sensors.

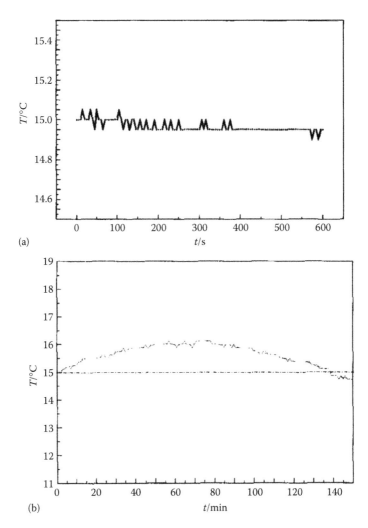

(a)

(b)

FIGURE 6.16 Bridge's temperature–time relationship monitored by the embedded FBG: (a) in 10 min and (b) in 150 min.

Concluding Remarks

1. Stability and durability of FBG is far better than that of traditional foil gauge when embedded in concrete structure.
2. FBG strain sensors can be used to monitor the fabrication process, especially the rebar deformation of box beam during post-tension process effectively, which can provide useful information for the design, fabrication process control, and construction technique.
3. FBG strain sensor-based bridge health monitoring system can be used to monitor the strain and temperature history of the bridge, which is essential

for fatigue degradation and deflection analysis. It greatly contributes to the health evaluation and service life prediction, so that catastrophic failure can be avoided and condition-based maintenance can be carried out, thereby increasing the safety and lowering the cost of the full life of the bridge.

4. FBG strain sensor-based bridge health monitoring system can be utilized to count vehicle numbers, load distribution in time domain, and passing speed of vehicles on the bridge, which are useful data for the management department to improve the efficiency of the bridge.

Distributed Optical Fiber Sensor Network

Introduction

A foil gauge is a traditional strain sensor, which is a kind of single point measurement and will need a large amount of metallic wires to connect with strain indicator in practical measurements on large-scale structures. It results in the requirements of an immense amount of time and manpower to paste foil gauges and weld wires, and multiple strain indicators should work in parallel predictively.

The development of optical technology brought about discrete optical fiber sensors, such as FBG sensors. The number of measuring points is limited by the bandwidth of light source [26]. A larger measuring range will lead to a smaller number of measuring points, which make it hard to achieve enough structural information for large-scale structures. During the beginning of the twenty-first century, the optical technology developed to a higher level. Researchers utilized swept-wavelength interferometry to interrogate distributed optical fiber sensors, which is sensitive to local strain and temperature through examining the fluctuation of the Rayleigh backscatter. The distributed optical fiber sensors enable distributed strain and temperature measurements with millimeter-range spatial resolution, which is an advantage over FBG sensors.

The Rayleigh backscatter can be measured as a function of position in the optical fiber [35], whereas the Rayleigh backscatter in the optical fiber is caused by random fluctuations in the index profile along the fiber length. The Rayleigh backscatter amplitude is a random but static property of a given fiber without any changes in fiber status. This character is called the inherent texture information of optical fibers [40]. By comparing the scattered light of a sensor to a reference measurement that was recorded with the fiber in a known state, one can determine the physical state of the fiber at the time of measurement.

In this research, measuring features and results between distributed optical fiber sensors and foil gauges are compared through the bending test of nonuniform thickness of the cantilever beam made of aluminum. Then, the surface strain field of the cantilever beam is rebuilt with the strain data measured by distributed optical fiber sensor, and changes in stiffness of the cantilever beam are identified. Also, the bending test of the nonuniform thickness of the cantilever beam is simulated on ANSYS. The results between the simulation and the test are compared as well.

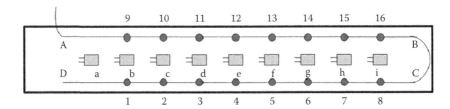

FIGURE 6.17   Test specimen.

System Description

The experimental subject is an aluminum with the size of $50 \times 4 \times 0.3$ cm, and a section of 10 cm length from the left side is fixed on the experiment table, whereas the right side is free. An aluminum tablet with the thickness of 0.3 cm is pasted on the lower surface at the position of 19–23 cm from the left side, which enhanced the local stiffness. A distributed optical fiber sensor and nine foil gauges are used during the measurement, as is shown in Figure 6.17. The distributed optical fiber sensor is connected with an interrogator through a standoff cable at point A. Pasted along the beam, the measuring sections are segment AB and segment CD, with eight foil gauges marked as "b–i" to measure the strain and one foil gauge marked as "a" for temperature compensation. The foil gauges marked as "b–i" are pasted on the beam uniformly spaced, whereas the foil gauge marked as "a" is contacted instead of being pasted.

Using the hot-spot method, the locations of points A, B, C, and D are recorded as starting points and ending points of the fiber segments, which is necessary during the postprocessing afterward. We keep a fiber segment with the length of about 20 cm, which is unstrained, to measure the temperature as the fiber's temperature compensation. A mass block with small size is used as a loading at the free side of the beam.

Monitoring Results

Measuring data of the distributed optical fiber sensor and foil gauges were recorded when the cantilever beam was free from external loads, which was treated as reference state. The cantilever beam was loaded by mass blocks with different weights as point loads at the location of 40 cm from the left side. When the cantilever beam became static, measuring data of the distributed optical fiber sensor and the foil gauges were recorded. Consider an example with the point load of 1.64 N to see the results. During the simulation with ANSYS, the surface strain along the two lines on which the distributed optical fiber sensor was exactly pasted was recorded and compared with measuring data of the distributed optical fiber sensor in three different gauge lengths and the foil gauges. The result is shown in Figure 6.18. Strain data of the distributed optical fiber sensor had already been got rid of the effect from the temperature through temperature compensation. The label on the x-axis expresses the location of sensors, and the label on the y-axis expresses the microstrain.

The image demonstrates that the measured strain data of the distributed optical fiber sensor in three different gauge lengths are consistent with numerical simulation analysis, although the measuring results of foil gauges fit well. The distinct wave trough signifies a sharp drop in strain due to the changes in thickness of the cantilever beam.

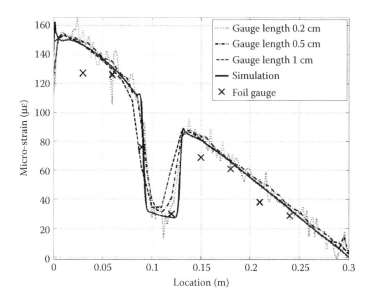

FIGURE 6.18 Comparison among results from distributed optical fiber sensor of three different gauge lengths, foil gauges, and simulation.

For the measuring results achieved at the 2 mm gauge length, the fluctuations of the strain data come from noise. In the region of wave trough, the strain data measured by the distributed optical fiber sensor depart somewhat from those of numerical simulation analysis, which could be judged as a difference between the actual structure and the numerical model. During the numerical simulation analysis, a Boolean operation was applied to glue an extra aluminum plate. For the actual structure, the extra aluminum plate was glued by a kind of a two-component epoxy adhesive, of which the stiffness is much smaller than that of aluminum. When the cantilever beam bends, the measured strain data are impacted by the strain transfer rate and elastic–plastic deformation, the result in the difference with the strain data received from the numerical simulation analysis. The strain data measured by foil gauges have a trend of moving downward generally, affected by the strain transfer ratio of adhesives. In a word, by observing the turning points of the strain data, one can accurately estimate that the length range of increased equivalent stiffness on the cantilever beam is 4 cm. In the test, the increase of the equivalent stiffness is caused by the increase in thickness. As the two segments of the distributed optical fiber sensor are pasted symmetrically, the measuring results are the same.

Figure 6.19 shows the surface strain contour along the length direction of the cantilever beam between 10 and 40 cm from the left side based on the numerical simulation results. When the cantilever beam bends, the strain near the clamped left side is the biggest, whereas the strain near the right side of tree is the smallest. The curve has a sudden drop in the region where the thickness of the beam increases.

Extracting the strain data of the two segments from the numerical simulation and actual measurements based on three different gauge lengths, respectively,

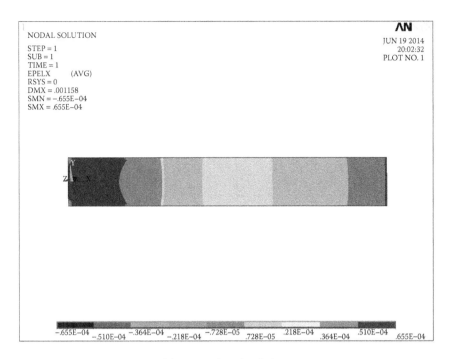

FIGURE 6.19   Strain contour of the numerical simulation.

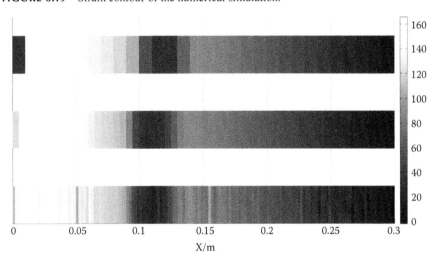

FIGURE 6.20   Comparison among the strain data of distributed optical fiber sensor with the gauge lengths of 1, 0.5, and 0.2 cm and the strain contour of the numerical simulation.

the strain field with a size of 30×2 cm between the two segments can be reconstructed using interpolation, as is shown in Figure 6.20. The horizontal axis expresses the length, and the different gray scales stand for different strain levels. The deeper the color, the smaller the strain. The unit of the strain is microstrain in this figure.

After comparison, the same sudden drop of the strain curve could be found in strain data based on results coming from numerical simulation and measurements of distributed optical fiber sensor in different gauge lengths, which indicated that there is an increase of the equivalent stiffness in the part of the structure. In this test, the increase of the equivalent stiffness is caused by the increase of the thickness. After contrasting, it can also be found that the range of thickness changes is almost the same as the result of numerical simulation, namely, 9–13 cm from the starting point of optical fiber is pasted. Among them, when the gauge length is 0.2 cm, the received data contained noises, thus leading to discontinuous changes in the reconstructed strain field. When the gauge length is 1 cm, the measurement is less sensitive to large changes of local strain gradient, resulting in a larger region that stands for thickness changes for the reconstructed strain field.

DISTRIBUTED PIEZOELECTRIC TRANSDUCER SENSOR NETWORK

In order to maximize the detection capability of a given SHM system for a hot spot in aerospace structures, a methodology for optimizing the layout of a piezoelectric transducer (PZT)-based actuator–sensor network is developed in this chapter. The proposed method simplified the optimal sensor network design by placing the sensors in arrays to ensure that the coverage rate of sensor network exceeds 95%.

Lamb Wave Damage Diagnosis Method

Damages (including flaw, delaminating, and cracking) appear in aging composite aircraft structure when the structures are subject to fatigue and impact. Conventional NDE for structural criticality, such as C-scan and radiographic inspection, or model-based methods, ranging from modal analyses to static parameters identification, are facing the challenge of compromise between satisfactory estimation accuracy and versatile applicability in practice. The damage extension diagnosis method could be used as the first stage for damage global detection. The results of damage extension diagnosis method will improve the aircraft maintenance efficiency by providing health reference information that includes the damage location and the size for NDE technology. With the aid of more sensitive NDE techniques, damage can be characterized and the actual shape of damage can be obtained. In primary structural components of the composite aircraft structure, Lamb wave techniques have been proven to provide more information about the damage type, severity, and location than previously tested methods (frequency response techniques), and may prove suitable for damage extension diagnosis method because they travel long distances. Therefore, Lamb waves–based technology was applied as the damage detection method in this research.

Lamb waves–based detection method utilizes the distributed sensor network to generate the Lamb wave. The sentence should be "Due to material properties changed, Lamb wave will be scattered. The scattered signal will be received by the sensor network. Using the diagnosis method, various features of scattering can be extracted from the captured signal, including time of flight (TOF), magnitude, and energy, which contain the essential information about the damage.

Features extracted from the captured signal that can be linked with damage at different states are termed damage indexes (DIs). Some literatures presented two representative relationships between DIs extracted from Lamb wave signals and a specific damage parameter. Because the DI utilizes the information of the forward scattered (or directly transmitted) wave, it is limited to provide a line interrogation and assess the damage only near, or in the line of, the actuator–sensor path.

The TOF defined as the time lag between the incident wave and the scattering wave is one of the most straightforward features of a Lamb wave signal for damage identification. In substance, it suggests the relative positions among the actuator, sensor, and damage. From the difference in the TOF between the damage-scattered and incident waves extracted from a certain number of signals, damage can accordingly be triangulated. The success of triangulating damage in terms of TOF is largely dependent on the accuracy of TOF extraction. The complexity of the composite aircraft structure reduces the accuracy of the TOF method.

Lamb wave tomography diagnosis has been developed based on the computed tomography using the X-ray radiography principle, which is widely used in clinical applications. In the diagnosis technique, a simple image concerning the damage information (the location, size, and shape of the damage) can be reconstructed using an appropriate image reconstruction technique. However, there is the requirement of a large number of rays to cover the entire inspection area for tomographic reconstruction, leading to either rotation of the object by very tiny increments or a large number of appropriately arranged transducers. Therefore, Lamb wave tomography diagnosis method is unfit for aircraft SHM.

Some researchers proposed some DI-based extension diagnostic methods to locate damage by analyzing the scattered waves. The process of damage diagnosis is shown in Figure 6.1. First, DI is generated with each pair of actuator–sensor to characterize the damage. Then, the monitored region is discretized into matrix, and each element (pixel) of the matrix corresponds to a structural point. Meanwhile, interpolation coefficients are estimated for every pixel in the monitored region based on the relative distance between the actuator–sensor paths and the pixels. Finally, a digital image is created to highlight the area of damage based on where propagating signals are affected by the local damage.

As shown in Figure 6.21, the damage extension diagnosis method is divided into three stages: (i) preprocessing stage, (ii) extension diagnosis stage, and (iii) postprocessing stage. In the first stage of the method, DIs of different paths in the sensor network were obtained by signal processing and feature extraction; in the second stage of the method, the probability–damage region was described by the probability-based diagnostic imaging (PDI) method combined with the empirical threshold value; in the last stage of the method, the damage location and severity size extension indicator were outputted as a result of the damage extension diagnosis method.

Using the sensor network-excited lamb wave signal, the monitored region was scanned. DIs were obtained by analyzing the captured signals of actuator–sensor paths.

The scattering signal energy transmitted along various actuator–sensor paths is shown as follows:

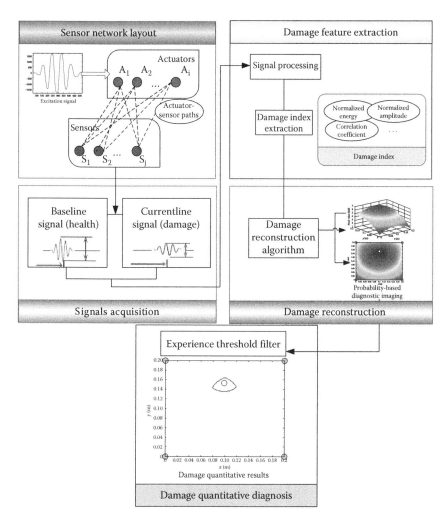

FIGURE 6.21 Extension damage diagnosis method.

$$\mathrm{DI}_{\mathrm{Path}(i)} = \sqrt{\mathrm{Eng}(\mathrm{scatter})/\mathrm{Eng}(\mathrm{base})} \qquad (6.10)$$

where Eng(scatter) and Eng(base) are the energies of the scatter signal and the baseline signal, respectively. The energy of signals is defined as the integration of the time-domain signal.

Following the classical damage imaging method, interpolation coefficients of various paths were utilized to draw the damage probability image.

In this work, the monitored region is discretized into a grid of pixels based on the required resolution in X and Y directions. The interpolation coefficient of Path(i) for pixel D is shown as follows:

$$Wn_D = \begin{cases} 1 - \dfrac{(L_{A-D} + L_{D-S})/L_{A-S}}{h}, & \dfrac{L_{A-D} + L_{D-S}}{L_{A-S}} < h \\[3mm] 0, & \dfrac{L_{A-D} + L_{D-S}}{L_{A-S}} > h \end{cases} \tag{6.11}$$

where:

L_{A-D}, L_{D-S}, and L_{A-S} are the distance between the pixel and the actuator of Path(i), the distance between the pixel and the sensor of Path(i), and the length of Path(i), respectively

$(L_{A-D} + L_{D-S})/L_{A-S}$ is referred to as damage relative distance

The scaling parameter h is employed to control the area influenced by the Path(i). The parameter h is specified as 0.5 in this chapter. The pixel value is shown as follows:

$$Pv = \sum_{i=1}^{n} DI_{Path(i)} Wn_D \tag{6.12}$$

where n is the amount of actuator–sensor paths. The probability of damage presence at node D is defined by Equation 6.3 and will be projected onto the X–Y plane in the form of a pixel value field, which can be illustrated in Figure 6.1. The peak of the pixel value field is counted as damage location. Furthermore, in order to quantify the damage, some development in this method has been carried out. An empirical threshold T is introduced as follows:

$$T = 0.2 \big[\max(\text{pixel value}) + \min(\text{pixel value}) \big] \tag{6.13}$$

The empirical threshold was installed for every imaging result.

Potential damage region is scaled out by the contour plot. As mentioned earlier, the monitored region was discretized into matrix, and each element (pixel) of the matrix corresponded to a structural point. The damage size can be quantified informally as follows:

$$S_{Damage} = \frac{NE_{Nonzero} S_{Region}}{NE_{Global}} \tag{6.14}$$

where:

S_{Damage} and S_{Region} represent the sizes of the damage and the entire monitored region, respectively

$NE_{Nonzero}$ and NE_{Global} denote the number of nonzero elements and the number of the entire elements in the processed matrix by the threshold alignment, respectively

The semiempirical size of the damage can be used as a simple size extension indicator. It is important to note that the relation between the semiempirical size estimation and the physical parameters (e.g., the delamination size) is not fully understood yet, but subsequent tests show that they have the same change trend. To determine the effectiveness of the method, simulated damage (added mass) location and severity extension test on the typical structure of aircraft (thin-walled structure) are presented. To illustrate the delamination growth monitoring capability of the damage extension diagnostic method, damage monitoring test in a typical reinforced component (T-type reinforced plate) of the composite aircraft during static load testing is presented.

The presented damage diagnostic method was employed to identify the damage in two common types of structures: thin-walled structure and carbon-fiber-reinforced polymer (CFRP) T-joint.

In order to validate the effectiveness of the damage diagnosis method in the composite aircraft structure, the damage location and the size evaluation of bonded added masses using the damage extension diagnosis method in a typical structure of the aircraft (thin-walled structure) were processed. The added masses bonded on the structure are proved to simulate delamination damage by changes in local stiffness.

An aluminum plate (430 × 500 × 3 mm) with a sensor network and a square mass (20 × 20 × 1 mm) made of steel is used as an experiment specimen. The sensor network consists of four PZT disks (APC 851) with a diameter of 6.35 mm and a thickness of 0.25 mm. The networked sensors were bonded on the specimen using quick-cure epoxy such that they were installed easily and quickly. The material properties of the aluminum plate and the epoxy layer are listed in Table 6.3, and the piezoelectric properties of PZT are listed in Table 6.4.

The locations of sensors and added mass are shown in Figure 6.22. The DIs were evaluated with the sensor measurements from 225 kHz input signals, which gave the highest signal-to-noise ratio. A five-peak sine wave modulated by a Gaussian envelope was used to drive actuators because of its narrowband signal. Signal was generated and captured by the damage diagnostic system.

TABLE 6.3
Mechanical Properties of Aluminum Plate and Epoxy Layer

	Young's Modulus E (GPa)	Poisson's Ratio υ	Density $\rho(\text{kg/m}^3)$
Plate	70	0.33	2700
Epoxy layer	14	0.31	1400

TABLE 6.4
Material Properties of PZT Transducer

Product Name	APC 851
Density	7.6
Electromechanical coupling factor	0.71
Relative dielectric constant	1950
Frequency constant	2040
Elastic constant E	63

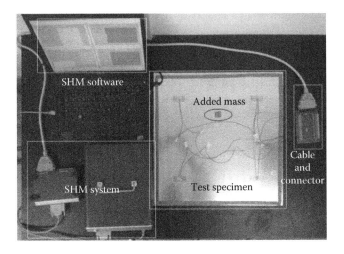

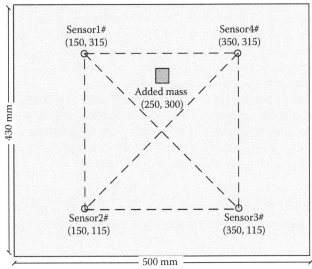

FIGURE 6.22   Experiment setup and test specimen.

In the experiment, PZTs were employed as both the actuator and the sensor. Each pair of the actuator and the sensor forms a path. PZT sensors were excited in a proper sequence; subsequently, signals in each path were obtained. Subtraction of the latter from the former measurement yielded the scattered wave due to the added mass. Then, the estimated interpolation coefficient for every pixel in the monitored region was based on the relative distance between the actuator and sensor paths to the pixel; finally, a digital image was created to highlight the area of damage based on where propagating signals were affected by the local damage.

Figure 6.23 shows the comparison of typical signals at different experiment stages.

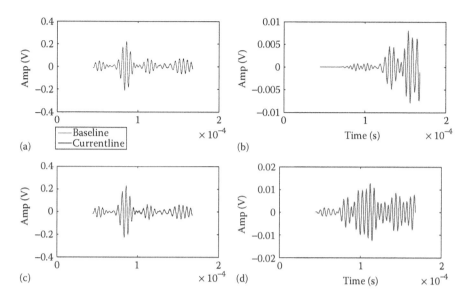

(a) Baseline / Currentline

(b) Time (s)

(c)

(d) Time (s)

FIGURE 6.23 Comparison of typical actuator–sensor paths: (a) signal of path 2–3, (b) scatter signal of path 2–3, (c) signal of path 1–4, and (d) scatter signal of path 1–4.

Figure 6.23a and c represent the signals of path 2–3/1–4, respectively; Figure 6.23b and d represent the scattering signals of path 2–3/1–4, respectively. The figure shows that the scattering signal of path 1–4 is stronger than that of path 2–3. The scattering signal energy transmitted along various actuator–sensor paths was used for monitoring the added mass location. The DIs are obtained through Equation 6.1. The normalized DI of each actuator–sensor path is shown in Figure 6.24.

DIs were extracted from various actuator–sensor paths for establishing different added masses. Figure 6.24 shows that the relationship between the DI and damage relative distance decreases. DIs and interpolation coefficients of the entire paths were combined with Equation 6.3, and probability-based diagnostic was achieved.

Applied with the pixel value function defined by Equation 6.3, PDI results for the added mass established by paths in the sensor network were fused, and the ultimate resulting image is shown in Figure 6.25.

After comparing with the empirical threshold, the diagnostic imaging results are shown as a contour plot in Figure 6.26.

In Figure 6.26, the contour plot represents the damage region result, the white circle represents the diagnosis location of the added mass, and the red square represents the real location of the added mass. The identified location (0.099, 0.153) matches well with the actual location (0.1, 0.15) of the added mass. The extension damage size could be obtained by Equation 6.4. The error in damage location identification is 2.5%.

The experiment result shows that the damage location and size were estimated accurately using the damage diagnosis method in the typical structure of the aircraft (thin-walled structure).

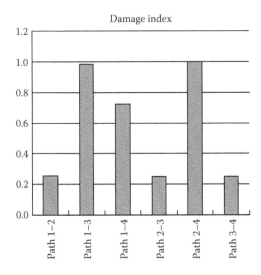

FIGURE 6.24 Comparison of actuator–sensor paths DIs.

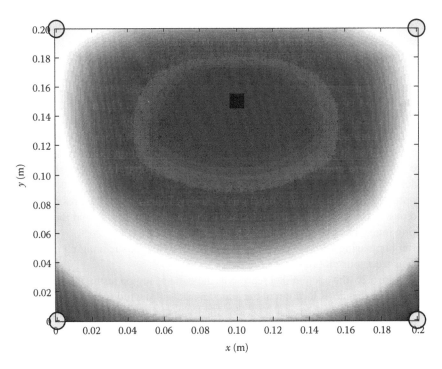

FIGURE 6.25 PDI for added mass.

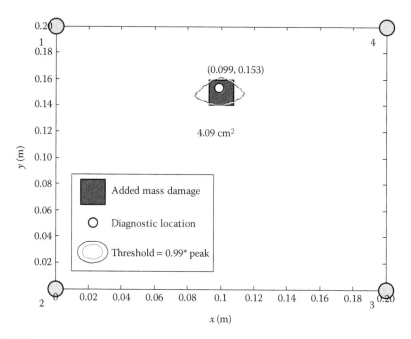

FIGURE 6.26 Damage diagnostic results in thin-walled structure.

Distributed PZT Sensor Network Design Method

The detection capability of a given SHM system does not only depend on its algorithm and the hardware sensitivities (signal-to-noise ratio), but also depends on the sensor network distribution. In order to maximize probability of detection (PoD) and reduce the influence on the monitored structure, damage diagnosis sensor network distribution should be optimized on the basis of the monitored region's structural features and the given SHM system.

Many attempts have been made toward PZT sensor network optimization design for SHM of a simple structure. Han and Lee [41] used genetic algorithms (GAs) to find efficient locations of piezoelectric sensors and actuators of a composite place considering controllability, observability, and spillover prevention. The experimental vibration control shows a significant vibration reduction for the controlled modes with little effect on the residual modes. Flynn and Todd [42] proposed an approach to optimize the sensor network by minimizing the Bayes risk. Guo et al. [43] implemented improved strategies for the GA-based technique to optimize sensor locations for truss structure. Vanli et al. [44] formulated the sensor network optimal placement problem as a minimax optimization in which the goal is to find the coordinates of a given number of sensors so that the worst (maximum) probability of nondetection of the sensor network is made as good as possible (minimized). Moore et al. [45] obtained the optimum sensor and impact locations for Bayesian model-based crack identification in plate by using a hybrid GA/steepest descent optimization method. Some researches focused on sensor network optimization for SHM of complex structures. Gao and Rose [46] presented the quantitative sensor placement optimization method with covariance matrix adaptation evolutionary strategy. The application of the evolutionary optimization sensor network in a multirib structure resulted in the minimum missed detection probability. The trade-off relationship between the optimized sensor network performance and the number of sensors was also presented in this chapter. Coelho et al. implemented the simulated annealing optimization framework in order to place sensors on complex metallic geometries such that a selected minimum damage type and size could be detected with an acceptable probability of false alarm. A lug joint component crack detection test verified the placement of sensors, which was able to interrogate all parts of the structure using the minimum number of transducers.

However, few studies have focused on efficiently placing the sensors on a complex practical structure. The purpose of this chapter is to design a distributed PZT sensor network for SHM of a full-scale composite horizontal tail with given SHM equipment and damage diagnosis algorithm in consideration of constraints and redundancy requirements.

On the basis of composite aircraft failure analysis, failure modes of composite horizontal tail under static include stiffener–airfoil interface delamination, fibrous fracture in upper airfoil, flat fracture in lower airfoil, and cracks in riveted joint of central wing box. In the static loading experiment, the airfoil near central wing box was chosen as a monitored region. The total area of the monitored region exceeds 4 m^2.

In order to guarantee diagnosis efficiency and capacity of sensor network, some constraints of sensor network design are presented.

The sensor distribution in the central wing box has been determined; placement of sensor arrays between stiffeners needs to be optimized. In this section, the sensor network optimal design was simplified to determine the number of sensor arrays to ensure that the coverage rate of sensor network exceeds 95%. The sensor network coverage rate is defined as follows:

$$\text{coverage rate} = \frac{\text{coverage of sensor network}}{\text{monitoring region area}} \tag{6.15}$$

where coverage of sensor network is the area of region which the superimposing of paths effective profile exceeds 2.

The sensor network is considered as a nonlinear system which is described by the following:

$$\dot{x}(t) = h\big(x(t), n\big)$$

$$y(t) = Cx(t), \quad t \in \big(0, t_f\big] \tag{6.16}$$

$$x(0) = x_0$$

where:

$x \in R^n, n \in R^m, y \in R^q$ are the state vector (SHM algorithm and equipment sensitivity), the control vector (placement of sensor arrays), and the output vector (coverage rate of sensor network), respectively

x_0 is the initial state vector

$h : C^{\infty}\left(R^n \times R^m\right) \to U_1 \subset R^n, h(0,0) = 0$

$C \in R^{q \times n}$ is a constant matrix

The objective is to find an optimal placement of sensor arrays n^* such that the coverage rate of the sensor network gives the desired output.

The flow chart of the sensor network successive approximation approach is shown in Figure 6.27.

Assuming that the state vector (SHM algorithm and equipment sensitivity) is time invariant in the experimental process. The number (n) of sensor arrays on the airfoils is initially identified. Then, the effective profile of single signal path is estimated for tolerable damage size of the monitored structure. The effective profile of the sensor network is determined by overlaying all effective profiles of signal paths, subsequently. If the effective profile of the coverage rate is below 95%, sensors are added in each sensor array and the process is iterated again.

In order to determine the effective profile of signal path for a tolerable damage (area = 2 cm²) for composite horizontal tail, Lamb wave propagation test was performed on composite stiffener panel with a simulation damage (stick sealant). The test specimen was a symmetric orthogonal laminated which material properties and structural style was same as the composite horizontal tail structure (1000 × 1000 × 7 mm) with sensor network.

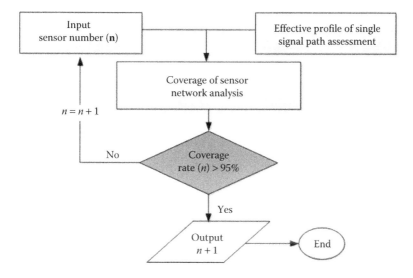

FIGURE 6.27   Flowchart of sensor network successive approximation approach.

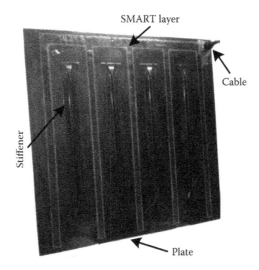

FIGURE 6.28 Lamb wave propagation test specimen.

The sensor network consists of 25 PZT disks (APC 851) with the diameter of 6.35 mm and the thickness of 0.25 mm; the piezoelectric characteristics of PZT are listed in Table 6.1. The sensor network was bonded on the specimen using quick-cure epoxy. The damage was simulated by a sealant placed nearby sensors 6# and 7#, as shown in Figure 6.28. The paths were divided into two types: cross-stiffener paths (CSPs) and uncross-stiffener paths (USPs).

The canonical correlation-based DI was extracted from signal paths and is described in Figure 6.29. In this study, the effective DI threshold was defined as 0.1, and its corresponding distance was the effective profile of signal path.

As shown in Figure 6.29, the effective profile of URP (p1) is estimated as 0.24 m, whereas the effective profile of CRP (p2) is estimated as 0.21 m.

The PoDs of different paths are defined as follows:

$$
\text{PoD}_{\text{URP}} =
\begin{cases}
1, & l \le 0.24 \text{ m} \\
0, & l > 0.24 \text{ m}
\end{cases}
\tag{6.17}
$$

$$
\text{PoD}_{\text{CRP}} =
\begin{cases}
1, & l \le 0.21 \text{ m} \\
0, & l > 0.21 \text{ m}
\end{cases}
\tag{6.18}
$$

where:

l is the Lamb wave distance

PoD_{URP} and $\text{PoD}_{\text{Stiffener}}$ and $\text{PoD}_{\text{Unstiffener}}$ PoD_{CRP} represent the damage detection probabilities of URPs and CRPs, respectively

As previously stated, the region between the load zone 2 and the central wing box of all the four tail airfoils was chosen as the monitoring region. The monitored region

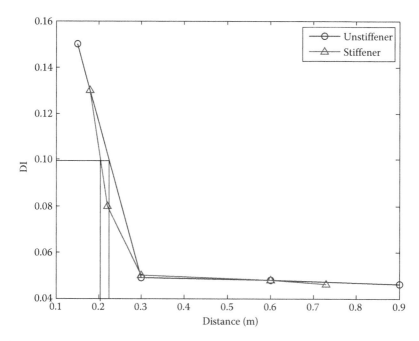

FIGURE 6.29 Effective profiles of different paths in the composite stiffener panel.

of horizontal tail was divided into four airfoils. The design of the horizontal tail optimal sensor network was performed on the upper-left airfoil.

As shown in Figure 6.27, the initial number of sensors of the successive approximation approach is defined as 4, and the placement is average along the direction of wingspan in the monitoring region. For each candidate sensor placement, the coverage rate of the sensor network is shown in Figure 6.30.

Based on the estimated effective coverage and the redundancy requirements, a seven-sensor array network was chosen as the optimized sensor network unit for each airfoil. To determine the optimal usage pattern, it is necessary to debug each unit of the SHM system before the load of the composite horizontal tail.

Experimental Procedure of the Full-Scale Composite Horizontal Tail

The static load of horizontal tail include symmetrical load and asymmetric load conditions. In the static load test, the proportions of the design load were defined as load levels in the static load test. A range of 40%, 67%, 100%, and 120% of the design load was performed in both the symmetrical and asymmetric load conditions. The load sequence of the full-scale composite horizontal tail is shown in Figure 6.31.

As shown in Figure 6.31, at the beginning of the static load experiment, 40% of asymmetric load case was performed to debug the load system; then, 67% and 100% of asymmetric load case were performed; after this stage, clamping is changed; later, a range of 67%, 100%, and 120% of symmetrical load case was performed.

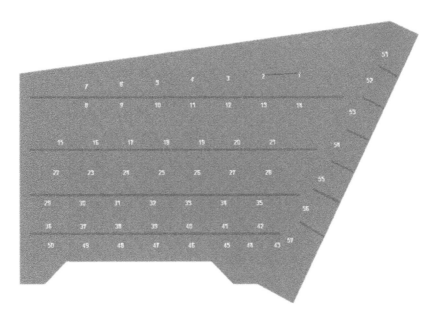

FIGURE 6.30 Result of the sensor network placement successive approximation approach.

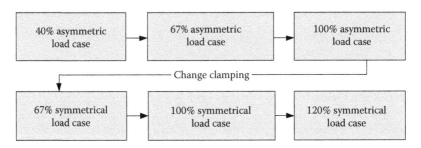

FIGURE 6.31 Load sequence of the full-scale composite horizontal tail.

In the 40%, 67%, and 100% load cases, 30 s of the load holding time was performed after each 20% load, respectively. In the 120% overload case, there is no load holding time. The 67% asymmetric load case, the 100% symmetrical load case, and the 120% symmetrical load case were chosen as representative from all load cases, and their SHM results were shown to determine the variation of horizontal tail health condition.

Damage imaging diagnosis was performed in all the tail airfoils after the unload of each load case with the optimized sensor network.

Damage diagnosis result shows that red zones occur in all airfoils, which means that delamination damage occurs in each airfoil; because of the accumulation of slight damage, the delamination distributions in different airfoils are unbalanced. On the basis of the distribution of red zones, most damages occur near the interface between the stiffeners and the skin; some damages occur near the joint location between the central wing box and the airfoil.

The damage diagnosis result showed that all the damage in the monitoring region was covered by the sensor network.

In this study, the distributional PZT sensor network was tailored for SHM of a full-scale composite horizontal tail. Based on the sensor network design constraints of the horizontal tail, the successive approximation approach was used to optimize the sensor network placement with given SHM equipment and algorithm. Then, the effective profiles of both kinds of paths were estimated by the Lamb wave propagation experiment in a multistiffener composite plate. The effective profile of URP (p1) was estimated as 0.24 m, whereas the effective profile of CRP (p2) was estimated as 0.21 m. Subsequently, based on the result of the successive approximation approach and the redundancy requirements, the seven-sensors array network was chosen as the optimization sensor network for the each airfoil. Finally, the efficient diagnosis of the full-scale composite horizontal tail is supported by the optimal sensor network. The result shows that all the damage in the monitoring region is covered by the sensor network; because of the redundancy consideration, damage in the location of invalid sensors is still detected.

Distributional PZT sensor network design can be applied to improve the efficiency of in-service SHM technology for full-scale composite aircraft structures.

LAMB WAVE–BASED MOBILE SENSING APPROACH

Unlike previous damage identification methods, which utilize a sensor network that is permanently attached to the structure, in the mobile sensing approach, a handmade mobile transducer set is used in order to generate and collect Lamb waves in the structure. The PDI technique based on the mobile sensing approach makes use of the handmade mobile transducer set to generate and collect Lamb waves from different locations of the structure under inspection. In this method, based on the current signals captured from the different locations of the structure, a diagnostic image was constructed, which highlights the most probable locations of the damages. Conventional point-scan NDE techniques can then be used once damage is found to identify the damage in detail.

METHODOLOGY

When an elastic wave such as Lamb wave is incident on damage, scattering happens in all directions. Such scattering waves can be utilized to investigate the damage. Usually, the damage scatter waves are obtained by comparing the current signal with a baseline. However, such a process is easily affected by environmental and operational variability, especially when signals are captured via a mobile transducer set. Therefore, a baseline-free damage identification algorithm is needed in order to define the damage scatter waves. In this study, the damage scatter wave separating process presented by Qiang and Shenfang [47] is adopted to define the damage scatter waves only by using the current signals.

Captured Lamb wave signals consist of three parts in time domain: the wave that directly comes from the actuator, the damage scatter wave, and the reflections from the structure boundaries. If the sensor is collocated with the actuator, because the distance from the actuator to the sensor is very short, the wave directly coming from the actuator only travels a very short time. Hence, it appears almost at the very beginning of the captured signal. Under this situation, based on the distance from the actuator to the sensor, the speed of the selected Lamb wave mode, and the length of the excitation signal, the end time of the wave directly coming from the actuator can be defined, and this wave can be separated from the damage scatter wave and the reflections from the boundaries.

However, the reflections from the boundaries still must be separated from the damage scatter wave. This can be achieved by calculating the time that the first boundary reflection is appeared in the captured signal. This time depends on the distance from the nearest structure boundary to the transducer set and the speed of the selected Lamb wave mode. Therefore, a time window function is utilized to separate the damage scatter wave from other waves in the captured signal. When the signal is multiplied by this function only, the damage scattered wave will be preserved. The time window function is expressed as follows:

$$f_{wg} = \begin{cases} 0 & t < t_1 - T_t \\ e^{-\frac{(t-t_1)^2}{2\sigma^2}} & t_1 - T_t \le t < t_1 \\ 1 & t_1 \le t < t_2 \\ e^{-\frac{(t-t_2)^2}{2\sigma^2}} & t_2 \le t < t_2 + T_t \\ 0 & t \ge t_2 - T \end{cases} \quad , \quad t_1 = t_0 + \frac{L_{AS}}{c_g} + L, \quad t_2 = t_0 + \frac{L_{Ab} + L_{Sb}}{c_g} \quad (6.19)$$

where:

t_0 is the beginning time of the excitation signal

t_1 is the end time of the wave directly coming from the actuator

t_2 is the beginning time of the reflections from the boundaries

c_g is the velocity of the selected Lamb wave mode

L_{AS}, L, L_{Ab}, and L_{Sb} represent the distance from the actuator to the sensor, the length of the excitation signal, the distance from the nearest structure boundary to the actuator, and the distance from the nearest structure boundary to the sensor, respectively

σ is the coefficient of the Gauss function

T_t is the length of the window transition

For a known structure and the excitation signal, with regard to the location of the transducer set, the times t_1 and t_2 can be calculated easily. Figure 6.32a and b illustrates the damage scatter wave separation process of a typical Lamb wave response.

If the damage happens near the structural boundaries, it is difficult to separate the damage scatter wave from the reflections from the boundaries because they may mix together. However, when the transducer set is placed along the structural boundaries, the first boundary reflection and the wave directly coming from the actuator mix

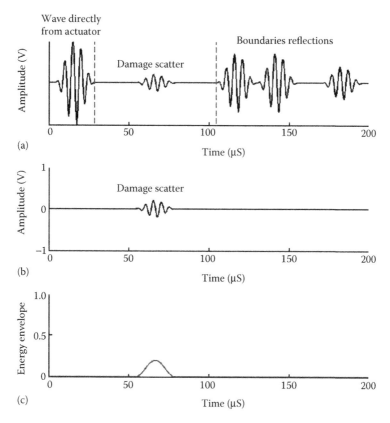

FIGURE 6.32 Typical Lamb wave response: (a) the signal separation process; (b) the separated damage scatter signal and (c) the energy envelope of the damage scatter signal.

together and appear at the first part of the signal. Under this situation, the damage scatter wave can be easily separated from such waves.

If the reference time is set at the peak of the excitation signal, the energy envelope of the damage scatter signal at that time corresponding to the wave traveling from the actuator to the damage and then from the damage to the sensor, is maximum (as shown in Figure 6.32c). Therefore, the energy envelope of the damage scatter signal is used to construct the diagnostic image. It was defined using Equation 6.19.

Assume that the data acquisition process has been done in N locations of the monitoring area. By using the damage scatter wave separation process, the damage scatter signal corresponding to each response was defined, and subsequently, the energy envelope of all damage scatter signals were obtained. For each response, a probability image can then be constructed as follows:

$$I_n(i,j) = e_n(t_{nij}), \quad n = 1, 2, \ldots, N \quad \text{and} \quad t_{nij} = \frac{L_n^{ad} + L_n^{ds}}{c_g} \quad (6.20)$$

where:

$I_n(i,j)$ and $e_n(t)$ represent the damage presence probability of the imaging pixel (i,j) and the energy envelope of the damage scatter signal, corresponding to the signal acquired at location n, respectively

t_{nij} is the time corresponding to the wave traveling from the actuator to the imaging pixel (i,j) and then from the imaging pixel (i,j) to the sensor

L_n^{ad} and L_n^{ds} denote the distance from the actuator located at location n to the imaging pixel (i,j) and the distance from the imaging pixel (i,j) to the sensor located at n, respectively

c_g is the velocity of the selected Lamb wave mode

Probability images highlight the region with the highest probability of damage presence in the monitoring area. The diagnostic image can then be defined by taking a summation of all I_n defined from different responses such that

$$P(i,j) = \sum_{i=1}^{N} I_n(i,j) \tag{6.21}$$

where $P(i,j)$ represents the probability of damage presence in the imaging pixel (i,j). The imaging pixels with the highest probability of damage presence highlight the most probable locations of the damages in the monitoring area.

EXPERIMENTAL SETUP

In order to generate and collect Lamb waves in the structure, a mobile handmade transducer set was designed. To design such a transducer set, its flexibility must be taken into consideration, because it should be able to easily move on the structure to excite and collect Lamb waves at different locations of the structure. In the transducer set, a sensor is collocated with the actuator to capture the wave reflected back from the damage. Piezoelectric disks (APC 851) are utilized in the transducer set whose diameter and thickness are 6.6 and 0.24 mm, respectively. The PZT sensor and the actuator are attached to the bottom of the separate holders. Two screws passing through a top square plate maintain the holders in balance. Figure 6.33 shows the designed transducer set.

To facilitate the signal separation process, only the A_0 mode of Lamb wave is generated in the structure. To achieve this, a kind of grease lubricant is applied under the bottom surface of the actuator. By employing the previous technique, a pure A_0 mode can be excited.

A five-cycle sinusoidal tone burst enclosed in a Hanning window at a central frequency of 50 kHz is generated using an arbitrary waveform generation unit (Agilent® 33220A). The analog signal is amplified to 85 V using a linear amplifier (T&C Power Conversion, Inc., Rochester, NY, AG series) to drive each PZT actuator in turn. Wave signals are captured using an oscilloscope (Agilent® DSO5032A) at a sampling rate of 10 MHz. Sampled signals are transmitted into the central processing unit for further analysis. The A_0 mode velocity is assumed constant in this research.

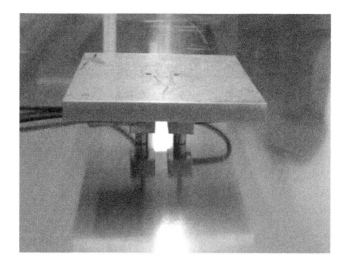

FIGURE 6.33   The designed handmade transducer set.

The length of the excitation signal (L) is 0.17 m. In the transducer set, the linear distance between the center of sensor and the center of actuator is 2 cm. Also in this study, the velocity of the A_0 mode is assumed constant.

In this section, two selected experimental studies are presented, which demonstrate the capability of the developed Lamb wave–based mobile sensing technique to identify (1) the through-thickness hole and crack in aluminum plates and (2) the added mass in composite plates.

1. *Through-thickness hole and crack in aluminum plates:* An aluminum plate (1000 × 1000 × 3 mm) is fixed along its four edges on a testing table. A through-thickness hole ($\varphi = 2$cm) is introduced into the plate, as seen in Figure 6.34.

   At the different locations on the plate (L1, L2,..., L24), Lamb waves are generated, and the corresponding responses (totally 24 responses) are collected. Such locations are considered for response collection because they cover the whole surface of the plate. Signal acquisition positions are listed in Table 6.5.

   All the captured signals are normalized to compensate the performance difference among the transducer set at the different locations on the plate. The typical responses collected by the transducer set at the locations L1, L7, L16, and L22 are shown in Figure 6.35.

EXPERIMENTAL RESULT

The damage scatter signal of all responses is defined using the time window function, and subsequently, the energy envelope of each damage scatter signal is defined.

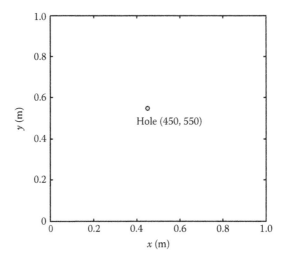

FIGURE 6.34   Identifying the through-thickness hole in the aluminum plate (unit: mm).

TABLE 6.5
Coordinates of Signal Acquisition Locations on Aluminum Plate

Position	Coordinates (x,y) (mm)	Position	Coordinates (x,y) (mm)
L1	(15,5)	L13	(700,500)
L2	(300,5)	L14	(985,500)
L3	(500,5)	L15	(15,700)
L4	(700,5)	L16	(300,700)
L5	(985,5)	L17	(500,700)
L6	(15,300)	L18	(700,700)
L7	(300,300)	L19	(985,700)
L8	(500,300)	L20	(15,995)
L9	(700,300)	L21	(300,995)
L10	(985,300)	L22	(500,995)
L11	(15,500)	L23	(700,995)
L12	(300,500)	L24	(985,995)

The diagnostic image was constructed by Equation 6.21. An average process is adopted to improve the imaging quality. To this end, the new value of one pixel is the average of the nearest 81 pixel values around it. Figure 6.36 demonstrates the diagnostic image. The damage presence probability values are normalized. The location of the actual hole is shown as a dark circle.

As seen in Figure 6.36, the region with the highest probability of damage presence coincident well with the actual location of the hole.

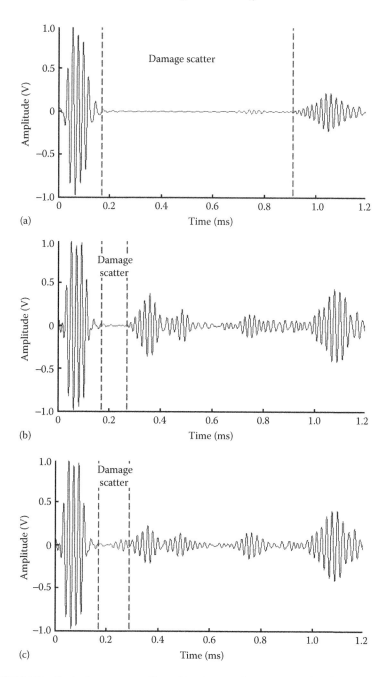

(a)

(b)

(c)

FIGURE 6.35 Typical responses collected by the transducer set at various locations: (a) L1; (b) L7 and (c) L16.   (*Continued*)

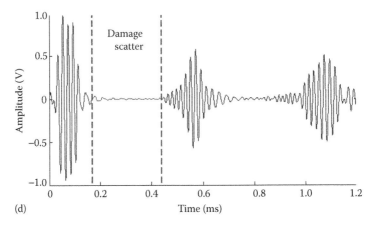

(d)

FIGURE 6.35 (Continued) Typical responses collected by the transducer set at various locations: (d) L22.

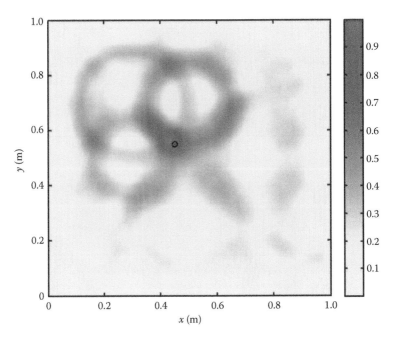

FIGURE 6.36 The diagnostic image defined for the aluminum plate with a through-thickness hole.

To identify the multidamage within the same inspection area is usually a bothersome task, due to the difficulty in isolating different waves scattered from the multidamage. To examine the feasibility of the developed PDI technique in identifying the multidamage, the aluminum plate bearing the hole used in the previous

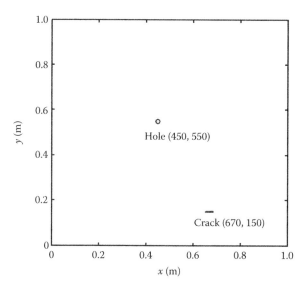

FIGURE 6.37 Identifying the through-thickness hole and crack in the aluminum plate (unit: mm).

case was introduced with a through-thickness crack (30 mm long and 1.5 mm wide) (see Figure 6.37).

In the developed PDI technique, each captured signal perceives the damage most sensitively near it only. Such a trait can be beneficial to detecting the multidamage. For example, the response collected at L12 in this application offers crucial information for the hole, whereas the response collected at L4 provides more information regarding the crack.

At the same locations (L1, L2, ..., L24), Lamb waves are generated, and the corresponding responses are collected. The diagnostic image is constructed and is displayed in Figure 6.38. The location of the actual crack is shown as a dark line.

As can be seen, both damage cases are identified accurately, matching well with the reality, which demonstrates the effectiveness of the developed method in highlighting the multidamage in the monitoring area.

2. *Added Mass on Composite Plates:* As an extension of the previous application, the approach is used to identify single and multiadded masses, which were bonded on the top surface of a glass/epoxy composite plate (1000 × 1000 × 3 mm) (see Figure 6.39). The added mass is a solid rectangular cube (20 × 40 × 3 mm) made of a kind of sealant.

The added mass simulates the wave scattering due to changes in local stiffness, which would represent that caused by a delamination. The composite plate is fixed along its four edges on a testing table.

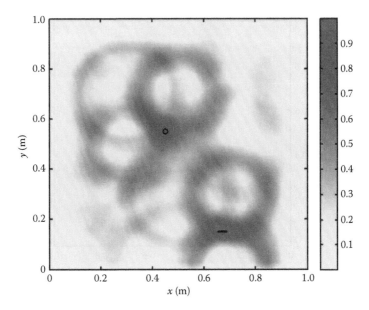

FIGURE 6.38 The diagnostic image defined for the aluminum plate with the through-thickness hole and crack.

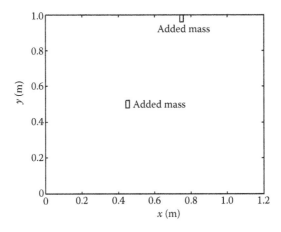

FIGURE 6.39 Identifying single and multiadded masses in the glass/epoxy composite plate.

At the different locations on the composite plate, totally 25 responses are collected. Signal acquisition positions are listed in Table 6.6.

The diagnostic images for composite plate with single and multiadded masses are constructed and shown in Figures 6.40 and 6.41, respectively. The actual location of the added mass is shown as dark rectangles.

TABLE 6.6

Coordinates of Signal Acquisition Locations on Composite Plate

Position	Coordinates (x,y) (mm)	Position	Coordinates (x,y) (mm)
L1	(15,5)	L14	(900,500)
L2	(300,5)	L15	(1185,500)
L3	(600,5)	L16	(15,750)
L4	(900,5)	L17	(300,750)
L5	(1185,5)	L18	(600,750)
L6	(15,250)	L19	(900,750)
L7	(300,250)	L20	(1185,750)
L8	(600,250)	L21	(15,995)
L9	(900,250)	L22	(300,995)
L10	(1185,250)	L23	(600,995)
L11	(15,500)	L24	(900,995)
L12	(300,500)	L25	(1185,995)
L13	(600,500)		

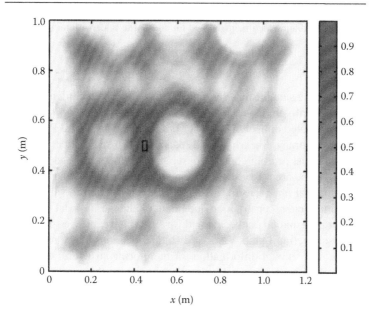

FIGURE 6.40   The diagnostic image defined for the composite plate with one added mass.

As can be seen in Figures 6.40 and 6.41, the regions with the highest probability of damage presence match well with the actual locations of the added masses. The mismatch of the results would be caused by the differences between the actual wave velocity and the assumed constant velocity.

In all the above applications, it can be seen that the developed PDI technique enables an approximate detection of the single damage and the multidamage in the plate-like structures.

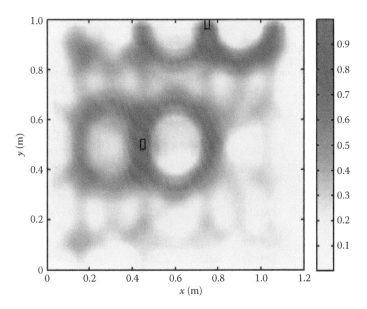

FIGURE 6.41 The diagnostic image defined for composite plate with multi added masses.

SUMMARY

In order to define the suitable data for damage identification techniques, a number of transducers are usually used as a sensor network.

First, the distributed optical fiber sensor based on the backward Rayleigh scattering owns good accuracy, linearity, and repeatability in strain measurement, and it has potential to assist or even completely replace traditional foil gauges in the field of general or specific area. Based on the high spatial resolution and high sensitivity of distributed optical fiber sensor, this research demonstrated the ability of surface quasi-static strain field reconstruction, judgment of changes in local equivalent stiffness, and other information connected with strain, which expresses structural status.

Second, by collecting Lamb waves at different locations of the monitoring area, sensors can provide enough information for monitoring structural damages. Using a sensor network instead of a single sensor will enhance the confidence of damage identification techniques and improve the tolerance to measurement noise.

The effectiveness of the sensor network depends on the density of sensors. However, a very dense network is impractical for some applications and will increase the cost. Therefore, practicability and desired resolution must take into consideration to design a suitable sensor network for a special application. In this chapter, three practical sensor networks were presented.

The PZT sensor network was tailored for SHM of the full-scale composite horizontal tail. Constraints of the optimal design were presented by analyzing the structural layout and load distribution of a composite horizontal tail. Then, the effective profiles of URPs and CRPs were estimated by the Lamb wave propagation experiment in a multistiffener composite plate. Based on the coverage rate and the redundancy requirements, the seven-sensor array network was chosen as the optimization sensor network for each airfoil. Finally, an efficient diagnosis of full-scale composite horizontal tail is supported by the optimal sensor network. The result shows that all the damages in the monitoring region are covered by the sensor network.

Finally, the PDI technique based on the mobile sensing approach was presented to highlight the most probable location of single and multidamages in the structures in an easily interpretable diagnostic image. In order to construct such a diagnostic image, the A_0 mode of Lamb wave was generated and collected at different locations of the structure using a mobile handmade PZT transducer set. Because the developed PDI technique was based on the mobile sensing approach, it can provide an area-scan inspection rather than a point-scan one, which is required in most of the conventional NDE techniques, and in this way minimize the time and cost of inspection. Conventional point-scan NDE techniques, such as ultrasonic C-scan and X-ray, can then be used once damage is found to characterize the damage in detail.

REFERENCES

1. Alleyne, D.N. and P. Cawley, The excitation of Lamb waves in pipes using dry-coupled piezoelectric transducers. *Journal of Nondestructive Evaluation*, 1996, 15(1): pp. 11–20.
2. Giurgiutiu, V., *Structural Health Monitoring: With Piezoelectric Wafer Active Sensors*. 2007, Academic Press, New York.
3. Wu, Z., X.P. Qing, and F. Chang, Damage detection for composite laminate plates with a distributed hybrid PZT/FBG sensor network. *Journal of Intelligent Material Systems and Structures*, 2009, 20(9): pp. 1069–1077.
4. Guo, Z., J.D. Achenbach, and S. Krishnaswamy, EMAT generation and laser detection of single lamb wave modes. *Ultrasonics*, 1997, 35(6): pp. 423–429.
5. Gao, W., C. Glorieux, and J. Thoen, Laser ultrasonic study of Lamb waves: Determination of the thickness and velocities of a thin plate. *International Journal of Engineering Science*, 2003, 41(2): pp. 219–228.
6. Kwun, H., S.Y. Kim, and G.M. Light, The magnetostrictive sensor technology for long range guided wave testing and monitoring of structures. *Materials Evaluation*, 2003, 61(1): pp. 80–84.
7. Grattan, K. and T. Sun, Fiber optic sensor technology: An overview. *Sensors and Actuators A: Physical*, 2000, 82(1): pp. 40–61.
8. Culshaw, B., Optical fiber sensor technologies: Opportunities and-perhaps-pitfalls. *Journal of Lightwave Technology*, 2004, 22(1): p. 39.
9. Giallorenzi, T.G. et al., Optical fiber sensor technology. *IEEE Transactions on Microwave Theory and Techniques*, 1982, 30(4): pp. 472–511.
10. Udd, E., Fiber optic smart structures. *Proceedings of the IEEE*, 1996, 84(6): pp. 884–894.
11. Lee, B., Review of the present status of optical fiber sensors. *Optical Fiber Technology*, 2003, 9(2): pp. 57–79.
12. Hill, K.O. et al., Photosensitivity in optical fiber waveguides: Application to reflection filter fabrication. *Applied Physics Letters*, 1978, 32(10): pp. 647–649.
13. Kawasaki, B.S. et al., Narrow-band Bragg reflectors in optical fibers. *Optics Letters*, 1978, 3(2): pp. 66–68.
14. Hill, K.O. and G. Meltz, Fiber Bragg grating technology fundamentals and overview. *Journal of Lightwave Technology*, 1997, 15(8): pp. 1263–1276.

15. Meltz, G., W.W. Morey, and W.H. Glenn, Formation of Bragg gratings in optical fibers by a transverse holographic method. *Optics Letters*, 1989, 14(15): pp. 823–825.
16. Morey, W.W., G. Meltz and W.H. Glenn, Fiber optic Bragg grating sensors. In *OE/FIBERS'89*, 1990, International Society for Optics and Photonics, Bellingham, WA.
17. Malo, B. et al., Point-by-point fabrication of micro-Bragg gratings in photosensitive fibre using single excimer pulse refractive index modification techniques. *Electronics Letters*, 1993, 29(18): pp. 1668–1669.
18. Froggatt, M. et al., Correlation and keying of Rayleigh scatter for loss and temperature sensing in parallel optical networks. In *Optical Fiber Communication Conference*, 2004, Optical Society of America, Washington, DC.
19. Soller, B. et al., High resolution optical frequency domain reflectometry for characterization of components and assemblies. *Optics Express*, 2005, 13(2): pp. 666–674.
20. Soller, B.J., M. Wolfe, and M.E. Froggatt, Polarization resolved measurement of Rayleigh backscatter in fiber-optic components. *OFC Technical Digest*, Los Angeles, CA, 2005.
21. Froggatt, M.E. et al., Characterization of polarization-maintaining fiber using high-sensitivity optical-frequency-domain reflectometry. *Journal of Lightwave Technology*, 2006, 24(11): pp. 4149–4154.
22. Soller, B.J. et al. High-resolution fiber reflectometry for avionics applications. In *IEEE Conference on Avionics Fiber-Optics and Photonics*, 2005, IEEE, Minneapolis, MN.
23. Gifford, D.K. et al., Optical vector network analyzer for single-scan measurements of loss, group delay, and polarization mode dispersion. *Applied Optics*, 2005, 44(34): pp. 7282–7286.
24. Grattan, K. and T. Sun, Fiber optic sensor technology: An overview. *Sensors and Actuators A: Physical*, 2000, 82(1): pp. 40–61.
25. Rao, Y., Recent progress in applications of in-fibre Bragg grating sensors. *Optics and Lasers in Engineering*, 1999, 31(4): pp. 297–324.
26. Kersey, A.D. et al., Fiber grating sensors. *Journal of Lightwave Technology*, 1997, 15(8): pp. 1442–1463.
27. Wu, Z. et al., Engineering approach to in-situ bridge health monitoring with fiber bragg gratings. *Journal-Harbin Institute of Technology* (English Edition), 2006, 13(5): p. 588.
28. Zhao, H. et al. Monitoring and controlling manufacturing for composite using fiber Bragg grating. In International Conference on Smart Materials and Nanotechnology in Engineering, 2007, International Society for Optics and Photonics, Bellingham, WA.
29. Zhang, X., Z. Wu, and B. Zhang, Strain dependence of fiber Bragg grating sensors at low temperature. *Optical Engineering*, 2006, 45(5): p. 054401.
30. Samiec, D., Distributed fibre-optic temperature and strain measurement with extremely high spatial resolution. *Photonik International*, 2012, 1: pp. 10–13.
31. Bos, J. et al., Fiber optic strain, temperature and shape sensing via OFDR for ground, air and space applications. In *Proceedings of 7th Nanophotonics and Macrophotonics for Space Environments Conference*, 2013, International Society for Optics and Photonics, Bellingham, WA.
32. Froggatt, M., D. Gifford, and J. Bos. Unaltered optical fiber as an absolute wavelength reference for OFDR systems. In *Proceedings of the 2011 Optical Fiber Communication Conference and Exposition and the National Fiber Optic Engineers Conference*, 2011, Los Angeles, CA.
33. Murayama, H. et al., Distributed fiber-optic sensing system with OFDR and its applications to structural health monitoring. *2nd International Conference on Smart Materials and Nanotechnology in Engineering*, SPIE, 2009, Harbin Institute of Technology, Weihai, China.

34. Murayama, H. et al., Distributed fiber-optic sensing system with OFDR and its applications to structural health monitoring. *2nd International Conference on Smart Materials and Nanotechnology in Engineering*, SPIE, 2009, Harbin Institute of Technology, Weihai, China.

35. Froggatt, M. and J. Moore, High-spatial-resolution distributed strain measurement in optical fiber with Rayleigh scatter. *Applied Optics*, 1998, 37(10): pp. 1735–1740.

36. Samiec, D., Distributed fibre-optic temperature and strain measurement with extremely high spatial resolution. Eds. E. Hering and R. Martin, Photonik International, Springer, Germany, 2012.

37. Ou, J.P., Damage accumulation and safety evaluation for important large infrastructures. 21st century's Chinese mechanics-9th science association reports of "forum for youth scientist". Tsinghua University Press, Beijing, China, 1996, pp. 179–189.

38. Ou, J.P. and X.C. Guan, State of art for intelligent structure and system in civil engineering. *Earthquake Engineering and Engineering Vibration*, 1999, 19(2): pp. 21–28.

39. Housner, G.W. et al., Structural control: Past, present, and future. *Journal of Engineering Mechanics*, 1997, 123(9): pp. 897–971.

40. Froggatt, M.E. and D.K. Gifford, Rayleigh backscattering signatures of optical fibers – Their properties and applications. Optical Fiber Communication Conference and Exposition and the National Fiber Optic Engineers Conference, March 17–21, 2013, IEEE Computer Society, Anaheim, CA.

41. Han, J. and I. Lee, Optimal placement of piezoelectric sensors and actuators for vibration control of a composite plate using genetic algorithms. *Smart Materials and Structures*, 1999, 8(2): p. 257.

42. Flynn, E.B. and M.D. Todd, A Bayesian approach to optimal sensor placement for structural health monitoring with application to active sensing. *Mechanical Systems and Signal Processing*, 2010, 24(4): pp. 891–903.

43. Guo, H.Y. et al., Optimal placement of sensors for structural health monitoring using improved genetic algorithms. *Smart Materials and Structures*, 2004, 13(3): p. 528.

44. Vanli, O.A. et al., A minimax sensor placement approach for damage detection in composite structures. *Journal of Intelligent Material Systems and Structures*, 2012, 23(8): pp. 919–932.

45. Moore, E.Z., K.D. Murphy, and J.M. Nichols, Optimized sensor placement for damage parameter estimation: Experimental results for a cracked plate. *Structural Health Monitoring*, 2013, 12(3): pp. 197–206.

46. Gao, H. and J.L. Rose. Ultrasonic sensor placement optimization in structural health monitoring using evolutionary strategy. In *AIP Conference Proceedings*, 2006, Portland, OR.

47. Qiang, W. and Y. Shenfang, Baseline-free imaging method based on new PZT sensor arrangements. *Journal of Intelligent Material Systems and Structures*, 2009, 20(14): pp. 1663–1673.

7 Damage Assessment Algorithms for Structural Health Monitoring

Viviana N. Meruane

CONTENTS

INTRODUCTION

The early detection of structural damage generates a wide interest in the civil, mechanical, and aerospace engineering fields. For the purpose of this chapter, damage is defined as changes to the material and/or geometric properties of a structure, which adversely affect its performance. Examples of structural damage are fatigue, crack, corrosion, and loosening of bolted connections among others. When the damage reaches a point where the structure is no longer functional, it is referred to as *failure*. The purpose of damage assessment is to detect and characterize damage at the earliest possible stage, and to estimate how much time remains before maintenance is required, the structure fails, or the structure is no longer usable. Damage assessment has a tremendous potential for life-safety and/or economic benefits: it reduces the maintenance cost and increases the structure safety and reliability. Damage assessment can be categorized into the following four levels:

Level 1: Detecting the presence of damage in the structure.
Level 2: Determining the geometric location of the damage.
Level 3: Quantifying the severity of the damage.
Level 4: Predicting the remaining lifespan.

The last step of damage assessment attempts to predict the system performance by assessing its current damage state, estimating the future loading environments, and predicting through numerical simulations and experience the remaining useful life of the system. Thus, damage prognosis requires the measurements of the current system state (damage location and quantification) and the prediction of the system deterioration when subjected to future loading. The prediction of the system deterioration requires knowledge in fracture mechanics, fatigue life analysis, and structural design assessment. According to Farrar and Lieven [43], damage prognosis qualifies as a "grand challenge" problem for engineers in the twenty-first century.

A global technique called vibration-based damage assessment [17] has been rapidly expanding over the last few years. The basic idea is that vibration characteristics (natural frequencies, mode shapes, damping, frequency response function [FRF], etc.) are functions of the physical properties of the structure. Thus, changes to the material and/or geometric properties due to damage will cause detectable changes in the vibration characteristics. Many studies have demonstrated that vibration measurements are sensitive enough to detect damage even if it is located in hidden or internal areas [158].

The field of damage identification is very broad and encompasses both local and global methods. This chapter is limited to global methods that detect damage from changes in the vibration characteristics of a structure. In general, these methods can be classified as linear or nonlinear. A linear method assumes that the structure's response can be modeled using linear equations of motion even after damage. Although damage is inherently a nonlinear phenomenon, low-level vibrations are typically used to identify the dynamic characteristics of the structures. For a wide variety of structures with different types of damage the measured response in vibration tests is linear.

Linear methods can be further classified as model-based or nonmodel-based. Nonmodel-based methods detect damage by comparing the measurements from the undamaged and damaged structures, whereas model-based methods locate and quantify damage by correlating an analytical model with experimental data of the damaged structure. Nonmodel-based methods usually provide the first two levels of damage assessment (detection and location), whereas model-based methods can achieve the third level (quantification). Additionally, model-based methods are particularly useful for predicting the system response to new loading conditions and/or new system configurations (damage states), allowing damage prognosis.

The most successful applications of vibration-based damage assessment are model updating methods based on global optimization algorithms. Model updating is an inverse method that identifies the uncertain parameters in a numerical model and is commonly formulated as an inverse optimization problem. In model updating-based damage assessment, the algorithm uses the differences between the models of the structure that are updated before and after the presence of damage to localize and determine the extent of damage. The basic assumption is that the damage can be directly related to a decrease of stiffness in the structure. Nevertheless, these algorithms are exceedingly slow and the damage assessment process is achieved via a costly and time-consuming inverse process, which presents an obstacle for real-time health monitoring applications.

Supervised learning algorithms are an alternative to model updating-based methods. The objective of supervised learning is to estimate the structure's health based on current and past samples. Supervised learning can be divided into two classes: parametric and nonparametric. Parametric approaches assumed a statistical model for the data samples. A popular parametric approach is to model each class density as Gaussian [91]. Nonparametric algorithms do not assume a structure for the data. The most frequently nonparametric algorithms used in damage assessment are artificial neural networks (ANNs) [5,55,131]. A trained neural network can potentially detect, locate, and quantify structural damage in a short period of time. Hence, it can be used for real-time damage assessment.

Building an accurate numerical model and defining an appropriate damage parameterization are crucial factors in model-based damage assessment. The following sections review and discuss these factors.

NUMERICAL MODEL

Numerical models are mathematical representations of a structure. A numerical model provides a means for predicting the response characteristics of the structure without actually building it. In most cases of practical interest, the model takes the form of a finite element (FE) model. In an FE model, the physical continuous domain of a complex structure is discretized into small components called FEs. The FE method is extensively used in research and industrial applications as it can produce a good representation of a true structure. An FE model of a linear structure is represented by the following $n \times n$ matrices: mass ($\mathbf{M}$), stiffness ($\mathbf{K}$), and damping ($\mathbf{C}$) matrices, where n is the number of degrees of freedom (DoFs). The motion of the linear system is described by

$$\mathbf{M}\ddot{\mathbf{x}} + \mathbf{K}\mathbf{x} + \mathbf{C}\dot{\mathbf{x}} = \mathbf{f}(t) \tag{7.1}$$

where:

$\mathbf{x}, \dot{\mathbf{x}}, \ddot{\mathbf{x}}$ are the displacement, velocity, and acceleration vectors

$\mathbf{f}(t)$ represents the time dependent external force

If the system is undamped or lightly damped, the natural frequencies (ω_i) and mode shapes (φ_i) are given by the solution of the eigenvalue problem.

$$\left(\mathbf{K} - \omega_i^2 \mathbf{M}\right)\varphi_i = 0 \tag{7.2}$$

Due to damage the system will change its characteristics, this change is expressed in the numerical model as matrix changes $\Delta \mathbf{M}$, $\Delta \mathbf{K}$, and $\Delta \mathbf{C}$. To describe the damage by physical parameters such as crack size, crack location, delamination, and/or loss of mass, the Δ-matrices are related to changes in physical parameters of the model, this is called damage parameterization.

The accuracy of the numerical model is of great importance. If the numerical model is not accurate enough, it becomes difficult to distinguish between numerical errors and actual changes due to damage. There are two approaches to overcome this problem. The first is to update the numerical model with the experimental data from the undamaged structure [45]. The changes to the model must be physically meaningful. However, even after updating there will still be differences between the numerical and experimental models. The second alternative is to define an objective function that considers the initial errors in the numerical model [46,62], thus the need for an accurate numerical model is avoided. This is achieved by correlating the data changes instead of their absolute value. The main assumption is that any change in the structure properties is caused by damage. Thus, any error in the undamaged model of the structure that is also present in the damaged model will be removed.

DAMAGE PARAMETERIZATION

Damage parameterization is a very important aspect of model-based damage assessment methods. The success of the algorithm relies on the quality of the parameterization used to model the damage. The type and complexity of the model depend on the type of structure, damage mechanism, requirements of the estimation procedure, and quality of the measured data.

LOCAL DAMAGE

The apparition of a crack is the most common case of local damage. There are several approaches to model cracks. Dimarogonas [32] and Ostachowitz and Krawczuk [116] give a comprehensive review of crack modeling approaches. According to Friswell and Penny [47], crack modeling falls in the three main categories illustrated in Figure 7.1: (1) local stiffness reduction, (2) discrete spring models, and (3) complex models in two or three dimensions.

Local stiffness reduction is the most widely used because of its simplicity. Damage is represented by an elemental stiffness reduction factor β_i, defined as the ratio of the stiffness reduction to the initial stiffness. The stiffness matrix of the damage structure $\mathbf{K}_d$ is expressed as a sum of element matrices multiplied by reduction factors.

$$\mathbf{K}_d = \sum_i \left(1 - \beta_i\right) \mathbf{K}_i \qquad (7.3)$$

where $\mathbf{K}_i$ is the stiffness matrix of the ith element. The value $\beta_i = 0$ indicates that the element is undamaged, whereas $0 < \beta_i \leq 1$ implies partial or complete damage. This is the simplest method to model damage, but suffers from problems in matching damage severity to crack depth and is affected by mesh density.

In the second class of methods, a rotational spring connecting two parts of the beam pinned at the crack location simulates the crack. These methods are commonly used when a continuous model of the beam is adopted. The equivalent spring stiffness is computed as a function of crack depth using fracture mechanics methods. This model has been successfully applied to crack identification of simply supported, cantilever and free–free beams [8,35,93,111,128]. However, their applicability is limited to beam structures.

(a)   (b)   (c)

FIGURE 7.1   Crack modeling in beam structures: (a) stiffness reduction; (b) discrete spring; and (c) crack model with 2D/3D elements.

The last class of methods is complex models with plate or brick elements. These approaches produce more detailed and accurate models than the methods previously discussed. However, these models are difficult to apply in structural damage assessment because they require a large number of DoFs, and they need to revise the mesh as the damage location changes. Furthermore, Friswell and Penny [47] demonstrated that at low frequencies, simple methods as the stiffness reduction and the spring model can correctly model a crack. The authors show that a more detailed model does not substantially improve the results from damage assessment.

DISTRIBUTED DAMAGE

Damage patterns in concrete beams are characterized by damaged areas with multiple cracks. One possibility is to update the bending stiffness of each element in the FE model. However, this requires a large number of updating parameters to describe properly the bending variation along the beam and a realistic damage pattern is not guaranteed. Wahab et al. [1] proposed to parameterize the damaged zones with a three-parameter damage function. With this parameterization not only the updating parameters are reduced, but also a realistic damage pattern is guaranteed. Teughels et al. [143] use a set of damage functions to identify the bending stiffness distribution. The updating parameters are the factors by which each of the damage functions is multiplied. They illustrate the algorithm with experimental data from a reinforced concrete beam with nonsymmetrical damage. The damage pattern detected is a smooth damage distribution that corresponds well with the experimental damage. On the other hand, they show that if the stiffness of each beam element is updated independently, the result is a nonsmooth damage distribution with several peaks.

OTHER TYPES OF DAMAGE

Damage assessment of composite materials and structures is a growing research area [108,167]. The aeronautical, naval, and automotive industries have increasingly introduced the use of composite materials as an alternative to conventional materials, due to their excellent mechanical properties, high strength, and low density. Nevertheless, failure modes of composite materials differ from conventional materials. Usual damage for composite materials are matrix cracking, fiber breakage, fiber-matrix debonding, and delamination between plies. Most of these damage mechanisms produce similar changes in the vibration response to that obtained for damage in metallic structures. However, delamination is one of the most important failures of laminated composite materials, and has no parallel to damage mechanisms in other materials.

Zou et al. [167] present a review of different methods to model delamination, most of the algorithms reviewed are based on the delaminated beam problem. Della and Shu [29] provide an extensive review of the different models for delaminated composite laminates (beams and plates). Particular attention is given to delaminated composites having piezoelectric sensors and actuators. Delamination appears as a debonding of adjoining layers in a composite material. Figure 7.2 illustrates the simplest delamination model for beam structures. This model assumes that delamination

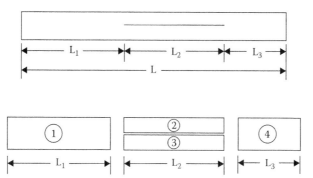

FIGURE 7.2 Simplified beam model of delamination.

divides the beam into four regions [146]. A Euler beam theory is applied to each region, and conditions of continuity couple the solutions. This simplified analytical model is capable of predicting the stiffness degradation observed in composite laminates. More accurate models are formulated through FEs using layer-wise theory [109]. Delaminations are modeled by discontinuities at the interfaces; the displacements on these interfaces remain independent, allowing for separation.

MODEL UPDATING METHODS

Model updating methods correlate a numerical model of the structure with the measured data to improve the model. In general, the numerical model is derived from FE analysis and the measurements are the vibration characteristics (natural frequencies, mode shapes, damping, FRF, etc.). The stiffness, mass, and damping distribution of the numerical model is updated to obtain the minimum difference between the numerical and experimental data, as is illustrated in Figure 7.3. The problem of detecting damage, is a constrained nonlinear optimization problem, where the damage indices of each element β_i are defined as updating parameters. An objective function $J(\beta)$ represents the error between the measured and numerical data.

$$\min \quad J(\beta)$$
$$0 \le \beta_i \le 1 \tag{7.4}$$

Setting up the objective function and using a robust optimization algorithm are the most important factors in model updating-based damage assessment.

CORRELATION TECHNIQUES

A proper selection of the objective function is crucial because it not only modifies the interpretation of the best correlation, but also influences the convergence of the optimization procedure. Several correlation techniques have been reported in the damage detection field, they can be categorized as follows.

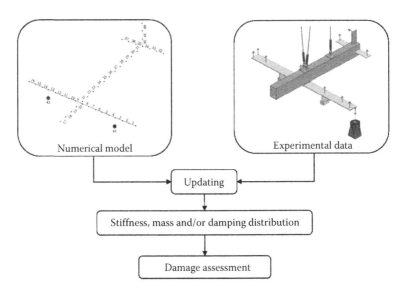

FIGURE 7.3   Scheme of model updating-based damage assessment method.

Natural Frequencies

Natural frequencies are the most popular damage indicators. The main reason for their popularity is that natural frequencies are easy to measure and are independent of the measuring position. In addition, natural frequencies are less contaminated by measured noise and can be identified more accurately than other modal parameters such as mode shapes and modal damping.

There is a large amount of literature regarding the use of natural frequencies in damage assessment. Salawu [132] presents an excellent review of methods using frequency shifts covering the period before 1997. Despite he emphasizes that damage detection using only natural frequencies is attractive because it is fast and economical, he concludes that it is not possible to locate and quantify correctly any damage situation by using only frequencies.

The most popular objective functions based on natural frequency changes are the Cawley–Adams criterion [19] and the damage location assurance criterion (DLAC) [101]. Cawley and Adams [19] were among the first to use frequency shifts in damage detection. They demonstrated that the ratio of the frequency changes in two modes is only a function of the damage location, although this only works for a single damage site. The measurement of a pair of frequencies allows us to predict possible damage locations. The Cawley–Adams criterion is defined as follows:

$$\varepsilon_{ij}(s) = \begin{cases} \dfrac{\partial\omega_{E,i}/\partial\omega_{E,j}}{\partial\omega_{A,i}(s)/\partial\omega_{A,j}(s)} - 1 & \text{if} \quad \dfrac{\partial\omega_{E,i}}{\partial\omega_{E,j}} \geq \dfrac{\partial\omega_{A,i}(s)}{\partial\omega_{A,j}(s)} \\[4mm] \dfrac{\partial\omega_{A,i}(s)/\partial\omega_{A,j}(s)}{\partial\omega_{E,i}/\partial\omega_{E,j}} & \text{if} \quad \dfrac{\partial\omega_{A,i}(s)}{\partial\omega_{A,j}(s)} \geq \dfrac{\partial\omega_{E,i}}{\partial\omega_{E,j}} \end{cases} \tag{7.5}$$

where:

$\partial\omega_{E,i}$ denotes the experimental change in the natural frequency for mode i

$\partial\omega_{A,i}(s)$ is the analytical prediction of the change in natural frequency for mode i with damage at location s

The total matching error is defined by

$$\varepsilon_s = \sum_{i=1}^{n-1}\sum_{j=i+1}^{n}\varepsilon_{ij}(s) \tag{7.6}$$

The predicted damage location is given by the minimum value using

$$E_s = \frac{\left(\varepsilon_s\right)_{min}}{\varepsilon_s} \tag{7.7}$$

Messina et al. [101] introduced the DLAC criterion to improve damage location using natural frequencies. The DLAC approach can only be used in single damage cases and is defined as follows:

$$\text{DLAC}(s) = \frac{\left|\partial\omega_E^T\partial\omega_A(s)\right|^2}{\left(\partial\omega_E^T\partial\omega_E\right)\left(\partial\omega_A(s)^T\partial\omega_A(s)\right)} \tag{7.8}$$

where:

$\partial\omega_E$ denotes the experimental change in the natural frequency vector

$\partial\omega_A(s)$ is the analytical prediction of the change in natural frequency vector with a damage at location s

$(.)^T$ indicates the vector transpose

A value of 0 indicates no correlation, whereas a value of 1 indicates a perfect correlation.

Messina et al. [102] proposed an extension of the DLAC criterion to identify damage in multiple locations. This method is referred to as multiple damage location assurance criterion (MDLAC). Using the same principle as DLAC, the predicted damage is given by a damage vector β that maximizes the MDLAC. They validated the algorithm with numerical and experimental data with good results. Ostachowicz et al. [115] applied a genetic algorithm (GA) approach to detect, locate, and quantify delamination and crack in a cantilever beam structure. They use the DLAC criterion to correlate the frequencies. According to the authors, the results are promising, particularly because the number of calculations needed for failure detection is much less than those required for classical search algorithms. Gomes and Silva [54] suggested a methodology based on GAs and frequency shifts for damage assessment. They used a modified MDLAC index. Toropov et al. [145] applied a genetic programming methodology based on natural frequencies for damage recognition of a steel portal frame. They formulate the objective function as follows:

$$J_\omega = \sum_j W_{\omega j} \left(\frac{\omega_{E,j} - \omega_{A,j}(\beta)}{\omega_{E,j}} \right)^2 \tag{7.9}$$

where:

$\omega_{E,j}$ and $\omega_{A,j}$ are the jth experimental and analytical natural frequencies
$W_{\omega j}$ is a weighting factor for mode j
β is a vector containing the damage indices

Friswell et al. [46] implemented a combined genetic eigensensitivity algorithm to locate and quantify structural damage. The authors proposed the following objective function:

$$J = W_\omega J_\omega + W_{ns} \delta_{ns} \tag{7.10}$$

where:

J_ω is the error in frequency
W_ω and W_{ns} are weighting factors

δ_{ns} is 0 if one site is damaged and is 1 if more than one site is damaged; this is a term to weight against more than one damage location. According to the authors, the damage penalization term is very important to avoid false damage assessment. The error in natural frequency J_ω is defined as

$$J_\omega = \sum_j W_{\omega j} \left(\frac{\delta\omega_{E,j} - \delta\omega_{A,j}(\beta)}{\omega_{E,j}} \right)^2 \tag{7.11}$$

where $\delta\omega_{E,j}$ and $\delta\omega_{A,j}$ are the changes due to damage in the jth experimental and analytical natural frequencies. This error function reduces the effect of modeling errors because it considers changes in frequency from the undamaged to the damaged structure.

Mode Shapes and Natural Frequencies

There are three issues to be addressed when experimental and analytical mode shapes are compared. First, analytical models have in general more DoFs than the number of measured DoF. For some correlation coefficients, the number of measured DoF have to be the same as the analytical DoF. In these cases, it becomes necessary either to expand the experimental modes or to reduce the analytical model; an overview of expansion and reduction techniques is given in the book of Heylen et al. [68]. In general, the use of reduction or expansion techniques introduces additional errors that affect the performance of the damage assessment algorithm. Therefore, correlation coefficients related only to the measured DoF are preferred, thus avoiding the need for model expansion or model reduction.

Second, natural frequencies and mode shapes of the experimental and analytical data must be paired correctly. This can become a difficult task in systems with high modal density. Arranging the natural frequency in ascending order is not sufficient, especially when two modes are closed in frequency. The modal assurance criterion (MAC) can be used to handle this problem. MAC is defined by Allemang and Brown [3] as

$$MAC_i = \frac{\left(\varphi_{A,i}^T \varphi_{E,i}\right)^2}{\left(\varphi_{A,i}^T \varphi_{A,i}\right)\left(\varphi_{E,i}^T \varphi_{E,i}\right)} \tag{7.12}$$

where:

φ_i is the ith mode shape

subscripts A and E refer to analytical and experimental, respectively

MAC is a factor that expresses the correlation between two modes. A value of 0 indicates no correlation, whereas a value of 1 indicates two completely correlated modes.

The third problem is mode shape normalization. Analytical mode shapes are often mass normalized, and measured mode shapes can also be mass normalized. However, if the mass matrices of the analytical and experimental models are different or the dynamic test is performed with ambient excitation, then the scaling of the mode shapes will be inconsistent. To avoid this, the measured mode shape may be scaled to the analytical mode shape by multiplying it by the modal scale factor (MSF) [45]:

$$\varphi_{E,i}^* = \varphi_{E,i} \cdot MSF_i \tag{7.13}$$

$$MSF_i = \frac{\varphi_{A,i}^T \varphi_{E,i}}{\varphi_{E,i}^T \varphi_{E,i}} \tag{7.14}$$

where:

$\varphi_{E,i}^*$ and $\varphi_{E,i}$ are the scaled and original ith experimental mode shape

$\varphi_{A,i}$ is the ith analytical mode shape

MSF_i is the modal scale factor for mode i

Multiplying the experimental mode shape by the corresponding MSF also solves the problem that the measured and analytical mode shapes could be 180° out of phase.

Most damage assessment methods make use of modal information as mode shapes and natural frequencies. In general, natural frequencies detect and quantify the presence of damage, while mode shapes locate it. Several damage assessment methods using mode shapes and natural frequencies have been proposed in literature; a review of the most representative ones is provided below.

Shi et al. [135] proposed an improved MDLAC criterion by including mode shape data. The damage sites are identified approximately using incomplete mode shapes first, and then the true damage sites and extents are determined using the natural frequencies. Results show that the inclusion of mode shape data improves the damage localization. The authors defined the MDLAC with mode shapes as

$$MDLAC_i(\beta) = \frac{\left|\partial\varphi_{E,i}^T \partial\varphi_{A,i}(\beta)\right|^2}{\left(\partial\varphi_{E,i}^T \partial\varphi_{E,i}\right)\left(\partial\varphi_{A,i}^T(\beta)\partial\varphi_{A,i}(\beta)\right)} \tag{7.15}$$

where:

$\partial\varphi_i$ is the ith mode shape change vector

β is the vector of damage indices

Subscripts A and E refer to analytical and experimental, respectively

$(.)^T$ indicates the vector transpose

The mode shape change vectors are defined at the measured DoFs, thus incomplete mode shapes can be used with no need of modal expansion or model reduction.

A well-established sensitivity-based algorithm is the inverse eigensensitivity method (IESM) [44]. This algorithm has been widely used in structural model updating. It correlates mode shapes and natural frequencies, and uses a gradient search optimization algorithm. In the IESM, the error functions are defined as

$$\varepsilon_{\lambda,r} = \frac{\lambda_{E,r} - \lambda_{A,r}}{\lambda_{E,r}}, \quad \varepsilon_{\varphi,r} = \varphi_{E,r} - \varphi_{A,r} \tag{7.16}$$

where λ_r, φ_r are the rth eigenvalue and mode shape, respectively. Subscripts A and E refer to analytical and experimental, respectively. In this case, the mode shape error can be computed at the measured DoF only, but a consistent mode normalization is required.

Wahab et al. [1] used the IESM to identify damage patterns of a reinforced concrete beam. They parameterized the damage patterns by a damage function to reduce the number of unknown parameters. Thus, the updating parameters are reduced to three and a realistic damage pattern is guaranteed. Fritzen and Bohle [51] described an application of the IESM and the modal kinetic energy criterion in the damage assessment of the Z-24 Bridge. The experimental modal data are extracted from output-only measurement data. Görl and Link [57] proposed an algorithm based on mode shape and frequency sensitivities to assess damage of a steel frame structure. They defined a reduced number of updating parameters to improve the condition of the sensitivity matrix. Although damage is successfully located and quantified, a careful selection of the regularization parameters was necessary to achieve satisfactory results. Wu and Li [155] proposed a two-stage eigensensitivity-based damage assessment algorithm. They applied the algorithm to experimental data of a steel frame structure.

Ruotolo and Surace [129] stated that the actual state of damage of the structure can be estimated only if some "fundamental" functions are properly combined and a weight constraint introduced that limits the total damage. They defined three fundamental functions:

$$\varepsilon_1(\beta) = \sum_i \left(1 - \frac{\omega_{E,i} / \omega_{E,i}^0}{\omega_{A,i}(\beta) / \omega_{A,i}^0}\right) \tag{7.17}$$

$$\varepsilon_2(\beta) = \sum_i \sum_j \left(\varphi''_{A,ij}(\beta) - \varphi''_{E,ij}\right)^2 \tag{7.18}$$

$$\varepsilon_3(\beta) = \sum_i \sum_j \left(\varphi_{A,ij}(\beta) - \varphi_{E,ij}\right)^2 \tag{7.19}$$

where:

φ_{ij} is the jth mode shape at a point i

φ_{ij}'' is the second derivative (curvature) of the jth mode shape at a point i

ω_i is the ith natural frequency

Subscripts A and E refer to analytical and experimental, respectively, and superscript 0 refers to the initial state (undamaged)

The first function computes the natural frequencies differences. The second, computes the modal curvatures difference, while the third computes the difference between normalized mode shapes. Ruotolo and Surace tested different combinations of the fundamental functions with simulated data of a cracked beam under five different damage scenarios. They found that the optimum combination to be maximized is the following:

$$J(\beta) = \left(v(\beta) + c_1\varepsilon_1(\beta) + c_2\varepsilon_2(\beta) + c_3\varepsilon_3(\beta) \right)^{-1} \tag{7.20}$$

where c_1, c_2, and c_3 are coefficients properly evaluated to ensure that each of the terms in the denominator have a similar value. The term $v(\beta) = \sum_i \beta_i$ is a weighting term that promotes the determination of damage at fewer sites, thus avoiding false damage assessment. The authors verified the methodology using a GA as optimization tool, and simulated and experimental data of a cracked beam.

Asce and Xia [62] defined an error function that includes the initial undamaged modal data. Thus, avoiding the need of a precise analytical model. They defined the error function as

$$J(\beta) = \sum_{i=1}^{nm} W_\lambda \left(\left[\frac{\lambda_{A,i}(\beta) - \lambda_{A,i}^0}{\lambda_{A,i}^0} \right] - \left[\frac{\lambda_{E,i} - \lambda_{E,i}^0}{\lambda_{E,i}^0} \right] \right)^2$$

$$+ \sum_{i=1}^{nm} W_\varphi \left(\left[\varphi_{A,i}(\beta) - \varphi_{A,i}^0 \right] - \left[\varphi_{E,i} - \varphi_{E,i}^0 \right] \right)^2 \tag{7.21}$$

where W_λ and W_φ are weighting factors. The authors used a GA to minimize the objective function. The results indicate that with proper weights, damage can be accurately detected, although the algorithm always detects false damages.

Au et al. [7] implemented a procedure to detect damage using noisy and incomplete modal data. First, the mode shapes are expanded with the system equivalent reduction expansion process. Then, they employed an elemental energy quotient difference to locate the damage, and last a micro-GA to quantify it. The micro-GA uses the following objective function:

$$J(\beta) = \sum_i \left(\frac{\omega_{E,i} - \omega_{A,i}(\beta)}{\omega_{E,i}} \right)^2 + \sum_i \frac{\left\| \varphi_{E,i} - \varphi_{A,i}(\beta) \right\|}{\left\| \varphi_{E,i} \right\|} \tag{7.22}$$

Simulated beams with multiple damage are used to verify the approach. The results show that noisy and incomplete data end up in a large amount of falsely detected damages. However, if the damage domain is reduced by a preliminary step as the elemental energy quotient difference, results are improved.

The difference between experimental and numerical modes can also be computed by indicators as the MAC [3], coordinate MAC (COMAC) [87], or the spatial MAC (SMAC) [67].

An advantage of MAC (see Equation 7.12) is that no normalization of the modes is necessary, thus operational modes can be used. If the number of measured DoFs is less than the numerical DoF, a partial MAC can be used. Hence, no mode shape expansion is needed. Brownjohn et al. [12] used the MAC values and frequency differences as correlation indices. They concluded that the algorithm provides quantitatively accurate information for damage assessment. Gao and Spencer [52] introduce the total MAC (TMAC) to determine the global correlation of the modes and not individually as with MAC.

$$\text{TMAC} = \prod_i \text{MAC}\left(\varphi_{A,i}, \varphi_{E,i}\right) \tag{7.23}$$

The modified total assurance criterion (MTMAC) is proposed by Perera and Torres [124]. This criterion is derived from the TMAC, but modifies it with the introduction of the frequency.

$$\text{MTMAC} = \prod_i \frac{\text{MAC}\left(\varphi_{A,i}, \varphi_{E,i}\right)}{\left(1 + \left|\dfrac{\lambda_{A,i} - \lambda_{E,i}}{\lambda_{A,i} + \lambda_{E,i}}\right|\right)} \tag{7.24}$$

They also proposed another criterion using residual forces, but this criterion fails in the presence of noisy or incomplete modal data. They implemented a GA to handle the optimization, and use simulated and experimental data of a beam to test the approach. The MTMAC approach gives acceptable damage predictions, although in the presence of experimental noise considerable false damages are detected.

The COMAC measures the correlation at each DoF averaged over a set of mode pairs. It identifies the coordinate at which two sets of mode shapes do not agree. It is defined by Lieven and Ewins [87] as

$$\text{COMAC}_i = \frac{\left(\sum_j \left|\varphi_{A,ij}\phi_{E,ij}\right|\right)^2}{\sum_j \varphi_{A,ij}^2 \sum_j \varphi_{E,ij}^2} \tag{7.25}$$

The COMAC takes a value between 1 and 0. A low value indicates discordance at a point and thus a possible damage location. Ndambi et al. [113] studied the possibilities of COMAC as a damage indicator in reinforced cracked beams. They concluded that with COMAC it is possible to detect and locate the damage, but is not possible to quantify it.

Heylen and Janter [67] introduced the application of the the SMAC to model updating. In analogy to the definition of the MAC two SMAC matrices can be defined:

$$\text{SMAC}_A = \left(\varphi_A^T \varphi_A\right)^{-1} \left(\varphi_A^T \varphi_E\right) \left(\varphi_E^T \varphi_E\right)^{-1} \left(\varphi_E^T \varphi_A\right) \tag{7.26}$$

$$\text{SMAC}_E = \left(\varphi_E^T \varphi_E\right)^{-1} \left(\varphi_E^T \varphi_A\right) \left(\varphi_A^T \varphi_A\right)^{-1} \left(\varphi_A^T \varphi_E\right) \tag{7.27}$$

In this case instead of comparing two vectors, the SMAC allows us to compare different vector spaces. From the diagonal elements it can be derived what vectors are well or badly described by the other vector space, if the vector spaces defined by φ_A and φ_E are identical, the SMACs become unity matrices.

Residual Forces

The modal characteristics of an undamaged structure are described by the eigenvalue equations:

$$\left(\mathbf{K} - \lambda_{A,i}\mathbf{M}\right)\varphi_{A,i} = 0 \tag{7.28}$$

where:
K and **M** are the stiffness and mass matrices of the numerical model
$\lambda_{A,i}$ and $\varphi_{A,i}$ are the ith analytical eigenvalue and mode shape

If the analytical modal parameters are replaced by the experimental ones, the residual force vector is defined as

$$\mathbf{R}_i = \left(\mathbf{K} - \lambda_{E,i}\mathbf{M}\right)\varphi_{E,i} \tag{7.29}$$

where:
$\lambda_{A,i}$ and $\varphi_{A,i}$ are the ith experimental eigenvalue and mode shape
$\mathbf{R}_i$ is the residual force vector for mode i

Excluding modeling errors and experimental noise, the residual force vector should equal to zero if the structure is undamaged. However, if the structure is damaged the stiffness matrix must be updated:

$$\mathbf{R}_i(\beta) = \left(\mathbf{K}(\beta) - \lambda_{E,i}\mathbf{M}\right)\varphi_{E,i} \tag{7.30}$$

where β is the vector of damage indices. This residue will be 0, if only a correct set of damage indices β are introduced. To compute the residual force vector the measured DoFs must be the same as the analytical DoFs. If not, it is necessary either to expand the experimental modes or to reduce the analytical model. In addition, experimental modes must be mass normalized.

Kosmatka and Ricles [83] used the residual force vector to locate potential damage elements, and then they quantified damage through an eigensensitivity algorithm. The results show that the residual force vectors are able to locate the regions of the structure containing damage. Fritzen and Bohle [50] compared the performance of the IESM algorithm and the modal force residual method. Similar results are obtained with both methods, but the modal force residual method is more sensitive to the amount of input data and to the experimental noise. A study of damage identification methods based on the residual force vector is performed by Yang and Liu [156]. They used the residual force vector to localize damages preliminarily, and then they applied three different damage quantification methods to identify damages more precisely; the first is the solution of the residual force equation, the second is the minimum-rank elemental update technique [34], and the third is the natural frequency sensitivity method. They concluded that the best combination is to localize damages with the residual force vector and quantify them with the frequency sensitivity method. Yang [157] extended this damage localization algorithm to locate damage at the element level rather than at the DoF and proposed a damage quantification algorithm using residual forces. The author concluded that the approach may be useful for structural damage identification, although he stated that the algorithm should be tested with experimental data and that further research it is necessary to tackle the problem of test/analysis DoF mismatch. Yung et al. [159] implemented a sensitivity-based algorithm with residual force vector residuals. They defined the residual as the difference between the damage and undamaged residual force vectors:

$$\varepsilon_R(\beta) = \mathbf{R}_{D,i}(\beta) - \mathbf{R}_{U,i} \tag{7.31}$$

where $\mathbf{R}_{D,i}$ and $\mathbf{R}_{U,i}$ are the damaged and undamaged residual force vectors for mode i. Mares and Surace [90] applied the concept of residual forces to build an objective function optimized by a GA. They defined the objective function to be maximized as

$$J(\beta) = \frac{c_1}{c_2 + \sum_i \mathbf{R}_i(\beta)^T \mathbf{R}_i(\beta)} \tag{7.32}$$

where:

c_1 and c_2 are constants
$\mathbf{R}_i$ is the residual force vector defined in Equation 7.30
β is the vector of damage indices
$(.)^T$ indicates the vector transpose

The authors were able to identify both the location and extent of the damage with a reasonable degree of accuracy. However, the results indicate that the algorithm is very sensitive to experimental noise. Similarly, Rao et al. [127] used the concept of residual force matrix to specify the objective function, the optimization procedure is attained by a GA. They defined the objective function as

$$J(\beta) = \frac{c_1}{c_2 + \sqrt{R(\beta)_{11}^2 + R(\beta)_{22}^2 + \cdots + R(\beta)_{nm}^2}} \tag{7.33}$$

where:

R_{nm} is the residual force value for mode n at the mth DoF
c_1 and c_2 are constants

Flexibility Matrix

An important group of structural damage assessment methods are derived from the change of flexibility [33]. The dynamic flexibility matrix is very sensitive to changes in the low-frequency modes. In addition, it can be built with a few low modes at any arbitrary set of DoFs [77,118]. These characteristics make flexibility-based methods very attractive. The flexibility matrix is basically defined as the inverse of the stiffness matrix, which can be written in modal form as

$$\mathbf{G} = \mathbf{\Phi}\mathbf{\Lambda}^{-1}\mathbf{\Phi}^T = \sum_i \frac{1}{\omega_i^2}\varphi_i\varphi_i^T \tag{7.34}$$

where:
  $\mathbf{\Phi}$ and $\mathbf{\Lambda}$ are the eigenvector and eigenvalue matrices
  φ_i and ω_i are the ith mode shape and natural frequency
  $(.)^T$ indicates the vector transpose

From Equation 7.34 we can see that the higher the frequency, the smaller the modal contribution to the flexibility matrix. According to Jaishi et al. [77] a good estimation of the flexibility matrix may be obtained from only few low-frequency modes at a reduced set of measured DoFs. Pandey and Biswas [118] showed that changes in the flexibility matrix can be used to detect and locate structural damage. They compute the experimental change in the flexibility matrix as

$$\Delta\mathbf{G} = \mathbf{G}_{E,U} - \mathbf{G}_{E,D} \tag{7.35}$$

where $\mathbf{G}_{E,U}$ and $\mathbf{G}_{E,D}$ are the experimental flexibility matrices for the undamaged and damaged cases. Similar nonmodel-based methods that use changes in the experimental flexibility matrices have been proposed; Doebling et al. [33] review most of them. Model-based methods that use the flexibility matrix are reviewed next.

Stutz et al. [139] proposed to minimize the difference between the experimental and analytical flexibility matrices to solve the inverse damage assessment problem:

$$J(\beta) = \left\| \mathbf{G}_E - \mathbf{G}_A \right\|_F^2 \tag{7.36}$$

where:
  $\mathbf{G}_E$ and $\mathbf{G}_A$ are the experimental and analytical flexibility matrices
  $\left\|\cdot\right\|_F$ stands for Frobenius norm

Jaishi and Ren [77] implemented a sensitivity-based algorithm using the modal flexibility residuals defined in Equation 7.36. The results show that the algorithm predicts damage with a good accuracy, although with an increment of the simulated noise the amount of falsely detected damages increases. A multiobjective damage identification method is studied by Perera et al. [123]. Two objective functions are used, the first is the MTMAC defined in Equation 7.24, and the second is formulated in terms of the modal flexibility. Similar to the MAC indicator they defined the MACFLEX as

$$\text{MACFLEX}_j = \frac{\left|\mathbf{G}_{A,j}^T\mathbf{G}_{E,j}\right|^2}{\left(\mathbf{G}_{A,j}^T\mathbf{G}_{A,j}\right)\left(\mathbf{G}_{E,j}^T\mathbf{G}_{E,j}\right)} \tag{7.37}$$

where:

$\mathbf{G}_{A,j}$ and $\mathbf{G}_{E,j}$ are the analytical and experimental flexibility vectors corresponding to the jth mode

$(.)^T$ stands for transpose

Strain Energy

The modal strain energy (MSE) of the ith mode is defined as

$$\text{MSE}_i = \frac{1}{2}\varphi_i^T\mathbf{K}\varphi_i \tag{7.38}$$

where:

φ_i is the ith mode shape vector
$\mathbf{K}$ is the global stiffness matrix
$(.)^T$ indicates the vector transpose

To compute the MSE the measured points must match the numerical DoFs. However, numerical models have in general more DoFs than the number of measurement points. Thus, it is necessary either to expand the experimental modes or to reduce the analytical model. Experimental modes must also be consistently normalized.

Changes in strain energy have been extensively used as an indicator to represent damage. Damage indices are determined through strain energy changes; this method is commonly known as the damage index method. The damage index has been widely used as a nonmodel method for simple structures. For example, in the case of a Bernoulli–Euler beam, the bending strain energy for the ith mode is

$$U_i = \frac{1}{2}\int_0^l \text{EI}\left(\frac{\partial^2\varphi_i}{\partial x^2}\right)^2 dx \tag{7.39}$$

where:

EI is flexural rigidity
l denotes beam length
$\varphi_i(x)$ is the ith mode shape
x is the coordinate along the span of the beam

Stubbs [138] showed that the change in bending rigidity at the kth location in the structure for the ith mode is given by

$$\frac{(\text{EI})_{ik}^D}{(\text{EI})_{ik}} = \frac{\displaystyle\int_{a_k}^{a_{k+1}}\left(\frac{\partial^2\varphi_i^D}{\partial x^2}\right)^2 dx \bigg/ \int_0^l\left(\frac{\partial^2\varphi_i^D}{\partial x^2}\right)^2 dx}{\displaystyle\int_{a_k}^{a_{k+1}}\left(\frac{\partial^2\varphi_i}{\partial x^2}\right)^2 dx \bigg/ \int_0^l\left(\frac{\partial^2\varphi_i}{\partial x^2}\right)^2 dx} \tag{7.40}$$

superscript D stands for damaged. The damage index at the kth location (DI_k) is defined as

$$DI_k = \sum_i \frac{(EI)_{ik}^D}{(EI)_{ik}} \tag{7.41}$$

Similar approaches have been introduced for bar [36] and plate [21] like structures without the need of a numerical model. In the case of structures that are more complex a numerical model is needed. Among nonmodel-based methods, the damage index has given the best results. Farrar and Jauregui [41,42] compared the performance of five damage assessment methods, in the damage assessment of the I-40 Bridge. In general, the damage index method performed the best, although for the most severe damage case all the methods accurately locate the damage. Humar et al. [73] studied the performance of several damage identification algorithms; they concluded that the damage index method appears to be most successful in predicting the damage location, and it was the most tolerant with experimental noise. Alvandi and Cremona [4] investigated the reliability of some usual damage identification techniques, such as, the strain energy, shape curvature, change in flexibility, and change in flexibility curvature methods. They concluded that the strain energy method provides the best stability in presence of noisy signals.

Shi et al. [134] defined a damage indicator that makes use of the change of MSE in each structural element before and after the occurrence of damage. This damage indicator called the modal strain energy change ratio (MSECR) is defined as

$$MSECR_{i,j} = \frac{\left| MSE_{i,j}^D - MSE_{i,j}^U \right|}{MSE_{i,j}^U} \tag{7.42}$$

where $MSE_{i,j}$ is the modal strain energy for element i and mode j, superscripts D and U refer to damaged and undamaged, respectively. If several modes are considered, the MSECR is defined as the average for all the modes. This algorithm was later improved to include damage quantification and the modal truncation errors [136]. Hsu and Loh [72] proposed a modified modal strain energy change (M-MSEC), the algorithm was applied to a full-scale 3D real frame structure. The results showed that the M-MSEC method is more reliable than the original MSEC method.

Previous discussed methods use the strain energy change as a damage indicator to detect and locate damage; these methods do not give information about the extent of the damage. Jaishi and Ren [78] proposed a model updating algorithm based on strain energy residuals. They use a multiobjective optimization technique based on strain energy and natural frequency changes. They defined the strain energy and frequency residuals as

$$\varepsilon_{MSE,i} = \left(\frac{\varphi_{A,i}^T \mathbf{K} \varphi_{A,i}}{\varphi_{E,i}^T \mathbf{K} \varphi_{E,i}} - 1 \right)^2 \tag{7.43}$$

$$\varepsilon_{\omega,i} = \left(\frac{\lambda_{A,i}}{\lambda_{E,i}} - 1 \right)^2 \tag{7.44}$$

Meruane and Heylen [96] studied the performance of five fundamental objective functions: (1) frequencies, (2) modes and frequencies, (3) MAC and frequencies, (4) strain energy and frequencies, and (5) modal flexibility. The optimization process is performed with a GA. The objective function based on MAC values and natural frequencies performed the best.

Frequency Response Function

The use of the FRF is an attractive alternative. The advantage is that no modal extraction is necessary, thus it avoids contamination of the data with modal extraction errors. However, FRFs have the disadvantage that they cannot be identified from output only modal analysis with ambient excitation, thus the excitation by an artificial force is always required.

Lim and Edwinds [88] developed the response function method. This is a sensitivity-based method that directly uses measured FRF data. They defined the error function as

$$\varepsilon_{\alpha,j} = \alpha_j^A - \alpha_j^E \tag{7.45}$$

where α_j represents the jth column of the FRF matrix α, superscripts A and E refer to analytical and experimental, respectively. Wang et al. [151] implemented an adaptation of this technique for structural damage detection. They weighted selected locations and frequencies to minimize the influence of errors. Although their results are consistent with the experimental damage, the algorithm is very sensitive to modeling errors, experimental noise, and incompleteness of the measured FRF data. They concluded that more accurate and comprehensive damage indicators are needed. Thyagaraja et al. [144] proposed a gradient-based damage assessment method that uses FRF measurements from a minimum number of sensors. The authors stated that a more robust optimization algorithm should be used and that an algorithm should be developed to automatically select points from the FRFs. Fritzen and Bohle [49] applied an FRF sensitivity based method to identify damage of the I-40 bridge. Good damage predictions are obtained for damage scenario 4 and 3, but the method failed in the first two damage scenarios.

Imregun et al. [74,75] conducted several studies using simulated and experimental data to test the effectiveness of the response function method. They found that complex and noisy FRF data can make the convergence process very slow and often numerically unstable. Furthermore, the success of the method is highly dependent on the selection of the frequency points. Lammens [84] addresses how a poor selection of the frequency points can lead to an unstable updating process and inaccurate results. According to Levin and Lieven [86] the reason for this bad performance is that the function is dominated by the contributions at the resonant peaks. To reduce the influence at the natural frequencies and to increase the contribution at the anti-resonances, they proposed the following objective function:

$$J_q(\beta) = \sum_k \sum_\omega \left| \log\left(\left\| \alpha_{jk}^E(\omega) \right\|_2 \right) - \log\left(\left\| \alpha_{jk}^A(\omega) \right\|_2 \right) \right|^q \tag{7.46}$$

where:

ω are the frequency points selected to apply the algorithm

j is the selected column of the FRF matrix (excitation location)

k is the DoF where the response is measured

After evaluating the performance with different values of q, they found that the most effective value is $q = 1$.

Pascual et al. [119] introduced the concept of frequency shift between experimental and analytical FRFs with the frequency domain assurance criterion (FDAC):

$$FDAC_j(\omega_1, \omega_2) = \frac{\left| \sum_i \alpha_{ij}^E(\omega_1) \alpha_{ij}^{A*}(\omega_2) \right|^2}{\sum_i \left(\alpha_{ij}^E(\omega_1) \alpha_{ij}^{E*}(\omega_1) \right) \sum_i \left(\alpha_{ij}^A(\omega_2) \alpha_{ij}^{A*}(\omega_2) \right)} \qquad (7.47)$$

where:

ω_1 and ω_2 represent the experimental and analytical frequencies where the FDAC is evaluated

$(.)^*$ denotes complex conjugate

This criterion is equivalent to the MAC in the frequency domain, thus a value of 0 indicates no correlation, whereas a value of 1 indicates two perfectly correlated operational deflection shapes. Pascual et al. [120,121] proposed an improved model-updating algorithm; the FRF residues are computed at different frequencies that were previously selected through the FDAC criterion.

Zang et al. [161] defined the global shape correlation criteria and the global amplitude criterion to correlate FRFs,

$$GSC(\omega) = \frac{\left| \alpha_j^E(\omega)^H \alpha_j^A(\omega) \right|^2}{\left(\alpha_j^E(\omega)^H \alpha_j^E(\omega) \right) \left(\alpha_j^A(\omega)^H \alpha_j^A(\omega) \right)} \qquad (7.48)$$

$$GAC(\omega) = \frac{2 \left| \alpha_j^E(\omega)^H \alpha_j^A(\omega) \right|}{\left(\alpha_j^E(\omega)^H \alpha_j^E(\omega) \right) + \left(\alpha_j^A(\omega)^H \alpha_j^A(\omega) \right)} \qquad (7.49)$$

where $(.)^H$ denotes the conjugate transpose. Zang et al. [163] implemented a sensitivity-based method using the average of these two indicators to identify damage. An FRF-based method that uses a hybrid-GA and hill climbing is implemented by Raich et al. [126]. To compute the difference between the analytical and experimental FRFs, they suggested the following error function:

$$J(\beta) = \sum_k \left(\int_{\omega_0}^{\omega_1} \left| \alpha_{jk}^E(\omega) - \alpha_{jk}^A(\omega) \right| \right) \qquad (7.50)$$

where ω_0 and ω_1 are the lower and upper bound of the frequency range.

Antiresonances

Recently, great attention has been given to the possible use of antiresonances in structural damage assessment. Antiresonances are an attractive alternative because they can be determined easily and with less error than mode shapes [23]. In addition, antiresonances are very sensitive to small structural changes, which makes them good damage indicators.

Antiresonances can be derived from point FRFs, where the response coordinate is the same as the excitation coordinate; or from transfer FRFs, where the response coordinate differs from the excitation coordinate. Point FRFs are preferred because matching problems arise when antiresonances from transfer FRFs are used. Moreover, the distribution of the transfer antiresonances can be significantly modified with small structural changes [23]. On the other hand, the procedure to obtain point FRFs differs from common modal testing, such as the excitation DoF is moved together with the response DoF. This may become not practical or too expensive.

Williams and Messina [153] introduced antiresonances from point FRFs to the MDLAC algorithm. They concluded that the incorporation of antiresonance data improves the accuracy of the damage predictions. Dilena and Morassi [31] studied the problem of crack detection in beams using natural frequencies and antiresonances. They found that the use of antiresonances helps to avoid nonuniqueness of the damage location that occurs when only natural frequencies are used. However, they also found that experimental noise and modeling errors are usually amplified when antiresonances are included. Bamnios et al. [8] used the shift in the first antiresonance versus the measuring position to detect and locate a crack in a beam. They stated that the method can be used to locate the crack roughly and then other methods can be employed to determine the crack characteristics more precisely. The changes in the antiresonances and natural frequencies are used by Inada et al. [76] to locate and quantify delamination of a composite beam. A two-step procedure is implemented, first the delamination domain is identified from the antiresonance changes and next the location and size is defined with the natural frequency changes. The method was effective in identifying delamination locations and sizes. Wang and Zhu [150] suggested a method to identify cracks in beams, which makes use of natural frequencies and antiresonances from point FRFs. The methodology is similar to the one proposed by Bamnios et al. [8], the shift in the first antiresonance versus the driving point location is used to locate damage.

A sensitivity-based method using natural frequencies and antiresonaces can also be implemented [22]. This method is related to the same principle of the IESM, but using antiresonances instead of mode shapes, the errors are defined as

$$\varepsilon_{\lambda,j} = \lambda_{E,j} - \lambda_{A,j}, \quad \varepsilon_{r,j} = \lambda_{r,j}^E - \lambda_{r,j}^A \qquad (7.51)$$

where:
λ_j is the jth eigenvalue
$\lambda_{r,j}$ is the jth antiresonant eigenvalue
superscripts and subscripts A and E refer to analytical and experimental, respectively

D'Ambrogio and Fregolent [22] discussed the possibilities of antiresonances in model updating. They showed that antiresonances from transfer FRFs are very sensitive to small changes, which makes it very difficult to keep track correctly of the errors between analytical and numerical antiresonances. Furthermore, while model updating with antiresonances from point FRFs gives very good results, no convergence is obtained when antiresonances from transfer FRFs are used. They stated that the actions required to use transfer FRF antiresonances without problems can be the subject for future research. D'Ambrogio and Fregolent [23] applied the antiresonance sensitivity approach to update the model of a frame structure. With antiresonances from point FRFs the method is robust and leads to good results. In contrast, with antiresonances from transfer FRFs the method is very unstable. Only with a careful selection of the updating parameter and a good match between experimental and numerical antiresonances they could reach results.

Antiresonances from transfer FRFs are used by Jones and Turcotte [80] to update a six-meter flexible truss structure. They studied the correctness of the updated model by using it to detect damage. The Group for Aeronautical Research and Technology in Europe (GARTEUR) structure is updated with an antiresonance-based method by D'Ambrogio and Fregolent [24]. The unmeasured point FRFs are synthesized through a truncated modal expansion. In a later work [25], they proposed the use of zeros from a truncated expansion of the identified modes; they refer to these zeros as "virtual antiresonances." They compared the results with an updating method using MAC and natural frequencies. The updated models using either true or virtual antiresonances were more accurate than with MAC. Nam et al. [110] extend the sensitivity-based method by adding additional spectral information. The basic spectral information corresponds to the natural frequencies and the additional information to antiresonances and static compliance dominant (SCD) frequencies. The SCD frequencies are the frequencies that yield the same static compliances of the structural system on the FRF. They defined the errors as

$$\varepsilon_{\lambda,j} = \frac{\lambda_{E,j} - \lambda_{A,j}}{\lambda_{E,j}}, \quad \varepsilon_{r,j} = \frac{\lambda_{r,j}^E - \lambda_{r,j}^A}{\lambda_{r,j}^E}, \quad \varepsilon_{s,j} = \frac{\lambda_{s,j}^E - \lambda_{s,j}^A}{\lambda_{s,j}^E} \tag{7.52}$$

where $\lambda_{s,j}$ is the jth SCD eigenvalue. They used antiresonances from point and transfer FRFs. A numerical spring mass system evaluates the algorithm. They concluded that the accuracy of the algorithm can be improved with the use of additional spectral information as antiresonances and compliance dominant frequencies.

Meruane and Heylen [97] demonstrated that antiresonances are a good alternative to mode shapes in damage assessment. However, they stated that further research is required for the identification of experimental antiresonances and for the matching of experimental and numerical antiresonances. Meruane [100] presented a model updating and damage assessment algorithm that uses antiresonant frequencies derived from transmissibility data. Antiresonance frequencies correspond to the dips in FRFs and consequently to the dips and peaks in transmissibility functions. Hence, it is possible to identify antiresonance frequencies using transmissibility information.

Genetic Algorithms

Classical optimization algorithms used in damage assessment are sensitivity-based methods. These algorithms obtain the optimum solution through a sensitivity-based search. Because of the nonlinear relation between the vibration data and the physical parameters, the algorithm performs an iterative optimization process. The book of Friswell and Mottershead [45] discuss these methods in detail. Nevertheless, these searching mechanisms are highly sensitive to the initial searching conditions and they usually lead to local minimums. Moreover, the ill-conditioned inverse problem leads to instabilities and the convergence is not always assured. According to Natke [112], the only solution to overcome ill-conditioning is the regularization of the equations. However, the regularization parameters must be carefully selected because the speed of convergence and the quality of the results strongly depend on them [165]. Weber et al. [152] studied the sensitivity-based damage detection in an ill-posed problem. Numerical simulations show that with an appropriate parameter selection, regularization significantly improves results.

The limitations of classical optimization techniques restrict their use and therefore favor the implementation of more recently developed global searching algorithms. Genetic algorithms (GAs) have become very popular in the damage assessment field because of their efficiency and easy implementation in damage assessment problems [158]. Unlike traditional gradient-based optimization methods, GAs use multiple points to search and the calculation of gradients is not required. The GA is a global searching process derived from Darwin's principle of natural selection and evolution. It was first developed by John Holland in 1975 [71], and since then it has been increasingly introduced in diverse areas such as music generation, genetic synthesis, strategy planning, machine learning, and damage assessment [137].

As shown in Figure 7.4, the GA starts with creating a random initial population. A set of possible solutions, referred to as chromosomes, form the initial population. A sequence of genes that represents the variables of the problem forms each chromosome. The fitness function evaluates the fitness of each chromosome. Next, the algorithm passes the initial population through a selection process. Chromosomes with a higher fitness have a higher probability of surviving in the next generation. After the selection process, the chromosomes are randomly paired. Each pair of chromosomes is referred to as parents. The algorithm uses the basic GA operators, crossover and mutation, to reproduce the parents. As a result, it creates new pairs of children. Crossover and mutation are applied randomly with a probability of p_c and p_m, respectively. After the process of selection, genetic operations, and replacement, the algorithm evaluates the new population. This process is iterated for a number of generations until a convergence criterion is achieved.

Representation

Any GA needs a chromosome representation. This representation determines how the problem is structured, and defines which genetic operators and parameters are used. A sequence of genes in a certain alphabet represents each chromosome. This alphabet can be binary digits (0 and 1), real numbers, and symbols (i.e., A,B,C), among others. In damage assessment problems each chromosome represents one possible damage distribution, and each gene is the stiffness reduction factor of an element.

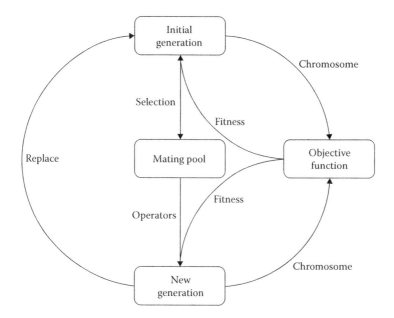

FIGURE 7.4   Working principle of a simple GA.

Binary-coded genetic algorithms (BCGAs) are the classic approach; they are preferred because of their simplicity and traceability. These methods first code the variables in a fixed-length binary string. For example, the following is a chromosome that represents an n variable problem:

$$\underbrace{1011111}_{x_1}\underbrace{10000}_{x_2}\underbrace{1000011}_{x_3}\ldots\underbrace{100001}_{x_n}$$

However, to solve problems where the values of the variables are continuous, it is more natural to represent the genes directly as real numbers. Thus, the representations of the solutions are closer to its natural formulation:

$$\underbrace{9.5}_{x_1}\underbrace{1.6}_{x_2}\underbrace{6.7}_{x_3}\ldots\underbrace{3.3}_{x_n}$$

Researchers of diverse areas suggest the application of real-coded genetic algorithms (RCGAs) [39,154]. RCGAs are more efficient than BCGAs when they work with continuous variables. They are inherently faster because decoding the binary variables before each evaluation is not necessary. In addition, RCGAs do not have a limited precision as do BCGAs; this allows a representation to the machine precision [64].

Initialization

The GA starts with an initial population of N chromosomes. GAs usually generate the initial population with random numbers. However, to reduce the searching time, some algorithms seed the initial population with potentially good solutions, while the rest of the population is randomly generated.

Fitness Function

The fitness function maps the objective function values into fitness values. A fitness value is a measure of the chromosome performance. A GA always maximizes the fitness. Thus, in a maximization problem, the fitness values are equal to the objective function values. On the other hand, in a minimization problem, the fitness function defines the fitness values as a suitable number subtracted by the objective function values.

To maintain uniformity and to prevent premature convergence caused by a dominant chromosome. The algorithm can apply fitness scaling methods, some of these methods are

- *Linear scaling*: In this method the fitness is scaled as

$$F_{i'} = a \cdot F_i + b \tag{7.53}$$

  where:
  F_i and $F_{i'}$ are the initial and scaled fitness values of chromosome i
  a and b are constants chosen to keep the fitness in a predefined range
- *Sigma truncation*: This method is an improvement of the linear scaling:

$$F_{i'} = F_i + (\overline{F} - c \cdot \sigma) \tag{7.54}$$

  where:
  $\overline{F}$ and σ are the population's mean fitness value and standard deviation respectively
  c is a small integer (usually in the range $[1-5]$)
- *Power law scaling*: In this method the initial fitness is scaled to some specific power:

$$F_{i'} = F_i^k \tag{7.55}$$

The value of the parameter k depends on the problem.

Selection Strategies

Selection of individuals to produce successive generations plays an important role in GAs. In general, selection strategies favor the selection of better individuals depending on the selection pressure. The higher the selection pressure the higher the probability of selecting the best individuals. The selection pressure determines the convergence speed of a GA. Thus, if the selection pressure is too low, the GA will take an unnecessarily long time. In contrast, if the selection pressure is too high, the GA algorithm will converge prematurely to a suboptimal solution. The selection strategy and its parameters define the selection pressure. Selection schemes are classified as proportional selection, ranking selection, and tournament selection. Ranking and tournament selections have the advantage of not needing fitness scaling. Roulette wheel [71] (proportional), normalized geometric [79] (ranking), and tournament selection [104] (tournament) are the most representative methods of each selection scheme.

- *Roulette wheel*: Roulette wheel, was the first selection method. The probability p_i of selecting an individual i is proportional to his or her fitness value,

$$p_i = \frac{F_i}{\sum_j F_j} \tag{7.56}$$

  where F_i is the fitness of individual i. In general, the roulette wheel selection scheme is inherently slow [27]. To counteract this, the algorithm can adopt an elitist strategy. An elitist strategycopies the best or a few of the best chromosomes into the succeeding generation, which speeds up the searching process. Even though, this elitist strategy may increase the speed of domination of a population by a super chromosome, it appears to improve the performance [37].
- *Normalized geometric*: Normalized selection assigns the probability of selecting an individual p_i based on his or her rank when all individuals are sorted. The probability of selecting the best individual (q) must be defined in advance:

$$p_i = q'(1-q)^{r-1} \tag{7.57}$$

where

$$q' = \frac{q}{1-(1-q)^p} \tag{7.58}$$

where:
    r is the rank of the individual
    p is the population size

The selection pressure is increased/reduced by increasing/reducing the probability of selecting the best individual q.
- *Tournament selection*: Tournament selection, selects n individuals randomly from the population, and inserts the best individual of the n into the new population. The number of individuals, n, is called the tournament size. An increment/reduction of the tournament size increases/reduces the tournament selection pressure. Rao et al. [127] show that this method provides a good selective pressure by holding a tournament competition among two individuals.

Many researchers have investigated the selection schemes. Golder and Deb [53] studied the convergence time and grow ratio of different selection schemes. They found that proportional selection is significantly slower than the other types. Nonlinear ranking and tournament selection with a tournament size larger than two provided the best performance. Zhang and Kim [166] compared the performance of different selection schemes. They evaluated the performance of each selection scheme in terms of convergence speed and solution quality. The results show strong evidence that ranking selection and tournament selection are a better choice than proportional selection. Elkamchouchi and Wagih [37] compared the performance of the three selection strategies for an antenna array optimization problem. The results show that normalized geometric selection gives a more reliable convergence while tournament comes next and last is the roulette wheel selection.

Crossover

Crossover is the most important operator of GAs. The development of effective crossover operators is the central research topic of GAs. Consequently, many types of crossovers have been proposed in literature [60]. Crossover is applied with a probability p_c, if no crossover takes place, the two children are exact copies of their respective parents. The main crossover operators used in BCGAs are the following.

Let us assume that $x = \{x_1, x_2, ..., x_n\}$ and $y = \{y_1, y_2, ..., y_n\}$ ($x_i, y_i \in [a_i, b_i]$) are two chromosomes selected for application of the crossover operator. Whereas, $x' = \{x'_1, x'_2, ..., x'_n\}$ and $y' = \{y'_1, y'_2, ..., y'_n\}$ are the two children created by the crossover operation.

- *Single-point crossover*: This is the simplest crossover, the parents cross over genes at a randomly chosen point to form two children [71]. If they are crossed after the kth position, the resulting children are

$$x' = \{ \quad x_1, \quad x_2, \quad ..., \quad x_k, | y_{k+1}, \quad ..., \quad y_n \quad \}$$
$$y' = \{ \quad y_1, \quad y_2, \quad ..., \quad y_k, | x_{k+1}, \quad ..., \quad x_n \quad \}$$

- *Two-point crossover*: Two-point crossover selects two random points and the parents swap genes between these points [38],

$$x' = \{ \quad y_1, \quad y_2, \quad ..., \quad y_k, | x_{k+1}, \quad ..., \quad x_m, | y_{m+1}, \quad ..., \quad y_n \quad \}$$
$$y' = \{ \quad x_1, \quad x_2, \quad ..., \quad x_k, | y_{k+1}, \quad ..., \quad y_m, | x_{m+1}, \quad ..., \quad x_n \quad \}$$

- *Uniform crossover*: A uniform crossover looks at each gene in the parents and randomly assigns the gene from one parent to one child and the gene from the other parent to the other child [140]. The operator generates first a random mask. This mask is a vector of random ones and zeros and is the same length as the parents. If $mask_i = 0$, then $x_{i'} = x_i$ and $y_{i'} = y_i$, otherwise if $mask_i = 1$, then $x_{i'} = y_i$ and $y_{i'} = x_i$:

$$mask = \{ \quad 1, \quad 0, \quad ..., \quad 0, \quad 1, \quad ..., \quad 0 \quad \}$$
$$x' = \{ \quad y_1, \quad x_2, \quad ..., \quad x_k, \quad y_{k+1}, \quad ..., \quad x_n, \quad \}$$
$$y' = \{ \quad x_1, \quad y_2, \quad ..., \quad y_k, \quad x_{k+1}, \quad ..., \quad y_n, \quad \}$$

Although these strategies work fine for binary representations, in real-coded algorithms there is a continuum of values, in which we are merely interchanging data points. The operators previously discussed rely totally on mutation to introduce new genetic material. Blending methods solve this problem; they combine values from the two parents into new variable values in the children. Recently, several sophisticated real-coded crossover operators have been developed. The book of Gwiazda [60] reviews some of them. The following are the most popular ones.

- *Flat crossover*: Introduced by Radcliffe [125], the flat crossover generates a child with random values between the two parents:

$$x' = \{x_1', x_2', ..., x_n'\}$$

where x_i' is a uniform random number in the interval $[\min(x_i, y_i), \max(x_i, y_i)]$.
- *Arithmetic crossover*: The arithmetic crossover produces two complimentary linear combinations of the parents [103]:

$$x' = r \cdot x + (1-r) \cdot y$$

$$y' = r \cdot y + (1-r) \cdot x$$

where r is a uniformly distributed number between 0 and 1.
- *Heuristic crossover*: The heuristic crossover produces a linear extrapolation of the two parents. This is the only operator that uses fitness information [154]:

$$x' = x + r(x - y)$$

$$y' = x$$

where:
  r is a uniformly distributed number between 0 and 1
  x is better than y in terms of fitness

If x' is infeasible, that is, one or more of its genes are outside the allowed range, then a new random number r is generated and x' is reevaluated. To ensure halting, after t failures, the children are let equal to the parents.
- *Blend crossover*: The blend crossover (BLX-α) [39] generates a child with random values between the two parents:

$$x' = \{x_1', x_2', ..., x_n'\}$$

where x_i' is a uniform random number in the interval $[x_{min} - I\alpha, x_{max} + I\alpha]$, with

$$x_{min} = \min(x_i, y_i)$$

$$x_{max} = \max(x_i, y_i)$$

$$I = x_{max} - x_{min}$$

The parameter α determines the distance outside the bounds of the two parent variables that the child variable may lie. The BLX-0.0 crossover is equal to the flat crossover.

- *Parent-centric crossover:* The parent-centric crossover (PBX-α) operator [89] defines a child as

$$x' = \{x'_1, x'_2, ..., x'_n\}$$

where x'_i is a uniform random number in the interval $[l_i, u_i]$, with

$$l_i = \min(a_i, x_i - I\alpha)$$

$$u_i = \max(b_i, x_i + I\alpha)$$

$$I = |x_i - y_i|$$

The values of α are usually in the interval $[0.5, 1]$.

Other crossover techniques for real-coded algorithms are CIXL2 [114], fuzzy recombination [149], and simulated binary crossover [26] among others. Each crossover technique directs the search in different areas near the parents, some of them use more exploration (or extrapolation) and others more exploitation (or interpolation). For example, flat and arithmetic crossover only use exploitation (Figure 7.5a), heuristic crossover only uses exploration (Figure 7.5b), and BLX-α (Figure 7.5c) and (PBX-α) (Figure 7.5d) combine exploitation with exploration.

The selection of an appropriate crossover technique depends on the optimization problem. Some forms of crossover operators are more suitable to tackle certain problems than others. An alternative is to combine different types of crossovers, this makes the approach suitable to most practical problems. Herrera et al. [66] show that by combining different types of crossovers, the effectiveness of the search is improved. After studying several crossover combinations, they found that the joined application of BLX-α and dynamic heuristic crossover performs the best.

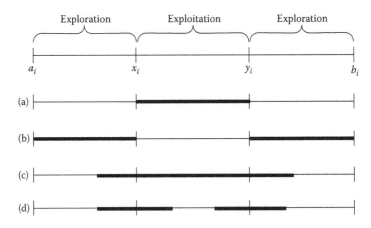

FIGURE 7.5 (a–d) Different action intervals of crossover operators.

Mutation

Mutation avoids losing potentially useful information. The mutation operation replaces a gene in a chromosome with one chosen randomly from the solution space. In the case of a BCGA, the mutation will switch one bit value at a randomly selected location. For RCGAs, several mutation have been developed, here two of them are considered: uniform and boundary mutation.

Let us assume that $x = \{x_1, x_2, ..., x_n\}$ $(x_i \in [a_i, b_i])$ is the chromosome selected for application of the mutation operator.

- *Uniform mutation*: Uniform mutation randomly selects one variable, x_j, and sets it equal to a uniform random number in the interval $[a_i, b_i]$ [103]:

$$x_i' = \begin{cases} U(a_i, b_i), & \text{if} \quad i = j \\ x_i, & \text{if} \quad i \neq j \end{cases} \tag{7.59}$$

- *Boundary mutation*: Boundary mutation randomly selects one variable, x_j, and sets it equal to either its lower or upper bound [103]:

$$x_i' = \begin{cases} a_i, & \text{if} \quad i = j, \quad r < 0.5 \\ b_i, & \text{if} \quad i = j, \quad r \geq 0.5 \\ x_i, & \text{if} \qquad i \neq j \end{cases} \tag{7.60}$$

where r is a uniformly distributed number between 0 and 1.

Other mutation operators that have been proposed for RCGAs are nonuniform mutation [103], time variant mutation [63], power mutation [28], and many more [61].

Termination Strategies

In general, the GA will stop once a convergence criterion is achieved. Some stopping criteria are

- The "target" fitness value is reached.
- The maximum number of generations is reached.
- The deviation of the population is smaller than some specified threshold, that is, the entire population has converged to the same solution.
- The best solution has not changed after a predefined number of generations.
- A combination of the above.

Selection of the GA Parameters

A correct selection of the GA parameters is crucial as they affect the solution and the algorithm runtime [37]. However, there is no general rule to select them; the right decision depends on the encoding, the number of genes, the objective function, and the application. The choice of the optimal parameters, such as population size and crossover and mutation probabilities, has been debated in several investigations.

The codependence among the GA operators and their strong dependence with the nature of the problem are the main difficulties to define an optimum set of parameters. However, we can outline some general rules:

- Increasing the crossover probability increases the recombination, but it also increases the disruption of good chromosomes.
- Increasing the mutation probability helps to introduce new information, but transforms the search in a random search.
- Increasing the population size increases diversity and reduces the probability of premature convergence, but it also increases the convergence time.

De Jong [30] performed the first investigation on how the parameter's variation affects the performance of GAs. He studied the performance of a simple BCGA tested over five cost functions. De Jong's results indicate that the best combination is a population size of 50–100 individuals, a crossover probability of 0.6, and a mutation probability of 0.001. Grefenstette [58] studied later the same test cases. He used a GA search to find the best parameters combination: 30 individuals, a crossover probability of 0.95, and a mutation probability of 0.001. These parameter settings become widely used, although some studies have shown that there are objective functions for which these parameter settings are not optimal [105]. Schaffer et al. [133] spend over a year testing different parameter combinations. Their results are similar to the ones obtained by Grefenstette: population size 20–30, crossover probability 0.75–0.95, and mutation probability 0.005–0.01. These parameter combinations have shown to work fine with BCGAs. However, in the case of RCGAs, the optimal combination changes. According to a study performed by Wright [154], RCGAs need a considerable higher mutation probability. Wright found that appropriate mutation rates for RCGAs are between 0.05 and 0.3. Haupt [65] investigated the optimum population size and mutation rate for a simple RCGA. Their results suggest that the best mutation rate lies between 5% and 20%, while the population size should be less than 16. According to Haupt and Haupt [64] the crossover rate, selection method, and type of crossover are not of much importance, but population size and mutation rate have a significant impact on the GA performance. They determined the optimum population size and mutation rate for a BCGA and an RCGA over five cost functions. The optimum population size is in the range 8–16 for the binary algorithm and 8–20 for the real coded algorithm. The optimum mutation rate is in the range 0.03–0.25 for the binary algorithm and 0.05–0.37 for the real-coded algorithm. These results indicate that a small population size is desirable for both the real-coded and binary-coded GAs. In general, the RCGA needs a higher mutation rate.

PARALLEL GAs

A GA search reevaluates the fitness of the entire population at each generation. Therefore, the algorithm calls the objective function several times during the optimization. In the case of complicated and time-consuming functions, this process may require high computational resources. There are several methods to increase the speed on these cases. One is to ensure that the algorithm does not evaluate identical chromosomes twice. It is then necessary to search in the population for identical twins. This is only worth the effort when the cost function evaluation takes longer than the search [64]. An alternative is to work with a hybrid-GA; it combines the power of the GA with the speed of a local optimizer. Thus, the GA finds the regions of the optimum, and then the local optimizer takes over to find the minimum [64].

Parallel genetic algorithms (PGAs) are also an attractive alternative; they are particularly easy to implement and provide a superior numerical performance. PGAs in addition to being faster than sequential GAs, lead to better results. The basic idea in parallel processing is to divide a large problem into smaller tasks. A group of processors solves these tasks simultaneously. Parallelization is applied to GA by different approaches, three main methods are distinguished: (1) global, (2) migration, and (3) diffusion. Figure 7.6 illustrates them. Multiple population GAs are the most popular parallel method and potentially the most efficient.

1. A global GA works in the same manner as a sequential GA, but the algorithm distributes the evaluation in different processors. It works with a master–slave configuration. The master stores the population, executes the GA operations and distributes individual to the slaves. The slaves only evaluate the fitness of the individuals. This method is only viable for time-consuming functions, where the time to evaluate the function is higher than the communication time between processors.
2. In migration GA, a number of populations are running in parallel. Each population runs a conventional GA individually. These populations exchange their individuals occasionally. This exchange is denominated migration. The separation into subpopulations prevents premature convergence because it allows each population to search in different zones. Migration GA is also known as multiple population, distributed, coarse-grained, or island GA.
3. Diffusion GA, also known as fine-grained or cellular GA, is suited for massively parallel computers. In this case, the subpopulations are very small, often composed of a single individual. The algorithm assigns each individual to a specific geographic location. Each individual only interacts with his or her closest neighbors. This method differs from a global GA because the genetic operators are decentralized.

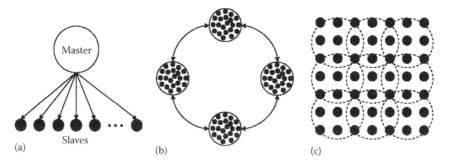

FIGURE 7.6 Three main parallelization methods of GAs: (a) global GA, (b) migration GA, and (c) diffusion GA.

The performance of multiple population GAs is mostly affected by the size and number of populations, migration rate, migration interval, migration topology, and the selection of the migrating individuals. The design of multiple population GAs is a difficult task, mainly because the parameters are intimately related and the effects of migration are not well understood. Cantú and Goldberg [13–16] have done extensive research in the theoretical and experimental parameter definition for parallel GAs. Their research helps to define appropriate values for the parameters and operators, without the need to guess or to do excessive experimentation.

Migration Rate

The migration rate defines the number of individuals that are exchanged in every migration. There is no general rule to select the migration rate. Mainly because there is a close relation among the migration rate, the migration interval, and the migration topology. Nevertheless, although it is not clear what the best value is, it seems that a low migration rate should be used (1%–10% of the population size) [2].

Migration Interval

The migration interval defines the number of generations between two successive migrations. Researchers recommend a moderate migration interval. Studies have demonstrated that a too frequent or too unfrequent migration degraded the algorithm performance [106,141]. The algorithm can perform a periodic migration or by a given migration probability p_m. Another technique is to perform the migration when the standard deviation of the population is small enough to make sense to exchange individuals. In this case, the purpose of migration is to restore diversity into the populations to prevent premature convergence.

Generally, the migration is synchronous, that is, the algorithm performs the migration and reception of individuals at the same time for all the populations. This means that the faster processors must wait for the slower ones to continue. Migration may also be asynchronous [56]. Here, each population independently decides when to send individuals, and accepts arriving individuals whenever they arrive. Thus, processors idle time are greatly reduced. However, the algorithm performance with asynchronous migration is difficult to predict or replicate because migration occurs at random times.

Topology

The topology determines the connections between populations. An important property of the topology is the degree of connectivity, which is the number of neighbors of each population. Cantú and Goldberg [16] demonstrated that the performance of a multiple population algorithm is mostly affected by the degree of connectivity and not by the topology type. They show that the solutions reached by different topologies with the same connectivity are almost identical. Ring migration topologies are the most popular. These topologies provide a good performance, and are easy to implement in different hardware platforms. Figure 7.7 illustrates three examples of ring migration topologies. Ring migration that transfers individuals directionally among adjacent populations. Neighborhood migration that transfers individuals in any direction among adjacent populations. Last, unrestricted migration that transfers individuals unrestricted to any population [142].

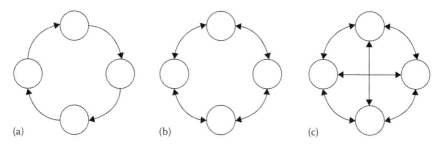

FIGURE 7.7   Ring topologies in multiple population GAs: (a) ring migration, (b) neighborhood migration, and (c) unrestricted migration.

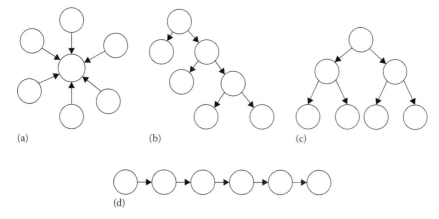

FIGURE 7.8   Tree topologies in multiple population GAs: (a) star, (b) sided binary tree, (c) balanced binary tree, and (d) line.

Tree topologies are an alternative to ring migration topologies. Tree topologies are a natural way of carrying the search in different levels. For example, in a first level, a GA roughly determines the optimum area. Then, GAs that are more specialized determine a more accurate solution inside this area. Figure 7.8 shows four typical tree topologies. Miyagi and Nakamura [106] performed a comparative study of these four tree topologies. They concluded that the line topology gives the best performance.

GENETIC ALGORITHMS IN DAMAGE ASSESSMENT

GAs are robust searching algorithms with an easy implementation to structural damage assessment problems. Several GA-based damage assessment algorithms have been proposed in literature; a summary is given in Table 7.1. The selection of operators and parameters is extremely important for the success of a GA. Despite this, there are not many investigations that address the GA parameter selection problem for the case of damage assessment. Furthermore, most of the papers summarized here do not give a clear explanation of why they selected those operators and parameters. The only work that shows the parameter tuning is the one of Ruotolo and Surace [129] and Meruane and Heylen [96].

TABLE 7.1
Summary of GA Applications for Structural Damage Detection

Author(s)	Mares and Surace [90]	Ruotolo and Surace [129]	Friswell et al. [46]	Chou and Ghaboussi [20]
Year	1996	1997	1998	2001
Objective function	Residual force	Mode shapes and frequencies	Frequencies	Static displacements
GA method	BCGA	BCGA	Hybrid BCGA	IRR BCGA
Initial population	Heuristic	Random	–	Random
Selection	Proportional	Roulette wheel	–	Roulette wheel
Crossover	Single point	Single point	Single point	Single point
Mutation	Binary	Binary	Binary	Binary
Structure(s)	Truss structure/ cantilever beam	Cantilever beam/ cantilever beam	Cantilever beam/ cantilever plate	Plane truss/truss bridge
Experimental (E), Simulated (S)	S/S	S/E	S/E	S/S
N variables	26/13	10/10	15/12	26/19
Population size	30	10	10	100/500
Crossover rate	0.65	0.8	0.6	1
Mutation rate	0.01	0.05	0.05	0.005/0.02

Author(s)	Asce and Xia [62]	Au et al. [7]	Koh et al. [81]	Rao et al. [127]
Year	2002	2003	2003	2004
Objective function	Mode shapes and frequencies	Mode shapes and frequencies	Acceleration	Residual force
GA method	RCGA	Hybrid micro BCGA	Hybrid BCGA	BCGA
Initial population	Random	–	–	Random
Selection	Roulette wheel	–	–	Tournament ($s = 2$)
Crossover	Single point	–	–	Double point
Mutation	Uniform	Jump and creep	–	Binary
Structure(s)	Cantilever beam/ portal frame	Simply supported beam/three span beam	10 DoF system/50 DoF system	Truss structure / cantilever beam/ portal frame
Experimental (E), Simulated (S)	E/E	S/S	S/S	S/S/S
N variables	20/30	16/12	12/52	11/10/6
Population size	40/50	–	50/100	40
Crossover rate	0.85	0.5	0.6	1
Mutation rate	0.1	0.06	0.001	0.01
Stopping criteria	Max. 500 generations	–	–	–

(Continued)

TABLE 7.1 (*Continued*)
Summary of GA Applications for Structural Damage Detection

Author(s)	Perera and Torres [124]	He and Hwang [69]	Vakilbaghmisheh et al. [147]	Raich and Liszkai [126]
Year	2006	2006	2007	2007
Objective function	Mode shapes and frequencies	Frequencies and static displacements	Frequencies	FRF
GA method	BCGA	Hybrid RCGA	Micro BCGA and RCGA	Hybrid BCGA
Initial population	Random	Random	–	–
Selection	Roulette wheel	Roulette wheel	50% of best chromosomes	Tournament $(s = 4/4/8)$
Crossover	Single point	Hybrid crossover	Single point/ arithmetic	Multiple point
Mutation	Binary	Boundary	Binary/uniform	Nonuniform and uniform
Structure(s)	Simply supported beam/simply supported beam	Curved beam/ cantilever beam/ clamped beam	Cantilever beam/ Cantilever beam	Cantilever beam/ two-span beam/ steel frame
Experimental (E), Simulated (S)	S/E	S/S/S	S/E	S/S/S
N variables	10/20	10/10/20	2/2	10/20/81
Population size	100	40	8	100/200/200
Crossover rate	0.8/0.7	0.7	–	0.9/0.9/0.9
Mutation rate	0.01/0.02	0.01	0.2/0.3	0.01/0.005/0.005
Stopping criteria	Max. 100 generations	Max. 100 generations	Max. 100 generations	Max. 30/20/20 generations

Author(s)	Borges et al. [11]	Gomes and Silva [54]	Kokot and Zembaty [82]	Meruane and Heylen [95]
Year	2007	2008	2009	2010
Objective function	Mode shapes and frequencies	Frequencies	Harmonic displacements	Mode shapes and frequencies
GA method	BCGA	RCGA	Hybrid RCGA	RC-PGA
Initial population	Heuristic	Random	Random	Random
Selection	Rank based	Roulette wheel	Stochastic universal sampling	Normalized geometric
Crossover	Double point	BLX-α	Intermediate/ linear	Arithmetic/heuristic
Mutation	Creep	Uniform	–	Boundary/uniform
Structure(s)	Plane frame/plane truss	Simply supported beam/portal frame	Two-span beam/3D frame/ bridge structure	Normalized geometric

(*Continued*)

TABLE 7.1 (*Continued*)
Summary of GA Applications for Structural Damage Detection

Author(s)	Borges et al. [11]	Gomes and Silva [54]	Kokot and Zembaty [82]	Meruane and Heylen [95]
Experimental (E), Simulated (S)	S/S	S/S	S/S/S	E
N variables	13/26	24/56	16/32/ 3	26/30
Population size	40/30	500	–	5 populations of 15
Crossover rate	[0.3–0.7]	1	–	0.85
Mutation rate	–	0.01	–	0.02
Stopping criteria	Max. 150/ 70 generations	Max. 2000 generations	–	5 populations reach the same solution

Author(s)	Meruane and Heylen [96]	Meruane and Heylen [97]	Villalba and Laier [148]	Meruane [100]
Year	2011	2011	2012	2013
Objective function	Different functions based modal data	Mode shapes, resonant and antiresonant frequencies	Mode shapes and frequencies	Antiresonant frequencies
GA method	Hybrid RCGA	RC-PGA	BCGA/RCGA	RC-PGA
Initial population	Random	Random	Heuristic	Random
Selection	Normalized geometric	Normalized geometric	Tournament $n = 3$	Normalized geometric
Crossover	Arithmetic	Arithmetic, heuristic, BLX-0.5, two point and uniform	BLX-0.5 and two point	Arithmetic, heuristic, BLX-0.5, two point and uniform
Mutation	Boundary	Boundary and uniform	Breep/jump	Boundary and uniform
Structure(s)	3D space-frame	Car exhaust system/space frame	Plane truss	8 DoF system/car exhaust system
Experimental (E), Simulated (S)	E	E	S	E
N variables	43	47/43	–	8/21
Population size	25	5 populations of 40	200	5 populations of 40
Crossover rate	0.72	0.85	0.7–0.9/0.8–0.95	0.85
Mutation rate	0.01	0.02	0.005–0.02/0.03–0.06	0.02
Stopping criteria	–	5 populations reach the same solution	Max. 400 generations or 50 generations with no improvement	5 populations reach the same solution

Note: The line – indicates that the paper does not give this information.

On the other hand, there have been few studies and improvements regarding the operators. Rao et al. [127] compared two selection-crossover strategies: roulette wheel with a single-point crossover versus tournament with a double-point crossover. They found that the second strategy gives a better performance. A novel crossover operator called "hybrid crossover" is proposed by He and Hwang [69]; this operator gives a better performance than the single-point, double-point, and uniform crossovers. Raich and Liszkai [126] suggested the use of an adaptive crossover that protects individuals with higher fitness. Borges et al. [11] introduced some modifications to the GA to improve its performance, one of the modifications is the introduction of the "A-mutation" operator. This is a mutation operator specifically designed for damage assessment, where the problem is restricted to a few damage elements. The "A-mutation" modifies the potential damage locations by changing the state of a random element from "passive" to "active" and vice versa.

Some studies have proposed the use of hybrid GA techniques to improve the efficiency of GAs. Friswell et al. [46] first located damage with a BCGA, and then they use an eigensensitivity method to quantify it. Koh et al. [81] applied a BCGA followed with a least-squares optimization. Au et al. [7] used the element quotient difference to locate the damage and a micro-GA to quantify the damage extend. He and Hwang [69] combined a RCGA with a simulated annealing algorithm. Raich et al. [126] used a BCGA followed by a hill climbing algorithm. Kokot and Zembaty [82] proposed an RCGA followed by a Levenberg–Margquardt algorithm. Meruane and Heylen [96] implemented a hybrid RCGA that considers a damage penalization term, which avoids false damage assessment caused by experimental noise or numerical errors.

There are only a few investigations on the implementation of PGAs to damage assessment problems. Meruane and Heylen [95] implemented a real-coded PGA to detect structural damage. Their results showed that PGAs give an important increase in the performance compared to sequential GAs. The parallel algorithm was not only much faster than the sequential algorithm, but it was also able to reach a better solution. This algorithm was later improved [97,98], making each population to work with a different crossover, being the following ones: arithmetic crossover, heuristic crossover, BLX-0.5 crossover, two-point crossover, and uniform crossover. This ensures an effective search with an adequate balance between exploration and exploitation.

ARTIFICIAL NEURAL NETWORKS

Recent studies have introduced ANNs as an alternative to model updating in damage assessment [5,55,131]. A trained neural network can potentially detect, locate, and quantify structural damage in a short period, and hence, it can be used for real-time damage assessment. Damage assessment by ANN has the advantage that it is a general approach, in principle, ANN can be applied to any correlation coefficient that is sensitive to damage. Additionally, ANN can be used with structures that exhibit a nonlinear response [94].

Figure 7.9 illustrates the principle of a damage assessment algorithm using ANN. The vibration characteristics of the structure act as the inputs to the neural network, and the outputs are the damage indices of each element in the structure.

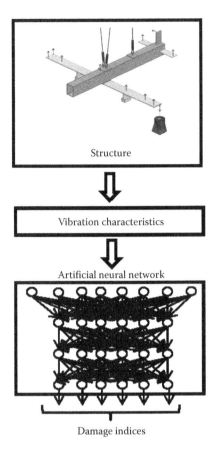

FIGURE 7.9 Damage assessment with ANNs.

ARCHITECTURE

An ANN is a data-processing algorithm that attempts to emulate the processing scheme of the human brain [6]. An ANN is formed by "neurons" that are interconnected to build a complex network. Knowledge is acquired by a learning process and stored in the interneuron connections known as the "synaptic weights." Different types of network architectures exist, and among them, the multilayer perceptron (MLP) is the most frequently used. An MLP network consists of an array of input neurons known as the input layer, an array of output neurons known as the output layer, and a number of hidden layers (see Figure 7.10). Each neuron receives a weighted sum from the neurons in the preceding layer and provides an input to every neuron of the next layer.

Multiple-layer networks are quite powerful. For instance, a network of two layers, where the first layer is sigmoid and the second layer is linear, can be trained to approximate any function (with a finite number of discontinuities) arbitrarily well.

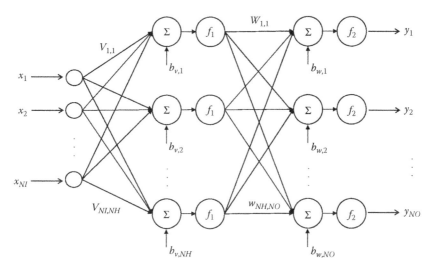

FIGURE 7.10   MLP network.

The outputs of a three-layer MLP are given by

$$y_i = f\left(\sum_{j=1}^{NH} \left[w_{i,j} f\left(\sum_{k=1}^{NI} v_{j,k} x_k + b_{v,k} \right) + b_{w,i} \right] \right)$$ (7.61)

where:

$\mathbf{x} = \{x_1, x_2, \ldots, x_{NI}\}$ and $\mathbf{y} = \{y_1, y_2, \ldots, y_{NO}\}$ are the input and output vectors

$w_{i,j}$ and $v_{j,k}$ are the interconnection weights

b represents the bias (or threshold) terms

$f_i(.)$ is the transfer function for ith layer

NI, NH, and NO are the number of input, hidden, and output nodes, respectively

The activation of each neuron is governed by a function known as the transfer function. As shown in Figure 7.11, typical selections for the transfer function are as follows:

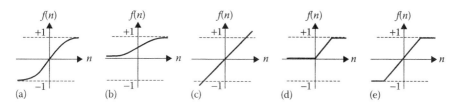

FIGURE 7.11   Transfer functions: (a) hyperbolic tangent sigmoid; (b) logarithm sigmoid; (c) linear; (d) saturating linear; and (e) symmetric saturating linear.

- *Hyperbolic tangent-sigmoid*:

$$f(n) = \frac{2}{1+e^{-2n}} - 1 \tag{7.62}$$

- *Logarithm sigmoid*:

$$f(n) = \frac{1}{1+e^{-n}} \tag{7.63}$$

- *Linear*:

$$f(n) = n \tag{7.64}$$

- *Saturating linear*:

$$f(n) = \begin{cases} 0 & n < 0 \\ n & 0 <= n <= 1 \\ 1 & 1 < n \end{cases} \tag{7.65}$$

- *Symmetric saturating linear*:

$$f(n) = \begin{cases} -1 & n < -1 \\ n & -1 <= n <= 1 \\ 1 & 1 < n \end{cases} \tag{7.66}$$

TRAINING

The values of the weights and bias are updated by training the neural network with data from the structure. The training problem consists of finding the weights that will minimize the mean square error E:

$$E = \sum_k \left\| \mathbf{y}^k - \mathbf{o}^k \right\|^2 = \sum_k E^k \tag{7.67}$$

where $\mathbf{y}^k$ and $\mathbf{o}^k$ are the actual and desired output vector for the kth training pattern, respectively.

The gradients of the error function are calculated by the backpropagation algorithm, which is described next.

Backpropagation Algorithm

In a general feedforward network, each node computes a weighted sum of its inputs as

$$a_j = \sum_i w_{j,i} z_i \tag{7.68}$$

where:

z_i is the node input

$w_{j,i}$ is the weight associated to that connection

The biases can be included as an extra input with a value equal to 1. Therefore, it is not necessary to deal with biases explicitly. The sum is transformed by an activation function to give the output of node j:

$$z_j = f(a_j) \tag{7.69}$$

where:

z_j is the node output

$f(.)$ is the transfer function

If this node is an output node, then z_j equals y_j. The derivative of E^k with respect to some weight $w_{i,j}$ is given by

$$\frac{\partial E^k}{\partial w_{j,i}} = \frac{\partial E^k}{\partial a_j} \frac{\partial a_j}{\partial w_{j,i}} \tag{7.70}$$

Using Equation 7.68, we can write

$$\frac{\partial a_j}{\partial w_{j,i}} = z_i \tag{7.71}$$

introducing the notation

$$\delta_j \equiv \frac{\partial E^k}{\partial a_j} \tag{7.72}$$

then we obtain

$$\frac{\partial E^k}{\partial w_{j,i}} = z_i \delta_j \tag{7.73}$$

Thus, to evaluate the derivatives, we need only to calculate the value of δ_j for each hidden and output node in the network. For the output node, the evaluation is straightforward:

$$\delta_j = \frac{\partial E^k}{\partial a_j} = f'(a_j) \frac{\partial E^k}{\partial y_j} \tag{7.74}$$

In the case of hidden nodes,

$$\delta_j = \frac{\partial E^k}{\partial a_j} = \sum_n \frac{\partial E^n}{\partial a_n} \frac{\partial a_n}{\partial a_j} \tag{7.75}$$

The sum runs over all nodes n to which node j sends connections. Substituting the definition of δ,

$$\delta_j = f'(a_j) \sum_n w_{n,j} \delta_n \tag{7.76}$$

Hence the value of δ for a particular hidden node is obtained by propagating the δs backward in the network. Because we already know the values of the δs for the output nodes, recursively applying Equation 7.76 we can evaluate the δs for all the hidden nodes in the network.

The backpropagation algorithm can be summarized as follows:

1. Take an input vector x^n and forward propagate through the network using Equations 7.68 and 7.69 to obtain the output of each node.
2. Evaluate δ_j for all the output nodes using Equation 7.74
3. Backpropagate the δs using Equation 7.76
4. Use Equation 7.73 to evaluate the derivatives

Optimization

The derivatives computed by the backpropagation algorithm can be used by an optimization algorithm to find the values of the weights and bias that minimize the total error. There are many optimization techniques available to address the network-training problem. The simplest method is gradient descent, also known as the steepest descent. This algorithm begins with an initial weight vector and iteratively updates it by moving in the direction of the greatest error decrease. The gradient descent rule can be written as

$$\Delta w_i^j = -\eta \frac{\partial E}{\partial w_i^j} \tag{7.77}$$

where Δw_i^j is the change for the weight w_i at step j. The convergence of this algorithm strongly depends on the learning rate η. If η is too large, the algorithm may overshoot, leading to divergent oscillations. However, if η is too small, the search might proceed quite slowly. Fang et al. [15] showed that an adaptive learning rate significantly improves the training performance. Another improvement to the gradient descent algorithm involves ensuring that each search direction is conjugate to all previous directions, thus avoiding unnecessary loops in the search process. The scale-conjugated gradient algorithm proposed by Moller [107] combines this concept and variable steps. Many additional optimization algorithms have been proposed to train neural networks, but the Levenberg–Marquardt algorithm has shown to be the most efficient [10]. This method is a gradient-based algorithm specifically designed to minimize the sum-of-squares error [92]. Let write the error function as

$$E = \frac{1}{2} \sum_k \left(e^k \right)^2 = \frac{1}{2} \|\mathbf{e}\|^2 \tag{7.78}$$

where e^k is the error for the kth pattern and $\mathbf{e} = \{e^1, e^2, \ldots, e^n\}$. The error vector $\mathbf{e}$ is a function of the weights, thus we can write a first-order Taylor approximation:

$$\Delta \mathbf{e} = \mathbf{Z} \Delta \mathbf{w} \tag{7.79}$$

where $\mathbf{w} = \{w_1, w_2, \cdots, w_m\}$ and $\Delta \mathbf{w} = \mathbf{w}^{j+1} - \mathbf{w}^j$ with $\mathbf{w}^j$ the current value of the weight vector and $\mathbf{w}^{j+1}$ the estimated weight vector. The matrix, $\mathbf{Z}$, contains the first derivatives of the errors $\mathbf{e}$ with respect to the weights $\mathbf{w}$:

$$\mathbf{Z}_{n,i} = \frac{\partial e^n}{\partial w_i} \tag{7.80}$$

The weight $\mathbf{w}$ is estimated by minimizing the following penalty function:

$$J(\mathbf{w}) = \left(\mathbf{Z}\Delta\mathbf{w} - \Delta\mathbf{e}\right)^T \left(\mathbf{Z}\Delta\mathbf{w} - \Delta\mathbf{e}\right) \tag{7.81}$$

where $(.)^T$ stands for transpose. The minimization problem is solved through least squares. The solution is given by

$$\Delta\mathbf{w} = \mathbf{Z}^+ \mathbf{e} \tag{7.82}$$

where $\mathbf{S}^+$ is the pseudoinverse of the matrix $\mathbf{S}$ [59]:

$$\mathbf{S}_{n\times m}^+ = \begin{cases} \mathbf{S}^{-1} & n = m \\ (\mathbf{S}^T\mathbf{S})^{-1}\mathbf{S}^T & n > m \\ \mathbf{S}^T(\mathbf{S}^T\mathbf{S})^{-1} & n < m \end{cases} \tag{7.83}$$

Equation 7.82 can be written in an equivalent form as

$$\mathbf{w}^{j+1} = \mathbf{w}^j + \mathbf{Z}^+(\mathbf{w}^j)\mathbf{e}(\mathbf{w}^j) \tag{7.84}$$

The final value of $\mathbf{w}$ is obtained through an iterative procedure. Equation 7.84 is iterated until convergence is obtained. The iterative problem may be ill-conditioned. Ill-conditioning leads to instabilities and the convergence is not always assured. A solution to overcome ill-conditioning is the regularization of the equations. In the Levenberg–Marquardt algorithm [92], this problem is handled by seeking to minimize the error function while trying to keep the step size small. This approach ensures that the linear approximation remains valid. The penalization function is now written as

$$J(\mathbf{w}) = \left(\mathbf{Z}\Delta\mathbf{w} - \Delta\mathbf{e}\right)^T \left(\mathbf{Z}\Delta\mathbf{w} - \Delta\mathbf{e}\right) + \lambda\Delta\mathbf{w}^T\Delta\mathbf{w} \tag{7.85}$$

where the parameter λ governs the step size. The solution of Equation 7.85 is

$$\mathbf{w}^{j+1} = \mathbf{w}^j + \left[\mathbf{Z}(\mathbf{w}^j)^T\mathbf{Z}(\mathbf{w}^j) + \lambda\mathbf{I}\right]^{-1}\mathbf{Z}(\mathbf{w}^j)^T\mathbf{e}(\mathbf{w}^j) \tag{7.86}$$

where $\mathbf{I}$ is the unit matrix. If λ is very small we recover the Newton formula, whereas for large values we recover the standard gradient descend.

ANNs in Damage Assessment

One of the main challenges is the selection of an appropriate measure of the system response, as input to the network, that is sufficiently sensitive to small damage. The idea of directly using the FRFs to train the neural networks has attracted many researchers. Among all of the dynamic responses, the FRF is one of the easiest to obtain in real-time because the in situ measurement is straightforward. However, the numbers of spatial response locations and spectral lines are overly large for neural network applications. The direct use of FRFs will lead to networks with a large number of input variables and connections, thus rendering them impractical. Hence, it becomes necessary to extract features from the FRFs and use these features as inputs to the neural networks. Castellini and Revel [18] presented an algorithm to detect and locate structural damage based on laser vibrometry measurements and a neural network for data processing. Using features extracted from the FRFs as inputs to the neural network, the authors were able to use the same network to detect and locate damage in three different experimental structures. To reduce the number of input variables, Zang and Imregun [162] applied a principal component analysis technique to the measured FRFs. The output of the neural network gives the actual state of the structure: undamaged or damaged. The algorithm was able to distinguish between the undamaged and damaged cases with satisfactory accuracy. Fang et al. [40] selected key spectral points near the resonance frequencies in the FRF data. These selected points were used as the inputs of a neural network, and the outputs were the stiffness reduction factors. The algorithm showed high accuracy in identifying damage to a simulated cantilever beam under different damage scenarios.

The input data can be further reduced if modal analysis is performed first. Thus, the input variables are the modal parameters of the structure. The natural frequencies and mode shapes are the most frequently used parameters. In general, the natural frequencies detect and quantify the presence of damage, whereas the mode shapes provide the location. Zapico and González [164] presented an algorithm designed to identify the damage to an experimental two-floor structure; they used the natural frequencies as the input variables and trained a neural network with data from a numerical model. The neural network contained two outputs representing the level of damage of each floor. The proposed methodology was able to predict the damage with an error of less than 8.6%. Yun et al. [160] used the natural frequencies and mode shapes as input data for a neural network used to detect damage in the joints of framed structures. The neural network was trained using a noise-injection learning algorithm to reduce the effects of experimental noise. The authors found that the algorithm could estimate damage with reasonable accuracy, but the performance strongly depended on the level of experimental noise. A disadvantage of using mode shape displacements is that mode shape sensitivity to damage is not significant. The mode shape curvatures are more sensitive to small structural modifications than the mode shape displacements [117]. Sahin and Shenoi [130] studied the effectiveness of different combinations of global (natural frequencies) and local (mode shape curvatures) vibration data as inputs for an ANN. They concluded that the best performance was obtained when the natural frequencies were used as the inputs to a first network to predict the severity of the damage and the mode shape curvatures were used as the inputs to a second network to predict the damage location.

Recently, researchers have proposed the use of antiresonant frequencies as an alternative to mode shapes in neural networks. Meruane and Mahu [99] present a methodology designed to assess experimental damage using neural networks and antiresonant frequencies, which are obtained from point FRFs. Although measurement of the point FRFs is time-consuming because the excitation point is moved together with the response location, this method offers several advantages for network-based damage assessment:

1. All antiresonances contain independent information because they correspond to the resonant frequencies of the system grounded at different DoFs.
2. For a given FRF, the number of antiresonant frequencies does not change from one damage scenario to the next; this is a valuable property if we require these frequencies to act as inputs to a neural network.
3. An antiresonant frequency always lies between two resonant frequencies. Hence, there is no doubt whether a dip in an FRF is an antiresonant frequency or a minimum.

A disadvantage of ANN is the need for large training sets. It is highly difficult and time-consuming to produce sufficiently large training data sets from experiments. An alternative to generating training samples is to use a numerical model of the structure. Castellini and Revel [18] showed that it is possible to produce correct damage predictions in an experimental structure using a neural network that was trained with samples generated by an FE model. Nevertheless, this approach is highly dependent on the accuracy of the numerical model. There are two approaches to overcome this problem. The first is to update the numerical model using experimental data from the undamaged structure. However, even after updating, differences will still remain between the numerical and experimental models. The second alternative is to define an input parameter that considers the initial errors in the numerical model, thus avoiding the need for an accurate numerical model. This goal is achieved using the changes in the data instead of their absolute values. The main assumption is that any change in the structural properties is caused by damage. Thus, any error in the undamaged model of the structure that is also present in the damaged model will be removed. Lee et al. [85] showed that natural frequency changes due to structural damage in a system without modeling errors are approximated as the same as those in a system with modeling error. Hence, changes in the natural frequencies are less sensitive to modeling errors than the natural frequencies themselves. This group demonstrated the applicability of a neural network trained with mode shape changes. The damage locations were estimated with reasonable accuracy, although false alarms were detected at several locations.

Simulated data derived from a numerical model are noise free, whereas actual measurements are never free from experimental noise. Noise in the measurements will cause the network to estimate parameters that are different from the actual properties of the structure. A solution is to introduce artificial noise into the numerical data used to train the network. This process is known as data perturbation scheme [70]. Yun et al. [160] used a noise-injection learning algorithm and a data perturbation scheme. They implemented a neural-network-based damage assessment algorithm to detect the damage in structural joints. The joints are modeled as semirigid connections using rotational springs, and accurate results are obtained in the cases of moderated noise. Sahin and Shenoi [130] trained a neural network with artificially added noise to detect single damage in beam-like composite laminates. The network contains two outputs: the location and the amount of damage. They studied different combinations of features extracted from the resonant frequencies and mode shape curvatures as inputs to the neural network, and their results show that feature selection plays a crucial role in the predictions accuracy. Zang and Imregun [162] proposed a different approach to address experimental noise and used principal component projection of the FRF data as an input to a neural network with two outputs of healthy or damaged. The authors stated that reconstructing the response using the highest principal components should not only achieve data compression but also remove a proportion of the noise. The principal component analysis compression acts as a noise filter, which can be useful in modal analysis as well.

The successful application of a neural network depends on the representation and the learning algorithms. Nevertheless, their selection is problem dependent and is usually determined by trial and error [9]. Sahoo and Maity [131] used a GA to automate the trial-and-error process. The network parameters (number of neurons, learning rate, etc.) are set as variables in an optimization problem handled by a GA. They used an MLP network with two hidden layers trained by a backpropagation algorithm. Meruane and Mahu [99] addressed the setup of an MLP network parameters and provided guidelines for their selection in similar damage assessment problems. Fang et al. [40] explored the use of a tunable steepest descent algorithm that dynamically adjusts the learning speed during the training process. They showed that this methodology significantly increases the training speed while maintaining the learning stability.

APPLICATION EXAMPLE

The structure consist of a steel beam with a rectangular cross section. The beam measures 1 m in length and has a cross-sectional area of 25×10 mm^2. As shown in Figure 7.12, soft springs suspend the structure to simulate a "free–free" boundary condition. A shaker excites the beam at one end, and the response is measured by 11 accelerometers. Both the excitation force and the measured responses are in the horizontal direction. In this direction, antiresonant frequencies are more sensitive to the experimental damage. Thirty-six antiresonant frequencies were identified from the transmissibility measurements using the algorithm presented in Reference [100].

The numerical model was built in MATLAB® with 2D beam elements. The model featured 20 beam elements and 40 DoFs, as shown in Figure 7.13. In the undamaged case, the maximum difference between the experimental and numerical antiresonances was 3.63%. Shadowed elements represent possible locations of damage, resulting in 18 damage locations and 144 patterns.

The structure was subjected to three different damage scenarios containing single and double cracks. Cracks were introduced into the structure by saw cuts of length l_c, as illustrated in Figure 7.14. Table 7.2 summarizes the two damage scenarios, indicating the distance from the left end to the cut, the corresponding element in the numerical model, and the cut length.

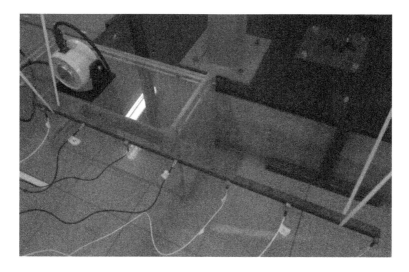

FIGURE 7.12   Experimental beam.

| 1 | 2 | 3 | 4 | 5 | 6 | 7 | 8 | 9 | 10 | 11 | 12 | 13 | 14 | 15 | 16 | 17 | 18 | 19 | 20 |

FIGURE 7.13   Numerical model of the beam and element numbering.

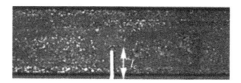

FIGURE 7.14   Saw cut introduced into the beam.

TABLE 7.2

Damage Scenarios Introduced to the Beam

Damage Scenario	Distance from the Left End (mm)	Element Number	Saw Cut Length (mm)	Distance from the Left End (mm)	Element Number	Saw Cut Length (mm)
1	685	14	9	–	–	–
2	360	8	12	697	14	6
3	360	8	5	810	17	17

Note: The experimental data are available for download at the following website: http://viviana.meruane. com/des_en.htm

Model Updating–Based Method

Formulation of the Optimization Problem

Damage is represented by elemental stiffness reduction factors, α_i, defined as the ratio of the stiffness reduction to the initial stiffness:

$$\alpha = \{\alpha_1, \alpha_2, \cdots, \alpha_m\} \tag{7.87}$$

where m is the number of structural elements. The stiffness matrix of the damaged structure, $\mathbf{K}_d$, is expressed as a sum of element matrices multiplied by reduction factors:

$$\mathbf{K}_d = \sum_{i=1}^{m} (1 - \alpha_i) \mathbf{K}_i, \tag{7.88}$$

where $\mathbf{K}_i$ is the stiffness matrix of the ith element. Thus, $\alpha_i = 0$ indicates that the ith element is undamaged, whereas $0 < \alpha_i < 1$ implies partial damage and $\alpha_i = 1$ complete damage.

Defining α_i as the ith updating parameter, the model updating problem is a constrained nonlinear optimization problem, where $\alpha = \{\alpha_1, \alpha_2, \cdots, \alpha_m\}$ are the optimization variables. The objective function correlates antiresonant frequencies. The error in antiresonances is represented by the ratio between the experimental and analytical antiresonances:

$$\varepsilon_{i,r}(\alpha) = \frac{\omega_{i,r}^{A}(\alpha)^2}{\omega_{i,r}^{E\,2}} - 1 \tag{7.89}$$

superscripts A and E refer to analytical and experimental, $\omega_{i,r}$ is the ith antiresonance of the rth FRF. The objective function is given by

$$J(\alpha) = \sum_{r} \sum_{i} \left\| \varepsilon_{i,r}(\alpha) \right\| \tag{7.90}$$

The optimization problem is defined as

$$\begin{aligned} \min \quad & J(\alpha) \\ \text{subject to} \quad & 0 \le \alpha_i \le 1 \end{aligned} \tag{7.91}$$

Configuration

Given the good performance of the parallel algorithm studied in Reference [100], most of its settings are used here:

- A multiple population GA with four populations
- Neighborhood migration

TABLE 7.3
Configuration of the PGA

Parallel Genetic Algorithm	Multiple population GA
Topology	Neighborhood migration
Number of populations	4
Migration rate	One individual
Migration selection	Best individual
Replacement policy	Worst individual
Type of migration	Delayed migration
Perform migration	After 20 generations with no improvement
Population size	40 individuals per population
Selection	Normalized geometric selection
Crossover	Arithmetic crossover, heuristic crossover, BLX-0.5 crossover, two-point crossover and uniform crossover
Mutation	Uniform mutation and boundary mutation
Crossover rate	0.8
Mutation rate	0.02

- Migration rate of one individual
- Selection of the best individual to migrate
- Replacement of the worst individual

The proposed damage assessment algorithm works with a delayed migration scheme. If a population has no improvement after a predefined number of generations, the GA stops and exchanges the individuals with their neighbors. This exchange of individuals is synchronous, that is, the algorithm waits until the five populations are ready to perform the migration. At each migration, each population sends its best individual, whereas the received individual replaces its worst individual. Before migration, the algorithm compares the best individuals from all populations, if they have all converged to the same solution it means no further improvement is possible and the algorithm stops. The PGA configuration is shown in Table 7.3.

NEURAL NETWORK–BASED METHOD

Input Vector

The input vector corresponds to the experimental changes in the antiresonant frequencies with respect to the intact case:

$$
\mathbf{X}^j = \left\{ \frac{\omega_{1,1}^D - \omega_{1,1}^U}{\omega_{1,1}^U}, \frac{\omega_{2,1}^D - \omega_{2,1}^U}{\omega_{2,1}^U}, \ldots, \frac{\omega_{n_1,1}^D - \omega_{n_1,1}^U}{\omega_{n_1,1}^U}, \ldots, \frac{\omega_{n_r,r}^D - \omega_{n_r,r}^U}{\omega_{n_r,r}^U} \right\}, \quad (7.92)
$$

Superscripts D and U refer to damaged and undamaged, respectively, and $\omega_{i,r}$ is the ith antiresonant frequency of the rth FRF.

To reduce the effects of experimental noise, the simulated data are polluted with random noise. As proposed by Hjelmstad and Shin [70], each set of perturbed data is created by adding a uniformly distributed random noise to the numerical data:

$$\omega_{i,r} = \omega_{i,r}\left(1 + \xi\right) \tag{7.93}$$

where ξ is a uniform random number with a specified amplitude. The variance of the perturbing noise should be the same as the variance of the measurement noise, which in this case was estimated to be 1.0%.

Output Vector

The network outputs are the damage indices defined in Equations 7.87 and 7.88.

Training and Validation Patterns

The distribution of training patterns plays a crucial role in the success of a neural network. The relationship between antiresonant frequencies and different damage levels is not linear. Therefore, a network might not be able to interpolate data. By combining multiple damages into the training, the number of training patterns is increased and a large number of training patterns could overwhelm the training procedure. In this study, training patterns were generated by considering up to two simultaneous damages, with 10 damage levels evenly distributed between 5% and 95%.

Network Parameters

The selected network parameters are the following [99]:

- *Network*: Three-layer MLP
- *Number of input nodes*: 20
- *Number of hidden nodes*: 80
- *Number of output nodes*: 18
- *Transfer function in the hidden layer*: Logarithm sigmoid
- *Transfer function in the output layer*: Symmetric saturating linear
- *Training method*: Levenberg–Marquardt

RESULTS

Figure 7.15 shows the results of the damage detected in the three experimental damage scenarios. The actual damage was estimated by using a full 3D model of the beam with the real cut. In the first two cases, damage is correctly located by both approaches, though the neural network approach detects a few false damages. In the last case, the model updating algorithm does not detect the small crack in element 8, this is because the larger cut hides the effect of the smaller cut. The neural network, on the other hand, detects both cracks but on the neighbor elements, such as elements 7 and 16 instead of 8 and 17, and it does not quantifies correctly the larger crack.

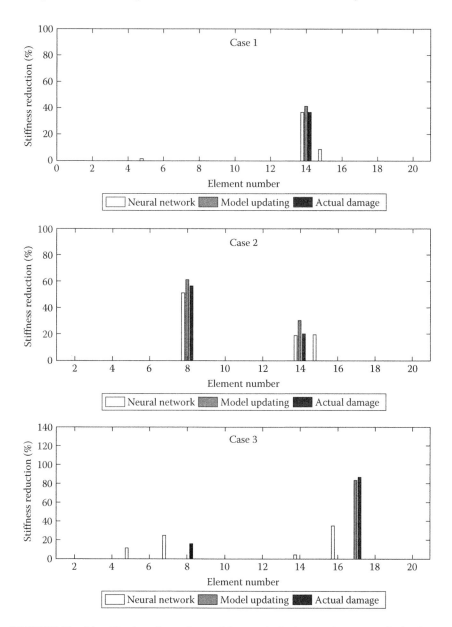

FIGURE 7.15   Identification of experimental damage in the beam using two methods: three-layer MLP neural network and model updating with parallel GAs.

In terms of time, the GA approach requires approximately 900 s to assess the experimental damage in each case, whereas the neural network approach requires only 0.17 s. It should be noted that the solution time does not include the time needed to train the network because the network requires to be trained once and then can assess any damage scenario. The time required to train the networks was about 4 h.

SUMMARY AND DISCUSSIONS

This chapter has given a review to the large literature available on damage assessment based on inverse methods. It provided a description of GAs and ANNs, and their application to structural damage assessment. The chapter explained their working principle and reviewed each factor that affects its performance.

The performance of two model-based algorithms to locate and quantify structural damage was demonstrated: a model updating approach based on PGAs and a three-layer MLP neural network. Both algorithms were successful in assessing the experimental damage. Although, the neural network approach is much faster than GAs, providing the possibility of continuously monitoring the state of a structure, the GA approach provides more accurate solutions. It is recommended that a neural network be used to yield rapid damage identification and that a GA approach be used to provide a more accurate solution.

Despite all the research that has been done, many difficulties remain in model-based damage assessment, such as modeling errors and experimental noise. Damage assessment algorithms in most cases interpret these differences as damage, which usually ends in a large amount of falsely detected damage. Additionally, these errors give a lower bound on the level of damage that can be assessed.

Another challenge is the effect of environmental conditions. Several structures have been instrumented with thermocouples, anemometers, and humidity sensor, the information gathered by these sensors has been used to eliminate environmental influences from the structure's dynamic response. Nevertheless, this information has not been directly applied to damage assessment methods. The most promising strategy is to include this information within the numerical model [48,98].

ACKNOWLEDGMENTS

The author acknowledges the partial financial support of the Chilean National Fund for Scientific and Technological Development (FONDECYT) under Grant No. 11110046.

REFERENCES

1. M.M. Abdel Wahab, G. De Roeck, and B. Peeters. Parameterization of damage in reinforced concrete structures using model updating. *Journal of Sound and Vibration*, 228(4):717–730, 1999.
2. E. Alba and J.M. Troya. A survey of parallel distributed genetic algorithms. *Complexity*, 4:31–52, 1999.
3. R.J. Allemang and D.L. Brown. A correlation coefficient for modal vector analysis. In *Proceedings of the 1st International Modal Analysis Conference*, pp. 110–116. Orlando, FL, 1982.
4. A. Alvandi and C. Cremona. Assessment of vibration-based damage identification techniques. *Journal of Sound and Vibration*, 292(1–2):179–202, 2006.
5. S. Arangio and J.L. Beck. Bayesian neural networks for bridge integrity assessment. *Structural Control and Health Monitoring*, 19(1):3–21, 2012.
6. M.A. Arbib. *The Handbook of Brain Theory and Neural Networks*. Bradford Book, Cambridge, MA, 2003.

7. F.T.K. Au, Y.S. Cheng, L.G. Tham, and Z.Z. Bai. Structural damage detection based on a micro-genetic algorithm using incomplete and noisy modal test data. *Journal of Sound and Vibration*, 259(5):1081–1094, 2003.

8. Y. Bamnios, E. Douka, and A. Trochidis. Crack identification in beam structures using mechanical impedance. *Journal of Sound and Vibration*, 256(2):287–297, 2002.

9. S.V. Barai and P.C. Pandey. Vibration signature analysis using artificial neural networks. *Journal of Computing in Civil Engineering*, 9(4):259–265, 1995.

10. M. Beale and H. Demuth. *Neural Network Toolbox*. The MathWorks, Inc., Version 7.0.3. Release 2012a, Natick, MA, 2012.

11. C.C.H. Borges, H.J.C. Barbosa, and A.C.C. Lemonge. A structural damage identification method based on genetic algorithm and vibrational data. *International Journal for Numerical Methods in Engineering*, 69:2663–2686, 2007.

12. J.M.W. Brownjohn, P.Q. Xia, H. Hao, and Y. Xia. Civil structure condition assessment by FE model updating: Methodology and case studies. *Finite Elements in Analysis and Design*, 37(10):761–775, 2001.

13. E. Cantú-Paz and D.E. Goldberg. Modeling idealized bounding cases of parallel genetic algorithms. In *Proceedings of the 2nd Annual Conference on Genetic Programming*, pp. 353–361, Morgan Kaufmann, San Francisco, CA, 1997.

14. E. Cantú-Paz and D.E. Goldberg. Predicting speedups of idealized bounding cases of parallel genetic algorithms. In *Proceedings of the 7th International Conference on Genetic Algorithms*, pp. 113–121, East Lansing, MI, 1997.

15. E. Cantú-Paz. *Efficient and Accurate Parallel Genetic Algorithms*. Kluwer Academic Publishers, Norwell, MA, 2000.

16. E. Cantú-Paz. Parameter setting in parallel genetic algorithms. *Intelligence (SCI)*, 54:259–276, 2007.

17. E.P. Carden and P. Fanning. Vibration based condition monitoring: A review. *Structural Health Monitoring*, 3(4):355–377, 2004.

18. P. Castellini and G.M. Revel. An experimental technique for structural diagnostic based on laser vibrometry and neural networks. *Shock and Vibration*, 7(6):381–397, 2000.

19. P. Cawley and R.D. Adams. The location of defects in structures from measurements of natural frequencies. *Journal of Strain Analysis*, 14:49–57, 1979.

20. J.H. Chou and J. Ghaboussi. Genetic algorithm in structural damage detection. *Computers and Structures*, 79(14):1335–1353, 2001.

21. P. Cornwell, S.W. Doebling, and C.R. Farrar. Application of the strain energy damage detection method to plate-like structures. *Journal of Sound and Vibration*, 224(2):359–374, 1999.

22. W. D'Ambrogio and A. Fregolent. Promises and pitfalls of antiresonance based dynamic model updating. In *Proceedings of the 2nd International Conference on Identification on Engineering Systems*, pp. 112–121, Swansea, UK, 1999.

23. W. D'Ambrogio and A. Fregolent. The use of antiresonances for robust model updating. *Journal of Sound and Vibration*, 236(2):227–243, 2000.

24. W. D'Ambrogio and A. Fregolent. Results obtained by minimizing natural frequency and antiresonance errors of a beam model. *Mechanical Systems and Signal Processing*, 17(1):29–37, 2003.

25. W. D'Ambrogio and A. Fregolent. Dynamic model updating using virtual antiresonances. *Shock and Vibration*, 11(3–4):351–363, 2004.

26. K. Deb and R.B. Agrawal. Simulated binary crossover for continuous search space. *Complex Systems*, 9(2):115–148, 1995.

27. K. Deb. Genetic algorithm in search and optimization: The technique and applications. In *Proceedings of the International Workshop on Soft Computing and Intelligent Systems*, pp. 58–87, Indian Statistical Institute, Calcutta, India, 1998.

28. K. Deep and M. Thakur. A new mutation operator for real coded genetic algorithms. *Applied Mathematics and Computation*, 193(1):211–230, 2007.

29. C.N Della and D. Shu. Vibration of delaminated composite laminates: A review. *Applied Mechanics Reviews*, 60(1):1–20, 2007.

30. K.A. De Jong. *An analysis of the behavior of a class of genetic adaptive systems*. PhD thesis, University of Michigan, Ann Arbor, MI, 1975.

31. M. Dilena and A. Morassi. The use of antiresonances for crack detection in beams. *Journal of Sound and Vibration*, 276(1–2):195–214, 2004.

32. A.D. Dimarogonas. Vibration of cracked structures: A state of the art review. *Engineering Fracture Mechanics*, 55(5):831–857, 1996.

33. S.W. Doebling, C.R. Farrar, and M.B. Prime. A summary review of vibration-based damage identification methods. *The Shock and Vibration Digest*, 30(2):91, 1998.

34. S.W. Doebling. Minimum-rank optimal update of elemental stiffness parameters for structural damage identification. *AIAA Journal*, 34(12):2615–2621, 1996.

35. E. Douka, G. Bamnios, and A. Trochidis. A method for determining the location and depth of cracks in double-cracked beams. *Applied Acoustics*, 65(10):997–1008, 2004.

36. T.A. Duffey, S.W. Doebling, C.R. Farrar, W.E. Baker, W.H. Rhee, and S.W. Doebling. Vibration-based damage identification in structures exhibiting axial and torsional response. *Transactions of the ASME-L-Journal of Vibration and Acoustics*, 123(1):84–91, 2001.

37. H.M. Elkamchouchi and M.M. Wagih. Genetic algorithm operators effect in optimizing the antenna array pattern synthesis. In *Proceedings of the 20th National Radio Science Conference*. Cairo, Egypt, 2003.

38. L.J. Eshelman, R.A. Caruana, and J.D. Schaffer. Biases in the crossover landscape. In *Proceedings of the 3rd international conference on Genetic algorithms*, pp. 10–19, San Francisco, CA, 1989.

39. L.J. Eshelman and J.D. Schaffer. Real-coded genetic algorithms and interval-schemata. *Foundations of Genetic Algorithms*, 2:187–202, 1993.

40. X. Fang, H. Luo, and J. Tang. Structural damage detection using neural network with learning rate improvement. *Computers and Structures*, 83(25–26):2150–2161, 2005.

41. C.R. Farrar and D.A. Jauregui. Comparative study of damage identification algorithms applied to a bridge: II. Numerical study. *Smart Materials and Structures*, 7(5):720–731, 1998.

42. C.R. Farrar and D.A. Jauregui. Comparative study of damage identification algorithms applied to a bridge: I. Experiment. *Smart Materials and Structures*, 7(5):704–719, 1998.

43. C.R. Farrar and N.A.J. Lieven. Damage prognosis: The future of structural health monitoring. *Philosophical Transactions of the Royal Society A: Mathematical, Physical and Engineering Sciences*, 365(1851):623–632, 2007.

44. R.L. Fox and M.P. Kapoor. Rates of change of eigenvalues and eigenvectors. *AIAA Journal*, 6(12):2426–2429, 1968.

45. M.I. Friswell and J.E. Mottershead. *Finite Element Model Updating in Structural Dynamics*. Kluwer Academic Publishers, the Netherlands, 1995.

46. M.I. Friswell, J.E.T. Penny, and S.D. Garvey. A combined genetic and eigensensitivity algorithm for the location of damage in structures. *Computers and Structures*, 69(5):547–556, 1998.

47. M.I. Friswell and J.E.T. Penny. Crack modeling for structural health monitoring. *Structural Health Monitoring*, 1(2):139–148, 2002.

48. M.I. Friswell. Damage identification using inverse methods. *Dynamic Methods for Damage Detection in Structures*, pp. 13–66, 2008.

49. C.P. Fritzen and K. Bohle. Identification of damage in large structures by means of measured FRFs—procedure and application to the I-40-highway-bridge. *Key Engineering Materials*, 167–168:310–319, 1999.

50. C.P. Fritzen and K. Bohle. Model-based damage identification from changes of modal data—a comparison of different methods. *Key Engineering Materials*, 204–205:65–74, 2001.

51. C.P. Fritzen and K. Bohle. Damage identification using a modal kinetic energy criterion and output-only modal data-application to the Z-24 Bridge. In *Proceedings of the 2nd European Workshop Structural Health Monitoring*, pp. 185–194, Munich, Germany, July 2004.

52. Y. Gao and B. F. Jr. Spencer. Damage localization under ambient vibration using changes in flexibility. *Earthquake Engineering*, 1:136–144, 2002.

53. D.E. Goldberg and K. Deb. A comparative analysis of selection schemes used in genetic algorithms. *Foundations of Genetic Algorithms*, 1:69–93, 1991.

54. H.M. Gomes and N.R.S. Silva. Some comparisons for damage detection on structures using genetic algorithms and modal sensitivity method. *Applied Mathematical Modelling*, 32(11):2216–2232, 2008.

55. C. Gonzalez-Perez and J. Valdes-Gonzalez. Identification of structural damage in a vehicular bridge using artificial neural networks. *Structural Health Monitoring*, 10(1):33–48, 2011.

56. M. Gorges-Schleuter. ASPARAGOS an asynchronous parallel genetic optimization strategy. In *Proceedings of the 3rd International Conference on Genetic Algorithms*, pp. 422–427, Morgan Kaufmann, San Francisco, CA, 1989.

57. E. Görl and M. Link. Damage identification using changes of eigenfrequencies and mode shapes. *Mechanical Systems and Signal Processing*, 17(1):103–110, 2003.

58. J.J. Grefenstette. Optimization of control parameters for genetic algorithms. *IEEE Transactions on Systems, Man, and Cybernetics*, 16(1):122–128, 1986.

59. T.N.E. Greville. The pseudoinverse of a rectangular or singular matrix and its application to the solution of systems of linear equations. *SIAM Review*, 1(1):38–43, 1959.

60. T.D. Gwiazda. *Genetic Algorithms Reference Volume I: Crossover for Single-Objective Numerical Optimization Problems*. Tomaszgwiazda e-books, Łomianki, Poland, 2006.

61. T.D. Gwiazda. *Genetic Algorithms Reference Volume II: Mutation Operator for Numerical Optimization Problems*. Tomaszgwiazda e-books, Łomianki, Poland, 2007.

62. H. Hao and Y. Xia. Vibration-based damage detection of structures by genetic algorithm. *Journal of Computing in Civil Engineering*, 16(3):222–229, 2002.

63. M.M.A. Hashem, M. Watanabe, and K. Izumi. Evolution strategy: A new time-variant mutation for fine localtuning. In *Proceedings of the 36th SICE Annual Conference*, pp. 1099–1104, Tokushima, Japan, 1997.

64. R.L. Haupt and S.E. Haupt. *Practical Genetic Algorithms*. Wiley-Interscience, Hoboken, NJ, 2004.

65. R.L. Haupt. Optimum population size and mutation rate for a simple real genetic algorithm that optimizes array factors. In *IEEE Antennas and Propagation Society International Symposium,* Vol. 2, Salt Lake City, Utah, 2000.

66. F. Herrera, M. Lozano, and A.M. Sanchez. Hybrid crossover operators for real-coded genetic algorithms: An experimental study. *Soft Computing—A Fusion of Foundations, Methodologies and Applications*, 9(4):280–298, 2005.

67. W. Heylen and T. Janter. Extensions of the modal assurance criterion. *Journal of Vibration and Acoustics*, 112:468–472, 1990.

68. W. Heylen, S. Lammens, and P. Sas. *Modal Analysis Theory and Testing*. Departement Werktuigkunde, Katholieke Universiteit Leuven, Leuven, Belgium, 2003.

69. R.S. He and S.F. Hwang. Damage detection by an adaptive real-parameter simulated annealing genetic algorithm. *Computers and Structures*, 84(31–32):2231–2243, 2006.

70. K.D. Hjelmstad and S. Shin. Damage detection and assessment of structures from static response. *Journal of Engineering Mechanics*, 123(6):568–576, 1997.

71. J.H. Holland. *Adaptation in Natural and Artificial Systems,* Vol. 31. University of Michigan Press, Ann Arbor, MI, 1975.

72. T.Y. Hsu and C.H. Loh. Damage diagnosis of frame structures using modified modal strain energy change method. *Journal of Engineering Mechanics,* 134:1000, 2008.

73. J. Humar, A. Bagchi, and H. Xu. Performance of vibration-based techniques for the identification of structural damage. *Structural Health Monitoring,* 5(3):215, 2006.

74. M. Imregun, K.Y. Sanliturk, and D.J. Ewins. Finite element model updating using frequency response function data II. Case study on a medium-size finite element model. *Mechanical Systems and Signal Processing,* 9(2):203–213, 1995.

75. M. Imregun, W.J. Visser, and D.J. Ewins. Finite element model updating using frequency response function data I. Theory and initial investigation. *Mechanical Systems and Signal Processing,* 9(2):187–202, 1995.

76. T. Inada, Y. Shimamura, A. Todoroki, and H. Kobayashi. Development of the two-step delamination identification method by resonant and anti-resonant frequency changes. *Key Engineering Materials,* 270–273:1852–1858, 2004.

77. B. Jaishi and W.X. Ren. Damage detection by finite element model updating using modal flexibility residual. *Journal of Sound and Vibration,* 290(1–2):369–387, 2006.

78. B. Jaishi and W.X. Ren. Finite element model updating based on eigenvalue and strain energy residuals using multiobjective optimization technique. *Mechanical Systems and Signal Processing,* 21(5):2295–2317, 2007.

79. J.A. Joines and C.R. Houck. On the use of non-stationary penalty functions to solve nonlinear constrained optimization problems with GA's. In *IEEE International Symposium Evolutionary Computation,* pp. 579–584. Orlando, FL, 1994.

80. K. Jones and J. Turcotte. Finite element model updating using antiresonant frequencies. *Journal of Sound and Vibration,* 252(4):717–727, 2002.

81. C.G. Koh, Y.F. Chen, and C.Y. Liaw. A hybrid computational strategy for identification of structural parameters. *Computers and Structures,* 81(2):107–117, 2003.

82. S. Kokot and Z. Zembaty. Damage reconstruction of 3D frames using genetic algorithms with Levenberg–Marquardt local search. *Soil Dynamics and Earthquake Engineering,* 29(2):311–323, 2009.

83. J.B. Kosmatka and J.M. Ricles. Damage detection in structures by modal vibration characterization. *Journal of Structural Engineering,* 125:1384–1392, 1999.

84. S. Lammens. *Frequency response based validation of dynamic structural finite element models.* PhD thesis, Katholieke Universiteit Leuven, Belgium, 1995.

85. J.J. Lee, J.W. Lee, J.H. Yi, C.B. Yun, and H.Y. Jung. Neural networks-based damage detection for bridges considering errors in baseline finite element models. *Journal of Sound and Vibration,* 280(3–5):555–578, 2005.

86. R.I. Levin and N.A.J. Lieven. Dynamic finite element model updating using simulated annealing and genetic algorithms. *Mechanical Systems and Signal Processing,* 12(1):91–120, 1998.

87. N.A.J. Lieven and D.J. Ewins. Spatial correlation of modes: The coordinate modal assurance criterion (COMAC). In *Proceedings of the 6th International Modal Analysis Conference,* pp. 1063–1070. Kissimmee, FL, 1988.

88. R.M. Lim and D.J. Ewins. Model updating using FRF data. In *Proceedings of the 15th International Modal Analysis Seminar,* Leuven, Belgium, pp. 141–163, 1990.

89. M. Lozano, F. Herrera, N. Krasnogor, and D. Molina. Real-coded memetic algorithms with crossover hill-climbing. *Evolutionary Computation Journal,* 12(3):273–302, 2004.

90. C. Mares and C. Surace. An application of genetic algorithms to identify damage in elastic structures. *Journal of Sound and Vibration,* 195(2):195–215, 1996.

91. M. Markou and S. Singh. Novelty detection: A review—Part 1: Statistical approaches. *Signal Processing,* 83(12):2481–2497, 2003.

92. D.W. Marquardt. An algorithm for least-squares estimation of nonlinear parameters. *Journal of the Society for Industrial and Applied Mathematics*, 11(2):431–441, 1963.

93. S. Masoud, M.A. Jarrah, and M. Al-Maamory. Effect of crack depth on the natural frequency of a prestressed fixed–fixed beam. *Journal of Sound and Vibration*, 214(2):201–212, 1998.

94. S.F. Masri, A.W. Smyth, A.G. Chassiakos, T.K. Caughey, and N.F. Hunter. Application of neural networks for detection of changes in nonlinear systems. *Journal of Engineering Mechanics*, 126:666–676, 2000.

95. V. Meruane and W. Heylen. Damage detection with parallel genetic algorithms and operational modes. *Structural Health Monitoring*, 9(6):481–496, 2010.

96. V. Meruane and W. Heylen. An hybrid real genetic algorithm to detect structural damage using modal properties. *Mechanical Systems and Signal Processing*, 25:1559–1573, 2011.

97. V. Meruane and W. Heylen. Structural damage assessment with antiresonances versus mode shapes using parallel genetic algorithms. *Structural Control and Health Monitoring*, 18(8):825–839, 2011.

98. V. Meruane and W. Heylen. Structural damage assessment under varying temperature conditions. *Structural Health Monitoring*, 11(3):345–357, 2012.

99. V. Meruane and J. Mahu. Real-time structural damage assessment using artificial neural networks and anti-resonant frequencies. *Shock and Vibration*, 2014:653279, 2014.

100. V. Meruane. Model updating using antiresonant frequencies identified from transmissibility functions. *Journal of Sound and Vibration*, 332(4):807–820, 2013.

101. A. Messina, A. Jones, and E.J. Williams. Damage detection and localisation using natural frequency changes. In *Proceedings of Conference on Identification in Engineering Systems*, pp. 67–76, Swansea, UK, 1996.

102. A. Messina, E.J. Williams, and T. Contursi. Structural damage detection by a sensitivity and statistical-based method. *Journal of Sound and Vibration*, 216(5):791–808, 1998.

103. Z. Michalewicz. *Genetic Algorithms + Data Structures = Evolution Programs*. Springer, Berlin, Germany, 1996.

104. B.L. Miller and D.E. Goldberg. Genetic algorithms, tournament selection, and the effects of noise. Technical Report 95006, Department of General Engineering, University of Illinois at Urbana-Champaign, IL, 1995.

105. M. Mitchell. *An Introduction to Genetic Algorithms*. MIT Press, Cambridge, MA, 1996.

106. H. Miyagi and M. Nakamura. A parallel genetic algorithm and its variance analysis for a new multiple knapsack problems. In *23rd International Technique Conference on Circuits/Systems, Computers and Communications*, Shimonoseki City, Japan, 2008.

107. M.F. Møller. A scaled conjugate gradient algorithm for fast supervised learning. *Neural Networks*, 6(4):525–533, 1993.

108. D. Montalvao, N.M.M. Maia, and A.M.R. Ribeiro. A review of vibration-based structural health monitoring with special emphasis on composite materials. *The Shock and Vibration Digest*, 38(4):295, 2006.

109. C.M.D. Moorthy and J.N. Reddy. Modelling of laminates using a layerwise element with enhanced strains. *International Journal for Numerical Methods in Engineering*, 43(4):755–779, 1998.

110. D. Nam, S. Choi, S. Park, and N. Stubbs. Improved parameter identification using additional spectral information. *International Journal of Solids and Structures*, 42(18–19): 4971–4987, 2005.

111. Y. Narkis. Identification of crack location in vibrating simply supported beams. *Journal of Sound and Vibration*, 172(4):549–558, 1994.

112. H.G. Natke. Problems of model updating procedures: A perspective resumption. *Mechanical Systems and Signal Processing*, 12(1):65–74, 1998.

113. J.M. Ndambi, J. Vantomme, and K. Harri. Damage assessment in reinforced concrete beams using eigenfrequencies and mode shape derivatives. *Engineering Structures*, 24(4):501–515, 2002.

114. D. Ortiz-Boyer, C. Hervas-Martnez, and N. Garca-Pedrajas. CIXL2: A crossover operator for evolutionary algorithms based on population features. *Journal of Artificial Intelligence Research*, 24:1–48, 2005.

115. W. Ostachowicz, M. Krawczuk, and M.P. Cartmell. Genetic algorithms in health monitoring of structures. Technical Report, Department of Mechanical Engineering, University of Glasgow, Glasgow, Scotland, 1996.

116. W. Ostachowicz and M. Krawczuk. On modelling of structural stiffness loss due to damage. *Key Engineering Materials*, 204–205:185–200, 2001.

117. A.K. Pandey, M. Biswas, and M.M. Samman. Damage detection from changes in curvature mode shapes. *Journal of Sound and Vibration*, 145(2):321–332, 1991.

118. A.K. Pandey and M. Biswas. Damage detection in structures using changes in flexibility. *Journal of Sound and Vibration*, 169(1):3–17, 1994.

119. R. Pascual, J.C. Golinval, and M. Razeto. A frequency domain correlation technique for model correlation and updating. In *Proceedings of the 15th International Modal Analysis Conference*, pp. 587–592, Orlando, FL, 1997.

120. R. Pascual, M. Razeto, J.C. Golinval, and Schalchli R. A robust FRF-based technique for model updating. In *Proceedings of the 27th International Seminar on Modal Analysis*, Leuven, Belgium, 2002.

121. R. Pascual, R Schalchli, and M. Razeto. Robust parameter identification using forced responses. *Mechanical Systems and Signal Processing*, 21(2):1008–1025, 2007.

122. R. Pascual. *Model based structural damage assessment using vibration measurements.* PhD thesis, University of Liège, Belgium, 1999.

123. R. Perera, A. Ruiz, and C. Manzano. Performance assessment of multicriteria damage identification genetic algorithms. *Computers and Structures*, 87(1–2):120–127, 2009.

124. R. Perera and R. Torres. Structural damage detection via modal data with genetic algorithms. *Journal of Structural Engineering*, 132(9):1491–1501, 2006.

125. N.J. Radcliffe. Equivalence class analysis of genetic algorithms. *Complex Systems*, 5(2):183–205, 1991.

126. A.M. Raich and T.R. Liszkai. Improving the performance of structural damage detection methods using advanced genetic algorithms. *Journal of Structural Engineering*, 133:449–461, 2007.

127. M.A. Rao, J. Srinivas, and B.S.N. Murthy. Damage detection in vibrating bodies using genetic algorithms. *Computers and Structures*, 82(11–12):963–968, 2004.

128. P.F. Rizos, N. Aspragathos, and A.D. Dimarogonas. Identification of crack location and magnitude in a cantilever beam from the vibration modes. *Journal of Sound Vibration*, 138:381–388, 1990.

129. R. Ruotolo and C. Surace. Damage assessment of multiple cracked beams: Numerical and experimental validation. *Journal of Sound and Vibration*, 206(4):567–588, 1997.

130. M. Sahin and R.A. Shenoi. Vibration-based damage identification in beam-like composite laminates by using artificial neural networks. In *Proceedings of the Institution of Mechanical Engineers, Part C: Journal of Mechanical Engineering Science*, 217(6):661–676, 2003.

131. B. Sahoo and D. Maity. Damage assessment of structures using hybrid neuro-genetic algorithm. *Applied Soft Computing*, 7(1):89–104, 2007.

132. O.S. Salawu. Detection of structural damage through changes in frequency: A review. *Engineering Structures*, 19(9):718–723, 1997.

133. J.D. Schaffer, R.A. Caruana, L.J. Eshelman, and R. Das. A study of control parameters affecting online performance of genetic algorithms for function optimization. In *Proceedings of the 3rd International Conference on Genetic Algorithms*, Morgan Kaufmann, San Francisco, CA, 1989.

134. Z.Y. Shi, S.S. Law, and L.M. Zhang. Structural damage localization from modal strain energy change. *Journal of Sound and Vibration*, 218(5):825–844, 1998.

135. Z.Y. Shi, S.S. Law, and L.M. Zhang. Damage localization by directly using incomplete mode shapes. *Journal of Engineering Mechanics*, 126(6):656–660, 2000.

136. Z.Y. Shi, S.S. Law, and L.M. Zhang. Improved damage quantification from elemental modal strain energy change. *Journal of Engineering Mechanics*, 128:521, 2002.

137. M. Srinivas and L.M. Patnaik. Genetic algorithms: A survey. *Computer*, 27(6):17–26, 1994.

138. N. Stubbs, J.T. Kim, and K. Topole. An Efficient and Robust Algorithm for Damage Localization in Offshore Platforms. In *ASCE 10th Structures Congress*, Vol. 92, pp. 543–546, San Antonio, TX, 1992.

139. L.T. Stutz, D.A. Castello, and F.A. Rochinha. A flexibility-based continuum damage identification approach. *Journal of Sound and Vibration*, 279(3–5):641–667, 2005.

140. G. Syswerda. Uniform crossover in genetic algorithms. In *Proceedings of the 3rd International Conference on Genetic Algorithms*, pp. 2–9, San Francisco, CA, 1989.

141. R. Tanese. Parallel genetic algorithms for a hypercube. In *Proceedings of the 2nd International Conference on Genetic Algorithms and Their Application*, pp. 177–183. Lawrence Erlbaum Associates, Inc. Mahwah, NJ, 1987.

142. K.S. Tang, K.F. Man, S. Kwong, and Q. He. Genetic algorithms and their applications. *Signal Processing Magazine, IEEE*, 13(6):22–37, 1996.

143. A. Teughels, J. Maeck, and G. De Roeck. Damage assessment by FE model updating using damage functions. *Computers and Structures*, 80(25):1869–1879, 2002.

144. S.K. Thyagarajan, M.J. Schulz, P.F. Pai, and J. Chung. Detecting structural damage using frequency response functions. *Journal of Sound and Vibration*, 210(1):162–170, 1998.

145. V.V. Toropov, L.F. Alvarez, and H. Ravaii. Recognition of damage in steel structures using genetic programming methodology. In *Proceedings of the 2nd International Conference on Identification in Engineering Systems*, pp. 382–391, Swansea, UK, March 1999.

146. J. Tracy and G. Pardoen. Effect of delamination on the natural frequencies of composite laminates. *Journal of Composite Materials*, 23:1200–1215, 1989.

147. M.T. Vakil-Baghmisheh, M. Peimani, M.H. Sadeghi, and M.M. Ettefagh. Crack detection in beam-like structures using genetic algorithms. *Applied Soft Computing*, 8(2):1150–1160, 2008.

148. J.D. Villalba and J.E. Laier. Localising and quantifying damage by means of a multi-chromosome genetic algorithm. *Advances in Engineering Software*, 50:150–157, 2012.

149. H.M. Voigt, H. Mühlenbein, and D. Cvetkovic. Fuzzy recombination for the breeder genetic algorithm. In *Proceedings of the 6th International Conference on Genetic Algorithms*, pp. 104–113, San Mateo, CA, 1995.

150. D. Wang and H. Zhu. Wave propagation based multi-crack identification in beam structures through anti-resonances information. *Key Engineering Materials*, 293–294:557–564, 2005.

151. Z. Wang, R.M. Lin, and M.K. Lim. Structural damage detection using measured FRF data. *Computer Methods in Applied Mechanics and Engineering*, 147(1–2):187–197, 1997.

152. B. Weber, P. Paultre, and J. Proulx. Structural damage detection using nonlinear parameter identification with Tikhonov regularization. *Progress in Structural Engineering and Materials*, 14(3):406–427, 2007.

153. E.J. Williams and A. Messina. Applications of the multiple damage location assurance criterion. *Key Engineering Materials*, 167–168:256–264, 1999.

154. A.H. Wright. Genetic algorithms for real parameter optimization. *Foundations of Genetic Algorithms*, 1:205–218, 1991.

155. J.R. Wu and Q.S. Li. Structural parameter identification and damage detection for a steel structure using a two-stage finite element model updating method. *Journal of Constructional Steel Research*, 62(3):231–239, 2006.

156. Q.W. Yang and J.K. Liu. Structural damage identification based on residual force vector. *Journal of Sound and Vibration*, 305:298–307, 2007.

157. Q.W. Yang. A numerical technique for structural damage detection. *Applied Mathematics and Computation*, 215:2775–2780, 2009.

158. Y.J. Yan, L. Cheng, Z.Y. Wu, and L.H. Yam. Development in vibration-based structural damage detection technique. *Mechanical Systems and Signal Processing*, 21(5):2198–2211, 2007.

159. G.J. Yung, K.A. Ogorzalek, S.J. Dyke, and W. Song. A parameter subset selection method using residual force vector for detecting multiple damage locations. *Structural Control and Health Monitoring*, 17(1):48–67, 2010.

160. C.B. Yun, J.H. Yi, and E.Y. Bahng. Joint damage assessment of framed structures using a neural networks technique. *Engineering Structures*, 23(5):425–435, 2001.

161. C. Zang, H. Grafe, and M. Imregun. Frequency-domain criteria for correlating and updating dynamic finite element models. *Mechanical Systems and Signal Processing*, 15(1):139–155, 2001.

162. C. Zang and M. Imregun. Structural damage detection using artificial neural networks and measured FRF data reduced via principal component projection. *Journal of Sound and Vibration*, 242(5):813–827, 2001.

163. C. Zang and M. Imregun. Structural damage detection and localization using FRF-based model updating approach. *Key Engineering Materials*, 245–246:191–200, 2003.

164. J.L. Zapico, M.P. González, and K. Worden. Damage assessment using neural networks. *Mechanical Systems and Signal Processing*, 17(1):119–125, 2003.

165. M.W. Zehn, O. Martin, and R. Onger. Influence of parameter estimation procedures on the updating process of large finite element models. In *Proceedings of the 2nd International Conference on Identification on Engineering Systems*, pp. 240–250, Swansea, UK, March 1999.

166. B.T. Zhang and J.J. Kim. Comparison of selection methods for evolutionary optimization. *Evolutionary Optimization*, 2(1):55–70, 2000.

167. Y. Zou, L. Tong, and G.P. Steven. Vibration-based model-dependent damage (delamination) identification and health monitoring for composite structures—A review. *Journal of Sound and Vibration*, 230(2):357–378, 2000.

8 Neural Networks and Genetic Algorithms in Structural Health Monitoring

Ratneshwar (Ratan) Jha and Sudhirkumar V. Barai

CONTENTS

INTRODUCTION

Structural Health Monitoring—Basic Concepts in SHM

Structural health monitoring (SHM) is aimed to determine health of a structure exposed to varying environmental and operational conditions as well as instrumentation noise (i.e., "real-world" conditions) while eliminating false indications. The basic premise of an SHM system is that damages alter stiffness, mass, or damping of a structure, and in turn cause changes in its dynamic response. The complete health state of a structure can be determined based on the presence, location, type and severity of damages (diagnostics), and estimation of remaining useful life (prognostics). The existing practice for ensuring structural safety (i.e., schedule-based maintenance) leads to expensive inspections, unnecessary downtime and retirement, and sometimes catastrophic failures without any warning. A reliable SHM system can inspect the structure continuously or periodically and provide the operator with up-to-date information about its health, which would provide tremendous benefits in terms of life-cycle costs by detecting damages early and allowing a much more efficient maintenance schedule (condition-based maintenance). Such a system would allow designers to relax the conservative designs and take full advantage of the benefits of advanced materials for engineered structures. Although SHM has experienced significantly increased research during the last decade, a damage detection method that can provide quantitative damage information anywhere in a complex structure, such as aircraft wings, wind turbine blades, or bridges, is still under development. Health monitoring of engineering structures is being pursued vigorously as evidenced by the proceedings of various conferences, journal publications, and recent books in the area.

Researchers are currently investigating both passive and active methods for damage diagnosis of structures. Passive methods analyze the structural vibrations from ambient excitations. Changes in vibrational characteristics due to the presence of damages (such as cracks or delaminations) allow for the detection of the presence of damage. Because structures are typically subjected to multiple sources of ambient vibrations during service, little equipment is needed to excite the structure. An extensive review of vibration-based SHM methods was presented by Doebling et al. (1998) and Sohn et al. (2003). Montalvo et al. (2006) have reviewed vibration-based SHM methods with emphasis on composite materials. Most of the vibration-based methods use modal properties (natural frequency, mode shapes, curvature of mode shapes, modal strain energy [MSE], etc.) of a structure for damage detection. Another class of methods is based on time series or frequency–time analysis of data (such as autoregressive models, signal processing using wavelet or Hilbert–Huang transform, etc.).

Active SHM methods analyze the health of a structure by exciting the structure and recording the corresponding response. The use of Lamb wave-based SHM has shown promise in recently published research. The books by Giurgiutiu (2008) and Su and Ye (2009) and the review article by Raghavan and Cesnik (2007) contain the fundamentals of Lamb wave-based structural diagnostics and extensive list of related publications. Lamb waves are of particular interest due to the similarity between their wavelength and the thickness of composite structures generally used and their ability to travel far distances. These two features allow for the detection of not only superficial but internal flaws and the ability to examine large areas. The multimodal and dispersive characteristics of Lamb wave propagation require careful signal processing for damage detection.

NEURAL NETWORK—DEFINITION, TYPES, AND TRAINING ALGORITHMS

A neural network (NN), also known as artificial NN (ANN), is a massively parallel interconnection of simple processors or neurons that has the ability to learn and generalize (Haykin, 1999). Properly formulated and trained NNs are capable of approximating any linear or nonlinear function to the desired degree of accuracy. NNs are extensively used in signal processing because they provide nonlinear parametric models with universal approximation power as well as adaptive training algorithms (Hu and Hwang, 2001). NNs have been applied in several other fields as well, including identification and control of dynamical systems (Narendra and Parthasarathy, 1990) and adaptive vibration suppression (Jha and He, 2004). Researchers have used NNs for SHM due to their ability to learn from the environment, store the acquired knowledge for future use, and identify underlying information from within noisy data (Reed, 2009). In SHM, NNs are used for mapping relationships between measured features and structural damage (physical) parameters and trained using known damage features and their corresponding physical parameters.

The NNs known as multilayer perceptrons (MLPs) are the most popular for SHM. The MLPs comprise a layer of input signals, one or more hidden layers of neurons, and an output layer of neurons wherein a layer consists of a single or multiple neurons. Neurons in each layer are connected to all neurons in adjacent layers. These networks are of feedforward type and the effects of the input signals are propagated through the networks layer by layer. Differences between the desired outputs (targets) and the network outputs give the errors. The connection strengths ("weights") and "biases" are updated during training (or, learning) such that the network produces the desired output for the given input. The MLPs trained with backpropagation (BP) are compact and provide excellent generalization (i.e., accurate outputs for inputs not encountered during training or learning).

Another type of NN often employed for SHM is known as radial basis function (RBF) network. RBF networks are nonlinear layered feedforward networks capable of providing universal approximations (Haykin, 1999). An RBF network has a single nonlinear hidden layer consisting of RBFs (such as Gaussian functions) that perform nonlinear transformation of the input vectors. Essentially, an RBF network solves a curve-fitting problem in a high-dimensional space. As such, learning is equivalent to finding a surface in a multidimensional space that provides a best fit to the training data. The linear weights of the output layer provide a set of adjustable parameters. Training an RBF network consists of two stages (Reed, 2009). The basis function parameters are first determined using an unsupervised method without reference to target data. The weights of the output layer are then obtained by solving a linear optimization problem. MLP and RBF networks are the most commonly used supervised networks.

GENETIC ALGORITHMS

Genetic algorithms (GAs) along with NNs and expert systems are the most widely used artificial intelligence tools for optimal solutions to engineering problems. GAs are a form of "evolutionary computation" methods that are based on Darwin's theory of natural evolution to conduct a search for optimization. These are most useful in cases where the derivative-based search techniques are not applicable. The three main steps of the basic GA are: initialization, selection, and generation (crossover–mutation). The first step randomly generates an initial population and each member of the population is compared with the other members for its fitness (i.e., ability to solve a problem). The next step selects members of the population for reproduction wherein fitter members are selected more times. Then crossover and mutation operators are applied to the intermediate population to create the next population. The above three steps form one generation in the evolution process of a GA and are continued till the optimization termination criteria are satisfied. GAs have been used in many disciplines such as design of communication networks, machine learning, signal processing, and combinatorial optimization. Examples of GA applications in SHM include condition monitoring of rotating mechanical systems (Saxena and Saad, 2007), identification of increasing levels of damage in airplane structure (Meruane and Heylen, 2010), and optimal sensor placement for damage detection (Yi et al., 2011).

THEORY

NEURAL NETWORKS

To set the scene it is useful to give a definition of what we mean by "Neural Net." However, the objective of the chapter is to make clear the terms used in this definition and to expand considerably on its content.

> A Neural Network is an interconnected assembly of simple processing elements, units or nodes, whose functionality is loosely based on the animal neuron. The processing ability of the network is stored in the inter-unit connection strengths, or weights, obtained by a process of adaptation to, or learning from, a set of training patterns.

The following will give insight of NNs as a computational tool.

Biological NNs and Physiological Aspects

A basic nerve cell or neuron is composed of a processing body, transmitting axons and receiving synapses and dendrites (Figure 8.1). Let us consider enormous numbers of such cells interconnected in an additive manner (Dayoff, 1990). If a processing cell receives a stimulus from say, an electrical or a chemical source or from a pressure or temperature change, this causes activation potential to form internally. This potential rapidly spreads along all axons. It is delayed by distance, electrochemical state, and synapse resistance at the terminus, where the axon joins another cell's body or its dendrites. The receiving cell is thus stimulated by all the transmitting axons connected to its body or its dendrite receptors. The signal received from any

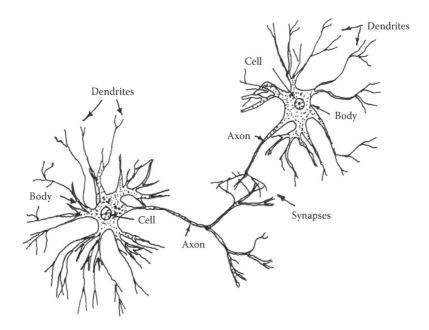

FIGURE 8.1   Biological neuron.

one axon or synapse or dendrite source is equivalent to the activation potential of the originating cell multiplied by a weight proportional to the time delay and resistance met along the axon, across the synapse and/or along the dendrite. The sum of all such weighted stimuli arriving within a short period becomes the input to the receiving nerve cell.

The state of that cell's internal processes determines the blending, squashing, and thresholding performed before the signal is transmitted out of all its axons to the temporally next layer on nerve cells. The recent history of the receiving cell thus determines its excitability and its activation function. The outgoing stimulus travels down more axons and, weighted by time and resistance, is received by the next round of processing cells.

Our brain is understood to possess 10^{10} to 10^{11} neurons massively interconnected with 10^3 to 10^4 synaptic connections resulting in 10^{13} to 10^{15} connections (Wasserman, 1989; Dayoff, 1990). Even when we assume the neurons to take binary forms (firing or not firing), the total number of degrees of freedom of the natural NN (i.e., the brain) is truly astronomical, reaching $2^{10^{15}}$! The modeling of brain function is an active area of neuroscience research and the actual function is not yet properly understood. However, the present state of knowledge has inspired and provided guidelines for developing ANN. In the next section attention has been given to the basic understanding of biological neurons and its extension to ANNs.

Artificial Neuron and ANNs

Mathematically, a single nerve cell can be modeled as an artificial neuron (computational unit) consisting of receiving sites (synapses), receiving connections (dendrites),

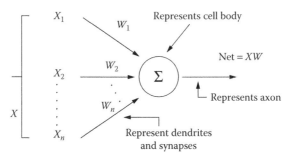

$$\text{Net} = X_1 W_1 + X_2 W_2 + \cdots + X_n W_n$$

FIGURE 8.2 Artificial neuron.

a processing element (cell body), and transmitting connections (axons). A typical architecture of an artificial neuron is shown in Figure 8.2.

A neuron (or node) receives input stimuli from other neurons if they are connected to it and/or to the external world. A neuron can have several inputs, but has only one output. This output, however, can be routed to the inputs of several other neurons. Each neuron has certain constant parameters associated with it. These are its threshold, transfer function, and the weights associated with its input. Each neuron performs a very simple arithmetic operation, that is, it computes the weighted sum of its inputs, subtracts its threshold from the sum, and passes the result through its transfer function. The output of the neuron is the result obtained from this function. The output of a neuron is, therefore, a mathematical function of its inputs. The most common transfer functions used in NN literature (Simpson, 1992) are the hard limiter, threshold, range, and sigmoid nonlinearities, such as shown in Figure 8.3.

ANNs are modeled on the concept of present understanding of the functioning of the brain thus comprising of a massively interconnected network of a large number of artificial neurons or computational units. NN models are specified by the net topology, node (artificial neuron) characteristics, and training or learning rules. The function of an NN is determined by these parameters. The architecture of the network determines the input of each node. The node characteristics (threshold, transfer function, and weights) determine the output of the node. The training or learning rules determine how the network will react when an unknown input is presented to it.

The large connectivity degree of the neurons, their massive parallelism, and their nonlinear analog response as well as learning capabilities are the basic factors that characterize the computational effectiveness of an NN. A computing device based on NN has a greater fault tolerance than a classical sequential computer due to the increased number of locally connected processing nodes. Thus, some neurons or links out of order do not diminish considerably the performance of the network.

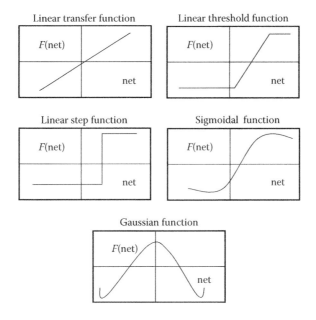

FIGURE 8.3   Activation functions.

NNs hold the promise of providing a fast, data-driven modeling system with desirable robustness properties because their strengths and weaknesses complement those of more traditional analytic techniques.

Characteristics of NNs

Parallel distributed processing, adaptive filters, connectionist/connectionism model, self-organizing systems, neurocomputers, neuromorphic systems, etc., are the various names under which the study of ANN has been carried out by researchers. The wide ranges of applications in fields have made neural nets, the center of attraction among researchers. The important characteristics of ANN are as follows (Dayoff, 1990).

Classification: They can extract classification (clustering) characteristics from a large number of input examples.

Pattern matching: They can produce the corresponding output patterns for given input patterns.

Pattern completion: For an incomplete pattern, networks can generate the missing portion of the input pattern.

Learning: Unlike expert systems, NNs learn many example patterns and their associations, that is, desired outputs or conclusions.

Generalization: They have associative memory. The network responds in an accretive or interpolative way to noisy, incompetent, or previously unseen data. An associative network, where input is equal to desired output, can produce a full output if presented with a potential input. This property is called "generalization."

Noise removal: A noise-corrupted input pattern presented to the network results in removal of some (or all) of the noise and produces a noise-free version of the input pattern as output.

Fault tolerance: In ANN, the memory is distributed and failure of some processing elements will slightly affect overall behavior of the network.

Optimization: For given initial values of a specific optimization problem, the networks help in arriving at a set of variables that represent a solution to the problem.

Control: Given the current state of a controller and the desired response for the controller as an input pattern, the networks generate proper command sequence to create the desired response.

Distributed memory: The connection weights are the memory units of the network. The values of weights represent the current state of knowledge of the network. A unit of knowledge, represented for example by an input/output pair is distributed across all the weighted connections of the network.

Storage memory: There is one set of network weights capable of representing a large space of stored patterns. Thus it provides an advantage of lesser amount of storage memory.

Historical Development of NNs

It will not be unfair to say that the study of ANN had a rather unsteady history. Early investigators into ANN provided a variety of simple demonstrations and based on the strength of these results, it was suggested that many previously unsolved problems might find their solution in ANN. The growth of ANN research has been tabulated in Table 8.1.

Various types of NN models, such as Hopfield net, Hamming net, Carpenter/ Grossberg net, single-layer perceptron, multilayer network, and so on. (Figure 8.4) have been described in the literature (Dayoff, 1990; Lippman, 1987; Wasserman, 1989). Here, a comparative study of these models has been summarized in Tables 8.2 and 8.3. The single-layer Hopfield and Hamming nets are normally used with binary input and output under supervised learning. The Carpenter/ Grossberg net, however, uses unsupervised learning. The single-layer perceptron can be used with multivalue input and output in addition to binary data. A serious disadvantage of the single-layer network is that complex decisions may not be possible. The decision regions of a single-layer network are bounded by hyperplanes whereas those of two-layer networks may have open or closed convex decision regions (Lippman, 1987).

MLP and Implementation Issues

MLP (Figure 8.5) consists of an array of input neurons, known as the input layer, an array of output neurons, known as the output layer, and a number of hidden

TABLE 8.1
NNs History

Present	Late 1980s to 2015	Interest explodes with conferences, articles, simulations, new companies, and government funded research
Late infancy	1982–	Hopfield at National Academy of Sciences
Stunned growth	1969–1981	1969—Some research continues
		Marvin Minsky and Seymour Papert's book Perceptrons condemning Rosenblatt's baby, the perceptron
		James Anderson (neurophysiologist) developed a model—Brain-State-in-a-Box
		Kunihiko Fukushima developed the model Neocognitron for visual pattern recognition
		Teuvo Kohonen developed a model similar to Anderson's but independently
		Grossberg, Rumelhart and McCelland; Marr and Poggio; Cooper and many more were busy in their labs
Early infancy	Late 1950s, 1960s	Excessive hype and research efforts expand
		1957: Frank Rosenblatt—The Perceptron—New model
		1959: Bernard Widrow and Marcian Hoff—ADALINE, MADALINE (Multiple ADAptive LINear Elements): Real world problem—Adaptive filters to eliminate echoes on phone line
		1960: Stephen Grossberg—Physiological research to develop NN model
Birth	1956–	Artificial Intelligence (AI) and Neural Computing fields launched—Dartmouth Summer Research Project
Gestation	1950–	Age of computer simulation
	1949–	Donald Hebb's The Organization of Behaviour highlights the connection between psychology and physiology pointing out that a neural pathway is reinforced each time it is used.
	1943–	Warren McCulloh and Walter Pitts' "How Neuron Might Work?" modeled a simple neuron network with electric circuit
	1936–	Turing uses brain as computing paradigm
Conception	1890–	A text by Williams James, Psychology (brief course) contains insight into brain activity and foreshadows current theories

Source: Nelson, M. M., and Illingworth, W. T. (1991) *A Practical Guide to Neural Nets*, Addison-Wesley. With permission.

layers. Each neuron receives a weighted sum from every neuron in the preceding layer and provides an input to every neuron of the next layer. The activation of each neuron is governed by a threshold function. To train the network, a popular training algorithm of the MLP is the BP or generalized delta rule where the error calculated at the output of the network is propagated through the layers of neurons to update the weights.

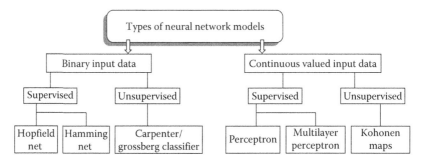

FIGURE 8.4   Classification of ANN models.

The implementation can be done at software and/or hardware levels. Software simulations generally allow use of various network models with a choice of transfer functions. The software can run on conventional computers (e.g., PCs, mainframe computers, etc.). With these conventional hardware simulations, the speed in training/testing cannot be achieved. So, for a more efficient implementation, recent research work has been diverted toward the hardware simulation. Basically, the hardware implementation leads to emulation within parallel architectures, neurocomputers, networks on chip, optical NNs, biological computers (Nelson and Illingworth, 1991), and so on. Here attention is focused on software implementation.

Phases of NNs Adaptation

The computational aspects of NN simulation are typically characterized by three main phases (Figure 8.6) as discussed in Moselhi et al. (1991).

Network design: Network topology and flow of data determine the interconnection pattern and the forward/backward flow of information inside the network. Having identified the problem and flow of data, a paradigm must be selected. Several innovatively developed NN paradigms have been outlined in Tables 8.2 and 8.3.

Network learning: Training/learning phase determines the weight updating function and error calculations. The task is to arrive at a unique set of weights that are capable of correctly associating all example patterns, used in learning, with their desired output patterns. There are two types of learning: supervised and unsupervised. A training algorithm is used and held responsible for specifying how weights adapt in response to a learning example (Tables 8.2 and 8.3).

Network recall: Recall phase establishes the network operation such as the propagating rule and threshold function. For the trained network, recall of a set of inputs to network produces outputs that are satisfactorily close to the desired set of outputs used in training.

TABLE 8.2
Commonly Used NN Paradigms

Paradigms	Hopfield Network	Hamming net	Bidirectional Associative Memory (BAM)	Boltzmann Machine	Adpative Resonance Theory (ART-2)	Perceptron	Backpropagation (BP)	Counterpropagation
Type of input	Binary, polar	Binary	Bipolar	Binary, continuous	Grayscale	Continuous	Continuous	Continuous
Type of output	Pattern, class	Pattern, class	Pattern, class	Pattern, real, class	Pattern, class	Real, class	Pattern, real, class	Pattern, real class
Transfer function	Hard-limiting	Hard-limiting	Clamped linear	Varies	Sigmoidal	Perceptron	Sigmoidal, hyperbolic tangent	Kohonen and sigmoid
Number of layers	1	2	1	2 or more	1	1	2 or more	2 or 4
Size of hidden layer	N/A	N/A	N/A	Small to medium	Increase by data category	N/A	Small to medium	Equal to number of data category
Connectivity	Fully interconnected	Fully interconnected	Fully interconnected	Fully interconnected, random	Fully interconnected	Random	Fully interconnected, random, slabs	Fully interconnected
Learning algorithm	Hopfield	Hamming	BAM	Boltzmann	ART-2	Perception convergence, delta rule	Generalized delta rule	Kohonen and Grossberg
Training method	Supervised	Supervised	Supervised	Supervised	Unsupervised	Supervised	Supervised	Supervised, Unsupervised
Network type	Recurrent	Recurrent	Recurrent	Nonrecurrent	—	Nonrecurrent	Nonrecurrent	Nonrecurrent

Source: Nelson, M. M., and Illingworth, W. T. (1991) *A Practical Guide to Neural Nets*, Addison-Wesley. With permission.
Bipolar: (−1 and +1) values as opposed to binary (0 and 1); Grayscale: Activation level opposed to Boolean.

TABLE 8.3
Performance Characteristics of Commonly Used Neural Paradigms

Network Paradigm	Hopfield Network	Hamming Net	BAM	Boltzmann Machine	ART-2	Perceptron	Backpropagation (BP)	Counterpropagation
Source	Hopfield (1982, 1984)	Lippman (1987)	Kosko (1987)	Hinton and Sejnowski (1986)	Carpenter and Grossberg (1987)	Rosenblatt (1961)	Rumelhart et al. (1986)	Hecht-Nielsen (1988)
Training time	Fast	N/A	Fast	Slow	Fast	Medium	Slow	Medium
Execution time	Medium	N/A	Fast	Slow	Medium	Fast	Fast	Fast
Information content	Low	N/A	Low	High	High	Low	High	High
Advantages	(1), (2), (3)	(4), (5), (6)	(1), (2), (3), (7), (8), (9)	(11),(12)	(13)	Very simple	(14), (15)	(15), (16), (17), (18)
Disadvantages	(1), (2), (3), (4), (5)		(2), (4), (5)	Very Slow, (5), (6)	—	(7)	(4), (5), (6)	(4), (5), (6), (8)
Utilization	High	Low	Low	High	High	Low	High	High

Advantages: (1) Provides dynamic (real time) performance. (2) Sufficiently responds to noisy, incomplete, or unseen inputs. (3) Can be used for optimization. (4) Implements the optimum minimum error classifier when bit errors are random and independent. (5) Requires fewer connections than the Hopfield net. (6) Does not suffer from spurious output patterns which can produce a "no-match" result. (7) Easier and more systematic training methodology than Hopfield. (8) Simpler and faster than Hopfield. (9) No stability problem. (10) Provides adaptive performance. (11) Alleviates local minima problem. (12) Uses probabilistic training method. (13) Facilitates learning new examples without destroying previous experience (incremental learning). (14) Powerful and accurate association. (15) Suitable for static environment (input does not change with time). (16) Good for rapid prototyping. (17) Can generate a function and its inverse. (18) Has powerful probabilistic capabilities.

Disadvantages: (1) Stability not guaranteed. (2) Memory limitations (stored patterns). (3) Difficult to formalize training method. (4) Could be trapped in local minima or paralyze. (5) No incremental learning. (6) Not suitable for real-time applications. (7) Limited representation capability. (8) Not as accurate as BP.

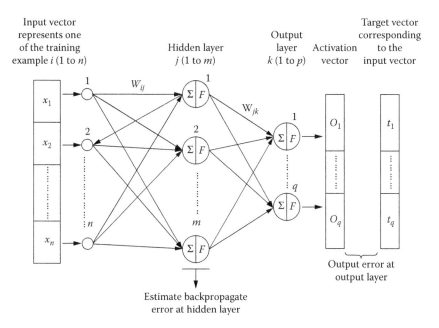

FIGURE 8.5 Typical MLP.

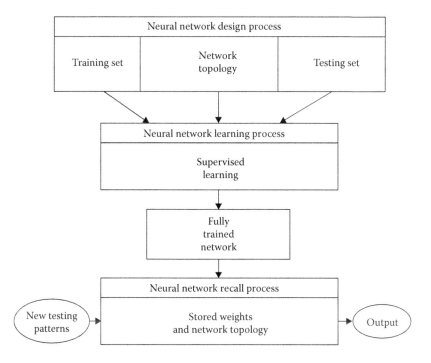

FIGURE 8.6 Phases of NN applications.

NNs Architectural Aspects

Choosing the correct topology of a network is a difficult task indeed. To design the architecture of a network, the following issues must be considered (Figure 8.7).

Choosing input–output nodes: Initially one has to decide upon the number of input nodes and output nodes. Number of facts per training example would determine the number of input nodes and corresponding desired output parameter would give the number of nodes in the output layer.

Training patterns: Presenting an appropriate training set to the network is a critical decision. If a very small percentage of the possible patterns are introduced in the training set, the resulting generalization may be poor while in the opposite case it is likely that higher oscillation may make it impossible to reach a state of global minima (Flood and Kartam, 1994a).

Normalization of training set: The input patterns must be normalized before being presented to the network. This is essential as the sigmoidal transfer function generates the output between 0 and 1. Various normalization or scaling strategies have been proposed in the NNs related literature (Brainmaker, 1989).

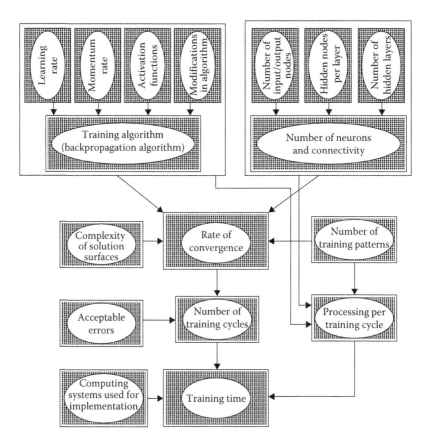

FIGURE 8.7 NN training paradigm.

The number of hidden layers in the network: In many investigations it has been mentioned that two to three hidden layers are sufficient for most problems. However the optimal number of layers is application dependent and is an open research topic. Keeping the number of neuron connections constant, whether the network that possesses more than one hidden layer may be superior to the network with single hidden layer is worth investigating as currently there is no convincing evidence either to support or to dispute the hypothesis in general. It is suggested in the literature (Brainmaker, 1989) that a multilayer network with linear neurons (neurons whose outputs are a linear function of their inputs) is equivalent to a two-layer network. Hence, the various weight matrices can be combined into a single matrix which serves the same purpose as the multilayer network with linear neurons. For this reason, using linear neurons in a multilayer network is considered rather unnecessary.

Number of neurons in hidden layers: The decision on how many hidden neurons should be used in a layer is rather arbitrary and has been usually decided by trial and error. A good guess would be to put an average of the number of input and output neurons. Another possibility would be to make the hidden layer of the same size as either the input or the output layer. Fewer hidden neurons mean fewer connections, hence less training capacity and faster training and running. Generally the hidden layer should not be the smallest layer in the networks nor should it be the largest (Brainmaker, 1989).

Choosing training parameters: The training parameters (learning parameter, momentum parameter), maximum error tolerance, epochs, etc. are again very difficult to decide upon and may be arrived at by considerably investigating the application domain. Though these parameters have generally been frozen in several investigations, it would be desirable to carry out further study of these parameters to assess their influence.

Choosing the activation function: Various types of activation functions are available in the literature (Brainmaker, 1989). Some of them are linear, linear threshold, step, sigmoid, and Gaussian activation functions (Figure 8.3). Among all these activation functions, the sigmoid function (also known as an S-shaped semilinear or squashing function) has been recommended in most of the BP applications. In sigmoid function the output is a continuous, monotonic function of the input. Both the function itself and its derivatives are continuous everywhere, and it asymptotically approaches low and high values.

Selecting an ANN architecture is a matter of further investigation in the domain of application. Here, several issues related to the design of suitable network need to be examined.

Most Widely Used BP Algorithm

BP (generalized delta rule) is the most widely used learning paradigm adopted in MLPs and has been applied successfully in several areas. The various applications include military pattern recognition, speech recognition, character recognition, sonar target recognition, image classification, signal encoding, knowledge processing, medical diagnosis, synthesis to robot autonomous vehicle controls, and other pattern analysis problems. In particular, in civil engineering, it has been adopted by various researchers (Table 8.4).

TABLE 8.4
Typical Examples of NNs Applications to Civil Engineering

Sr.	Domain	Problem Identification	Learning Paradigm	References
1.	Analysis	1. Simulation of structural analysis (optimization—beam test problem)	BP algorithm	Rogers (1994)
		2. Structural analysis problem (bilateral and unilateral structures)	Hopfield neural models and Hopfield–Tank neural models	Kortesis and Pangiotopoulos (1993)
		3. Earthquake selection in earthquake true time-history analysis for structural dynamics	BP algorithm	Cheng and Popplewell (1994)
		4. Nonlinear analysis of plates	BP algorithm	Pandey and Barai(1994)
2.	Design	1. Learning in engineering design (beam design problem)	Perceptron learning	Adeli and Yeh (1989)
		2. Optimum design of aerospace components	BP algorithm	Berke et al. (1993)
		3. Automated conceptual design of structural system (beam and frame structure)	Adaptive resonance theory (ART)-based neural model	Hajela et al. (1991)
		4. Structural preliminary design (gridworks structure)	Modified BP algorithm	Gan and Liu (1991)
		5. Structural analysis and design (truss, wing-box structure)	Flat networks, single hidden layer network, and functional link nets	Hajela and Berke (1991)
		6. Computational simulation of composite ply	NETS-BP-based algorithm	Brown et al. (1991)
3.	Damage assessment	1. Detection of damage in structural system (example of truss, portal frame, spatial truss)	Modified counterpropagation algorithm CPN-architecture	Szewczyk and Hajela (1994)
		2. Vibration signature analysis (five-storey three-dimensional steel frame)	BP algorithm	Elkordy et al. (1993, 1994)
		3. Detection of structural damage (three-storey frame subjected to earthquake base acceleration)	BP algorithm	Wu et al. (1992)
		4. Parameter identification (unilateral/bilateral structures)	Perceptron algorithm for supervised learning	Kortesis and Pangiotopoulos (1993)
		5. Identification and control of large structures	BP algorithm	Yen (1994)
		6. Seismic event identification	BP algorithm	Perry and Baumgardt (1991)
				(Continued)

TABLE 8.4 (*Continued*)
Typical Examples of NNs Applications to Civil Engineering

Sr.	Domain	Problem Identification	Learning Paradigm	References
		7. Minimizing distortion in truss structures	Hopfield networks	Fu and Hajela (1993)
		8. Identification of nonlinear dynamics systems (damped duffing oscillator under deterministic excitation)	BP algorithm	Masri et al. (1993)
		9. Building KBES for diagnosis PC pile	BP algorithm	Yeh et al. (1993)
		10. Priority rating of highway maintenance needs	BP algorithm	Fwa and Chan (1993)
		11. Seismic liquefaction potential assessment	BP algorithm	Goh (1994)
		12. Assessment of seismic liquefaction	BP algorithm	Tung et al. (1993)
4.	Other applications	1. Problem of determining truck attributes (such as velocity, axle spacing, and axle loads)	Layered feedforward network architecture radial–Gaussian incremental learning network systems	Gagarin et al. (1994)
		2. River flow prediction	Cascade-correlation algorithm (CASCADE network)	Karunanithi et al. (1994)
		3. Modular construction decision making	Kohonen self-organizing NNs	Murtaza and Fisher (1994)
		4. Representation of material behavior	BP algorithm	Wu et al. (1990), Ghaboussi et al. (1991)
		5. Optimum markup estimation in construction industry	BP algorithm	Moselhi et al. (1991)
		6. Analogy-based solution to markup estimation problem	BP algorithm	Hegazy and Moselhi (1994)
		7. Vertical formwork selection	BP algorithm	Kamarthi et al. (1992)
		8. Equation renumbering (profile front minimization PFM)	BP algorithm	Hoit et al. (1994)
		9. Material models for composites	BP algorithm	Pidaparti and Palakal (1993)
		10. Problems of (a) load location (b) concrete beam design and (c) simply supported plate	BP algorithm	Vanluchene and Sun (1990)

To understand implications of learning and implementation of the network, a simplified derivation of the generalized delta rule (Rumelhart et al., 1986) has been given in the following paragraph.

The net input to a node in layer j is given by (Figure 8.5):

$$\text{net}_j = \sum_{i=1}^{n} w_{ji} o_i \tag{8.1}$$

and the output of node j will be

$$o_j = f(\text{net}_j) \tag{8.2}$$

Here, f is the activation function and in this study a sigmoidal function has been used, which is given as

$$o_j = \frac{1}{1 + e^{\frac{-(\text{net}_j + \theta_j)}{\theta_0}}} \tag{8.3}$$

Now the input to the nodes of layer k is

$$\text{net}_k = \sum_{j=1}^{m} w_{kj} o_j \tag{8.4}$$

and its respective output is

$$o_k = f(\text{net}_k) \tag{8.5}$$

In the training process of the network, for the input pattern $x_p = i_{pi}$, the weight adjustment will take place in the links of NN for desired output t_{pk} at the output nodes. After achieving this first adjustment, the network will pick up another pair of x_p and t_{pk}, and will again adjust weights for new pair. In a similar way the process will go on till all the input–output pairs get exhausted. Finally the network will have a single set of stabilized weights satisfying all the input–output pairs.

Usually the outputs o_{pk} will not be the same as the desired output values t_{pk}. For each input–output pattern, the square of the error is given by

$$E_p = \frac{1}{2} \sum_{k=1}^{q} (t_{pk} - o_{pk})^2 \tag{8.6}$$

and the average system error by

$$E = \frac{1}{2P} \sum_{p=1}^{P} \sum_{k}^{q} (t_{pk} - o_{pk})^2 \tag{8.7}$$

Avoiding the p subscript in the Equation 8.6 for convenience, the expression will be

$$E = \frac{1}{2} \sum_{k=1}^{q} (t_k - o_k)^2 \tag{8.8}$$

In a true gradient search for a minimum system error, one has to compute the derivative of the error function E, with respect to any weight in the network and then change the weights according to the rule:

$$\Delta w_{kj} = -\eta \frac{\partial E}{\partial w_{kj}} \tag{8.9}$$

Here, η is the learning parameter, which gives

$$\frac{\partial E}{\partial w_{kj}} = \frac{\partial E}{\partial net_k} \frac{\partial net_k}{\partial w_{kj}} \tag{8.10}$$

Using Equation 8.4:

$$\frac{\partial net_k}{\partial w_{kj}} = \frac{\partial}{\partial w_{kj}} \sum_{j=1}^{m} w_{kj} o_j = o_j \tag{8.11}$$

Now δ_k can be defined by

$$\delta_k = -\frac{\partial E}{\partial net_k} \tag{8.12}$$

therefore,

$$\Delta w_{kj} = \eta \delta_k o_j \tag{8.13}$$

The weights on each connection should be changed by an amount proportional to the product of the term δ_k, available to the unit receiving input along that line and the activation o_j along that line. The determination of δ_k is a recursive process. To compute $\delta_k = -(\partial E/\partial net_k)$, the chain rule can be used to express δ_k in terms of two factors: first, the rate of change of error with respect to the output o_k and second, the rate of change of the output of the node k with respect to input to that same node. Therefore,

$$\delta_k = -\frac{\partial E}{\partial net_k} = -\frac{\partial E}{\partial o_k} \frac{\partial o_k}{\partial net_k} \tag{8.14}$$

Now these factors can be computed as

$$\frac{\partial E}{\partial o_k} = -(t_k - o_k) \tag{8.15}$$

$$\frac{\partial o_k}{\partial net_k} = f_k(net_k) \tag{8.16}$$

Using expressions (8.15) and (8.16), we have

$$\delta_k = (t_k - o_k)f_k(\text{net}_k) \tag{8.17}$$

For any output layer k, Δw_{kj} will be given by

$$\Delta w_{kj} = \eta(t_k - o_k)f_k(\text{net}_k)o_j = \eta\delta_k o_j \tag{8.18}$$

Similarly for the internal units:

$$\Delta w_{ji} = \eta\delta_j o_j \tag{8.19}$$

$$\delta_j = f_j(\text{net}_j)\sum_{k=1}^{q}\delta_k w_{kj} \tag{8.20}$$

The application of the BP algorithm involves two phases. In the first phase, the input is presented and propagated forward through the network to compute the output value of each unit. In the backward phase, the δs for all the units are computed. Once these two phases are complete, one can compute for each weight the Δws.

In summary, adding one more subscript p to denote the pattern number, we have

$$\Delta_p w_{ji} = \eta\delta_{pj}o_{pi} \tag{8.21}$$

If j are the nodes of the output layer, then

$$\delta_{pj} = (t_{pj} - o_{pj})f_j(\text{net}_{pj}) \tag{8.22}$$

or if j are nodes of internal or hidden units, then

$$\delta_{pj} = f_j(\text{net}_{pj})\sum_{k=1}^{q}\delta_{pk}w_{kj} \tag{8.23}$$

The BP is basically a gradient descent algorithm. In multilayer networks, the error surfaces will be complex with several local minima. It is possible that the gradient descent procedure may not reach the global minimum, but get trapped in one of the many local minima.

One way to increase the learning rate without leading to oscillation is to modify the BP algorithm by including the momentum term α as below.

$$\Delta w_{ji}(n+1) = \eta(\delta_j o_i) + \alpha\Delta w_{ji}(n) \tag{8.24}$$

Here, n is the presentation number and α a constant that determines the effect of previous weight changes on the current direction of movement in the weight space. This provides a kind of momentum in the weight space that effectively filters out high frequency variations of the error surface in the weight space. The advantages, disadvantages, and further improvement of BP algorithm are discussed below.

Advantages:
1. *Associative capability*: It can learn a variety of pattern mapping relationships. It does not require any prior knowledge of mathematical functions that maps the input pattern to the output patterns. BP merely needs examples of the mapping to be learned.
2. *Large number of design options*: In BP algorithm, one has options for the number of layers, interconnections, processing units, the learning parameters, and data representation. This facilitates the user to attack broad spectrum of applications.

Disadvantages:
1. *Convergence time*: For relatively simple problems, training session can require hundreds and thousands of iterations. In real applications, thousands of training examples may take days of computing time to train the network.
2. *Local minima problem*: Sometimes the network never converges to a point where it has learned the training set, that is, the network becomes stuck in a local minimum. At that stage, additional training does not improve the performance.
3. *Not suitable for real-time problem*: BP training takes enormous amount of time during training, thus it is not suitable for real-time problems.
4. *No incremental learning*: Once trained the network cannot be trained with additional training patterns. In such a case, one has to retrain the network with complete training patterns and this is a time consuming process.

Possible Improvements (Taylor and Mannion, 1989):
1. *Random walk*: Random walk helps by adding a small nudge to the gradient descent to avoid convergence. On time increment, these nudges diminish in size to zero. This is a kind of simulated annealing.
2. *Inertia*: Addition of small percent of the previous change in the weight to currently calculated changes in weights to give the actual change that is applied to the network helps in achieving the speed of convergence into the basic gradient descent.
3. *Second-order derivative*: In the derivation of learning rule, the consideration of second-order effects improves the performance of the NNs.

Hegazy and Moselhi (1994) have given further information on BP performance.

GENETIC ALGORITHMS

Computer algorithms based on the process of natural evolution have the capability to produce very powerful and robust search mechanisms, although the similarity between these algorithms and the natural evolution is based on crude imitation of biological reality (Mitchell, 1996). A number of examples are found in literature of this nature in which such algorithms have been successfully applied to solve various engineering problems quite effectively. GAs are such a tool that finds wide application in almost every field. Although being easy to comprehend and apply, they have been useful even in cases where normal search techniques and optimization models are found to be too tedious and inefficient. With advancements in computer technologies and development of more powerful and distributed computers, their efficiency is ever increasing. They are most efficient when applied to the parallel/distributed computer setup.

Three Main Steps of the Basic GA

1. Step 0 *initialization*: The first step in the implementation of any GA is to generate an initial population. In most cases the initial population is generated randomly. After creating an initial population, each member of the population is evaluated by computing the representative objective and constraint functions and comparing it with the other members of the population.
2. Step 1 *selection*: A selection operator is applied to the current population to identify the mating pool that will participate in the crossover to create the next generation population.
3. Step 2 *generation (crossover–mutation)*: To create the next generation, crossover and mutation operators are applied to the intermediate population to create the next population. Crossover is a reproduction operator, which forms a new child by combining each of the two parents. Mutation is a reproduction operator that forms a new child by making (usually small) alterations to the values of parent. The process of going from the current population to the next population constitutes one generation in the evolution process of a GA. If the termination criteria are satisfied, the procedure stops, otherwise, it returns to step 1. Figure 8.8 shows the working algorithm of a normal GA.

Coding in GA

Coding in GA is the form in which chromosomes and genes are expressed. There are mainly two types of coding: binary and real. In binary coding, chromosome is expressed as a binary string, whereas in real coding chromosomes are expressed as real numbers. Binary coding is still believed to be an ideal type of coding generally. However, the real coding is more applicable and easy in programming. Moreover, it seems that the real coding fits the continuous optimization problems better than the binary coding. From economic point of view also, real coding proves to be better because it requires less storing space and is also efficient in calculations.

Encoding of a Chromosome

An individual, or solution to the problem to be solved, is represented by a list of parameters, called chromosome or genome. The most common way of encoding chromosomes is a binary string but in real parameter-based GA they are represented

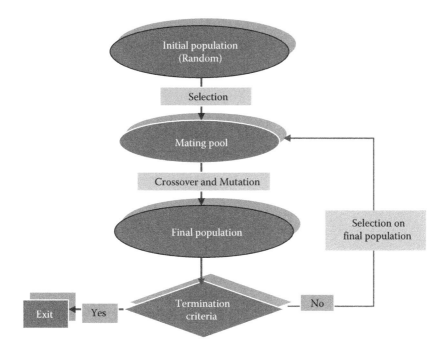

FIGURE 8.8   Working Algorithm of GA.

as integers. A wide variety of other data structures for storing chromosomes can also be used depending on the requirement. The encoding depends mainly on the solved problem. For example, one can encode directly integer or real numbers, sometimes it is useful to encode some permutations, and so on.

Simulated Binary Crossover

After we have decided what encoding we will use, we can proceed to crossover operation. Simulated binary crossover (SBX) operates on two parents and creates two new offspring. As the name suggests, the SBX operator simulates the working of the single-point crossover operator on binary strings. Figure 8.9 illustrates the SBX operator. A probability distribution is used around parent solutions to create two children solutions. Probability distribution is as follows:

$$P(\beta) = \begin{cases} 0.5(\eta_c + 1)\beta^{\eta_c} & \text{if} \quad \beta \leq 1 \\ 0.5(\eta_c + 1) / \beta^{\eta_c + 2} & \text{otherwise} \end{cases}$$

where

$$\beta = \left| \frac{y^{(2)} - y^{(1)}}{x^{(2)} - x^{(1)}} \right| \tag{8.25}$$

Parent 1	Parent 2
20	30
Child 1	Child 2
22	29

FIGURE 8.9   Simulated binary crossover.

The children solutions are then calculated as follows:

$$y^{(1)} = 0.5\left[(x^{(1)} + x^{(2)}) - \bar{\beta}\left|(x^{(1)} - x^{(2)})\right|\right] \tag{8.26}$$

$$y^{(2)} = 0.5\left[(x^{(1)} + x^{(2)}) + \bar{\beta}\left|(x^{(1)} - x^{(2)})\right|\right] \tag{8.27}$$

Crossover probability indicates how often crossover will be performed. If there is no crossover, offspring are exact copies of parents. If there is crossover, offspring have a share of parents' values. If crossover probability is 100%, then all offspring are made by crossover. If it is 0%, whole new generation is made from exact copies of chromosomes from old population (but this does not mean that the new generation is the same!). Crossover is made in hope that new children will contain good parts of old parents, and therefore, the new children will be better. However, it is good to leave some part of old population to next generation.

Mutation

After a crossover is performed, mutation takes place. Mutation is intended to prevent falling of all solutions in the population into a local optimum of the solved problem. Mutation operation randomly changes the offspring resulted from crossover. In case of real encoding, a polynomial probability distribution is used to create a solution y in the vicinity of a parent.

$$\bar{\delta} = \begin{cases} (2u)^{1/(\eta_m+1)} - 1, & \text{if} \quad u \le 0.5 \\ 1 - [2(1-u)]^{1/(\eta_m+1)}, & \text{otherwise} \end{cases} \tag{8.28}$$

$$y = x + \bar{\delta}\Delta_{\max} \tag{8.29}$$

where:
  Δ is the perturbance allowed in the solution
  u is a random number between 0 and 1

Mutation is illustrated in Figure 8.10.

Parent	30
Child	10

FIGURE 8.10   Mutation.

Mutation probability indicates how often members of population will be mutated. If there is no mutation, offspring are generated immediately after crossover (or directly copied) without any change. If mutation is performed, the offspring change. If mutation probability is 100%, parent is surely to be changed; if it is 0%, nothing is changed. Mutation generally prevents the GA from falling into local extremes. Mutation should not occur very often, because then GA will in fact change to random search.

Selection

Chromosomes are selected from the population to be parents for crossover. The problem is how to select these chromosomes. According to Darwin's theory of evolution, the best ones survive to create new offspring. There are many methods for selecting the best chromosomes. Examples are roulette wheel selection, Boltzmann selection, tournament selection, rank selection, steady-state selection, and so on.

GAs—Sequential and Parallel

Figure 8.11 gives the basic types of GAs. They are

1. *Generational GA*: Here the population evolves generation after generation through the process of selection, crossover, and mutation.
2. *Steady-state GA*: Here each offspring upon creation replaces the worst individual in parent population.
3. *Master–slave parallel GA*: Here one processor acts as the master and has several processors under it acting as slaves. The master sends the individual members of the population to his slaves who evaluate them and

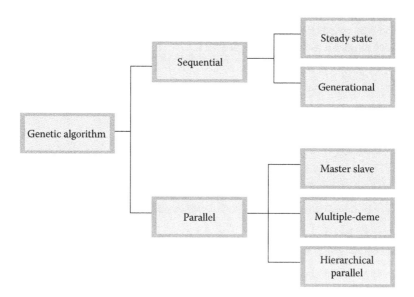

FIGURE 8.11   Types of GAs.

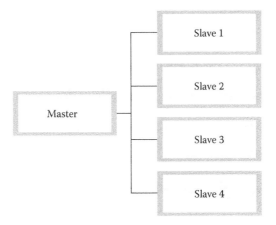

FIGURE 8.12 Master–slave parallel GA.

return them to the master. This method is more effective when we have very complex objective functions but can get extremely slow at times due to large number of communications between the processors and consequentially large communication time. Figure 8.12 shows a typical master–slave GA.

4. *Multiple Deme GA*: Here entire populations thrive on each processor and some communication is established between them. Some population members migrate from one processor to the other at predefined intervals. This form of GA has been intended to be employed in this project in future.

5. *Hierarchical parallel GA*: This combines the qualities of both master–slave and Multiple-Deme GAs. Each deme acts as a master and has some slave nodes under it as shown in Figure 8.13.

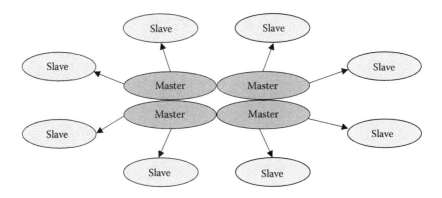

FIGURE 8.13 Hierarchical parallel GA.

NN APPLICATIONS IN SHM

Extensive literature for applications of NN in SHM exists. This section presents a detailed survey of applications related to civil infrastructure. For other application areas of composites, aerospace, and mechanical systems, only a sample of recently published results are presented as illustrative examples.

Civil Infrastructure

There has been an enormous amount of interest in the research and development of NN applications to civil infrastructure systems in last one decade. Researchers have proposed the use of NN in innovative applications to structural analysis, design, and diagnostics as given in Table 8.4, earthquake engineering (Barai, 1999), hydrology (ASCE, 2000a, b), mechanics (Zeng, 1998). Flood and Kartam (1994a, b) have given a review of NN application in civil engineering. An excellent reading for the novice could be publications by ASCE (Kartam et al., 1997, Flood and Kartam, 1998).

NNs have the capability of establishing the relationship between causes and effects of damage in structure. They can supplement alternative approaches of conventional structural identification, which are based on rigorous finite element methods (FEMs). An NN must be trained with sufficient training data to diagnose damage correctly. Training data can be actual damage states, which have been experienced by a structure, or they can be based on analysis studies. NNs are promising tools in damage detection. The compiled bibliography by Barai (2000) is an index of NNs activities in the field of damage detection. Following paragraph gives brief review on NNs in damage detection. It is worthwhile to mention that review of published all the papers in this field are out of the scope of this chapter. Hence, here we have made an attempt to review some of the representative papers that appeared during 1991–2000.

Kudva et al. (1991) have used a BP-based NN to identify damage in a stiffened plate. The plate was divided into a 4 × 4 array elements grid. Damage was introduced by cutting holes of various diameters in the plate at the centers of grid elements. The grids elements were of size 305 mm (12 in.) × 203 mm (8 in.) and the hole diameter ranged from 12.7 mm (0.5 in.) to 63.5 mm (2.5 in.). Under static uniaxial load, the structural strain readings were taken from elements in the grid elements. NNs were used to identify the map from the strain data to the location and size of the hole. In different experiments 8, 20, and 40 strains were used as input. The structure of the network was chosen to be two hidden layers, each with the same number of hidden nodes as the number of inputs. The network was trained with 3, 12, or 32 patterns, depending on which experiment was being tested. The authors claimed that the networks converged in less than 10 minutes on a 386 PC, depending on the example. It should be noted that in one example the NN failed to converge, and the authors were forced to modify their procedure to a two-step algorithm, which first predicted the hole quadrant, then the correct bay within the quadrant. The authors found that the NN was able to predict the location of the damaged bay without an error but that predicting hole size was more difficult with sometimes erratic results. In the cases where the NN successfully identified the hole size, typical errors were on the order of 50%.

Szewczyk and Hajela (1992, 1994) have demonstrated the application of ANN in the inverse problem of damage detection of truss and frame structures. They modeled the damage as a reduction in the stiffness of structural elements, which were associated with observed static displacements under prescribed loads. To generate this reverse mapping between stiffness of individual members of the structure and global static displacements, an improved *counterpropagation neural (CPN) network* was used. In the improved CPN, a control parameter referred to as a network solution was introduced to determine the size of networks and to control the accuracy of approximations. Both layers were trained simultaneously, and using an averaging operator, the outstar weights were computed during the training to eliminate a learning rate from the original network. In addition to this, a nonlinear interpolation scheme was introduced to increase the accuracy of network estimates. They performed simulation of a frame structure with nine bending elements and 18 degrees of freedom (displacements and rotation at each node). In training the network, 3600 randomly generated damage patterns were used. The size of the network architecture was governed by a control parameter, also called resolution parameter. In this investigation, the resolution parameter used was 0.9. They observed from the exercise that the network performance was generally precise but suffered with gradual deterioration in the presence of noisy and incomplete measurements data. They also concluded that the resolution parameter plays an important role in designing the size of the network and is highly problem dependent.

Wu et al. (1992) used BP algorithm in a multiperceptron architecture with one and two hidden layers to simulate damage states in a three-storey frame. The structure was subjected to earthquake base acceleration and the transient response was computed in time domain. The Fourier spectra of the computed relative acceleration time histories of the top floor for various members were used in training the NNs. The member damage was defined as a reduction in the member stiffness. An NN architecture with one hidden layer consisting of 200 input units, 3 output units, and a hidden layer with 10 nodes was selected by trial and error. The input represented the amplitudes of the Fourier spectra and the output represented the damage condition of each member. They used a total of 43 training cases consisting of 42 frequency spectra of relative accelerations recorded at top floor level along with the information that indicates which member is damaged and the extent of the damage. They observed that the performance of network was not satisfactory due to network not being provided with enough information during training to distinguish between various damage states. To enhance the performance of networks, they trained the NN with Fourier spectra recorded at two floor levels. The network had two 200-node input sets and two hidden layers with eight nodes per hidden layer. They concluded from this study that in the presence of adequate measurement data and one or two hidden layers, the performance of two-hidden-layer network was superior to that of one hidden layer network. Moreover, it was observed that additional input data further improve the detection capability of network. Obviously the performance of the network that depends on the hidden layer is problem dependent and needs to be researched further.

Elkordy et al. (1993, 1994) adopted BP NNs to model damage states of a five-storey steel frame. Three NNs were used. All the networks were designed with four input nodes, two output nodes, and 14 hidden nodes with two hidden layers chosen by trial and error. These networks were trained with analytically generated damage states, and tested with damage states obtained experimentally from a series of shaking tests. Although the results were promising, they concluded that the relation between the number of damage patterns required for training the network and those needed to perform satisfactorily and further the degree of simplification of the model needs to be investigated.

Leath and Zimmerman (1993) used an MLP NN based on a training method they developed to identify damage in a four-element cantilevered beam. Most of the paper illustrated the development of the training algorithm, which was subsequently used in the damage identification. The training algorithm was designed to fix the architecture of the network. The idea involved the creation of a network that exactly fits the data with a minimal number of hidden nodes. This criterion fixed the number of hidden nodes to be one less than the number of data points (the bias to the output neuron was the final adjustable parameter). There are two issues that raise questions about the intended structure of the NN. The first issue is that the number of training data points is essentially arbitrary, and one should not need to change architecture to incorporate more data. The second and much more important issue is that exactly fitting the data can cause problems in general. If there is any noise in the data, then an exact fit will model the noise as well. Additional measurements give no noise rejection and instead serve only to confuse the algorithm. This problem occurs, for example, when repeated measurements are made at the same input. The training algorithm to compute the weights for the hidden nodes was based on the principles of computational geometry and as such could only be applied with input dimensions of one or two. The training was intended to put the thresholds for the hidden layer sigmoids for maximum discrimination between the training data. The damage in the beam was modeled by reducing Young's modulus up to 95%. The NN was used to identify the map from the first two bending frequencies to the level of damage in each member. It should be noted that only the first two frequencies were used because the training algorithm could handle no higher dimension. The algorithm was able to identify damage to within 35%.

In Spillman et al. (1993), the authors used a feedforward NN to identify damage in a steel bridge element. The element was roughly 4.5 m (14.75 ft) long. Damage was introduced by cutting the element and bolting a plate reinforcement over top of the cut. With the plate attached, the element was considered to be undamaged. With the bolts loosened, the element was considered to be partially damaged; with the plate removed, the element was considered to be fully damaged. There were three sensors mounted on the element: two accelerometers and a fiber optic modal sensor. The beam was struck in four different locations with a calibrated impact. A total of 11 tests were performed. The time-history signal from each sensor was Fourier transformed, and the height and frequency of the first two modal peaks were used as inputs to the NN. The impact intensity and location were also provided as inputs. A network configuration was selected with the 14 inputs already mentioned, a hidden layer with 20 neurons, and 3 outputs, one for each of the possible damage states. The body of training data was cycled through the training algorithm until the self-prediction error converged to a minimum. Generally, convergence took less than 100 epochs. Other network configurations were also tried that used fewer of the inputs to see if, for example, the fiber optic sensor was providing any useful information. The results were moderately successful. Using all three sensors, the authors found the proportion of correct diagnoses to be 58%. The authors credited this number by citing the small size of the training data set.

Worden et al. (1993) used a BP NN to identify damage in a 20-member framework structure. Damage was modeled by removing one of the structural members completely. The NN was used to identify the map from static strain data to a subjective measure of the damage. The strain in eight members was used for input, and there was one output for each of the eight members where damage was to be identified. The subjective scale was between 0 and 1. A three-hidden-layer design with 12, 12, and 8 hidden nodes, respectively, was chosen. The network was trained on data generated by an FEM and tested on an experimental model of the same geometry. There were 192 training patterns, and the NN took approximately 50,000 epochs to converge. When applied to experimental data, the system was mostly able to identify the location of the damage. However, there were frequent misclassifications owing partially to the large size of the test set.

Yen and Kwak (1993) used a cerebellar model articulation controller (CMAC) network to identify sensor failures in damage detection. The CMAC network is a nearest-neighbor-type network. Interpolating the nearest data points approximates the function. This paper was not really a damage-identification paper because triply redundant sensors, which were perfect indicators of damage, were used. The network was used to determine if there were any sensor failures in the triply redundant sensing system. The basic idea of this network is that if one sensor's signal is sufficiently different from the other two, it has failed. If all three sensors produce different signals, then two sensors have failed. The inputs to the NN are the sensor readings, and the output is a binary value for each of five damage states: no damage, damage in one of the three sensors, or multiple damage. In the case of multiple damage, no attempt was made to determine which sensor had failed. The network was able to discriminate single failures but had difficulties with multiple failures.

Kirkegaard and Rytter (1994) used a BP NN to identify damage in a 20-m steel lattice mast subject to wind excitation. Damage was modeled by replacing lower diagonals with bolted joints of diminished thickness. The NN was used to identify the map from the first five fundamental frequencies to the percent damage in member stiffness. One output was used for each element of interest. The authors chose network architecture with two hidden layers of five nodes each. There were four outputs corresponding to four of the diagonals. The network was trained with 21 examples generated from an FEM. The training set used data with 0% to 100% reduction in diagonal cross-sectional area. The network was able to reproduce the training data, but it had less success on the test data. At 100% damage, the NN was able to locate and quantify the damage. At 50%, the network was able to predict the existence of damage, but not the magnitude. Damage less than 50% was not identified. This result was consistent for test data from both the FEM and the actual experiment. Manning (1994) used BP NNs to identify damage in a 10-bar truss structure and a 25-bar transmission tower with active members. Damage was modeled as changes in member cross-sectional area. Pole and zero locations were extracted from the frequency response functions (FRFs) between the member actuator and the two piezoceramic sensors on the same member. A measure of the member stiffness was also extracted from each FRF. The imaginary part of the pole and zero locations and the stiffness information were the data given to the NN. The NN identified the map from pole–zero location and member stiffness to member cross-sectional area. For the 10-bar truss, a network architecture with a single hidden node with nine hidden neurons was used. For the 25-bar transmission tower, an architecture with 40 inputs, two hidden layers with seven and five nodes, respectively, and 4 outputs was used. The author does not discuss how large the training sets were but claims that convergence occurred within 8000 epochs for the transmission tower. The networks were tested on three examples not in the training sample and predicted member area within 10% for most of the members.

Povich and Lim (1994) used a BP NN to identify damage in a 20-bay planar truss composed of 60 struts. Damage was modeled by removing struts from the structure. The structure was excited by a shaker in the 0–50 Hz frequency range. Two accelerometers were placed on the structure to provide input data. The frequency range studied contained the first four bending modes. The NN was used to identify the map from the Fourier transform of the acceleration history to damage in each desired member. The network had 394 inputs corresponding to the acceleration fast Fourier transforms at the frequencies of interest for two points and 60 outputs, one for each strut in the structure. The authors chose a two-hidden-layer network with 125 and 40 nodes in the respective layers. There were 61 training examples consisting of the removal of each strut and an undamaged case. The network took approximately 1000 epochs to converge. The NN was able to correctly identify the missing strut in 21 cases and was able to localize the damage to two adjacent struts in 38 cases. No cross-validation or testing of the network was done; the reported results are simply a measure of how well an NN could fit the data. The network's generalization capabilities were not tested.

Rhim and Lee (1994) used a BP NN to identify damage in a composite cantilevered beam. The damage was modeled as delamination in an FEM of the beam. The simulations were dynamic with both the force input and the measured output located at the beam tip. Before the NN was applied, a preprocessing step was done on the data. Autoregressive system identification was performed on the transfer function of the beam from the force input to the displacement output. The denominator of the transfer function or characteristic polynomial (equivalent to the poles) was then used in subsequent damage identification. The advantage of doing the system identification first was that it reduced the body of data to a smaller number of physically meaningful parameters. The NN was used to identify the map from the characteristic polynomial to an empirical damage scale. Each of the four outputs represented a different level of damage, where zero indicated no damage at that level, and one indicated total damage at that level. Ideally, there would be at most one nonzero output at a time. The authors chose a network architecture with 13 inputs, corresponding to a fixed 12th-order characteristic polynomial (the maximum number of resonances seen), one hidden layer with 30 nodes, and 4 outputs. The network was trained with 10 training patterns, and the network took 330,000 iterations to converge. The network generalization capability was tested on three examples and correctly identified the damage in those cases.

Stephens and VanLuchene (1994) used a BP network to identify damage in multi-storey buildings. The body of data was accumulated on a one-tenth scale reinforced concrete structure. The damage was modeled by introducing actual cracks into a concrete model of the structure. The network was used to identify the map from three empirical damage indices to a qualitative scale of damage in the structure. The three indices were measures of maximum displacement, cumulative energy dissipated in the building, and stiffness degradation. The output was a number between zero and one with the former being no damage, the latter total collapse. The NN was trained on 60 data points and tested on a training set of 32 data points. The authors experimented with different network topologies and settled on a single hidden layer with seven hidden neurons. The number of training epochs depended on the network architecture but was always less than 10,000. The NN correctly identified the damage index for about 25 of the 32 test data points. Interestingly, the researches applied the NN to the Imperial County Services Building, the one building to suffer a major earthquake while adequate sensors were in place to measure the damage indices. The network correctly identified the building as being lightly damaged. The NN result was shown to be superior to a linear regression model of the data by about 25%.

Szewczyk and Hajela (1994) used a counterpropagation NN to identify damage in truss structures. CPN builds what is essentially an adaptive look-up table from the data. The look-ups were keyed by the position of the input vector. During training, nodes were moved closer to adjacent input nodes. If no node was sufficiently near the input pattern, then a new node was added. Interpolating adjacent nodes made predictions. The advantage of CPN is that the body of data does not have to be cycled through more than once, as there is no error criterion to limit the convergence. Another advantage of the network is that the architecture is selected by the data, not user specified. The disadvantage of this network is that it may take a very large number of training points to adequately sample the desired function. In the paper, the authors show that the size of the training set does not seem to be a problem in their case. Damage was modeled by reducing Young's modulus in the truss members up to 100% (complete removal). The NN was used to identify the map from static deformation under load to Young's modulus of the members. The analysis is completely static, and no modal or frequency analysis is required. The damage identification algorithm was run on three structures of increasing complexity: a two-dimensional, 6-degree of freedom (DoF) system; a two-dimensional, 18-DoF system; and a 3-dimensional, 12-DoF system. The network was trained with 200, 3600, and 3000 examples respectively. The NN was then verified separately on test data and found to have a maximum error of approximately 30%.

Tsou and Shen (1994) used BP NNs to identify damage in two spring–mass systems: a 3-DoF system and the 8-DoF Kabe system, which has widely spaced eigenvalues. Damage was modeled by changing spring constants from 10% to 80%. For the 3-DoF system, the NN was used to identify the map from the change in modal frequencies to the percent change in the spring stiffnesses. In the 8-DoF problem, the NN was used to identify the map from the residual modal force, to the percent change in each spring constant. In the 8-DoF problems, the damage map was first broken up into a binary determination of whether damage was present in each spring. This body of data was then used to train another NN (via a look-up able) to estimate the extent of damage. This procedure led to a combinatorial explosion in the number of training patterns, which had to be stored. The architecture of the NNs was a single hidden layer with 40 hidden nodes for the 3-DoF problems. For the 8-DoF problems, different architectures were tried including 100, 60, and 40 hidden nodes. The networks for the two systems were trained with 27 and 105 patterns, respectively. The 3-DoF problems took 80,000 epochs to converge. The authors do not state how many epochs were required for the NN to converge in the 8-DoF case, but the computation was performed on a Cray supercomputer. The authors were able to identify the extent of damage to within 5% accuracy, provided the body of data was in an interpolation of existing data. However, data extrapolation produced errors up to 30%.

Pandey and Barai (1995) studied an application of multiplayer perceptron in the damage detection of steel bridge structures from static response. The issues relating to the design of network and learning paradigm were addressed and networks architectures were developed with reference to trussed bridge structures. The training patterns were generated for multiple damaged zones in a structure and performance of the networks with one and two hidden layers were examined. It was observed that the performance of the network with two hidden layers was better than that of single-layer architecture in general. The engineering importance of the whole exercise was demonstrated from the fact that measured input at only a few locations in the structure was needed in the identification process using the ANN. Further, Barai and Pandey (1995a) examined the performance of the network with reference to hidden layers and hidden neurons. Some heuristics were proposed for the design of ANN for damage identification in structures. These were further supported by an investigation conducted on five bridge configurations.

Barai and Pandey (1995b) used BP NNs to identify damage from dynamic response in a truss structure simulating a bridge. The authors used the NN to identify the map from various nodal time histories to changes in stiffness. To train the NN, the authors used a finite element simulation of the truss with a moving point force to simulate a vehicle being driven at constant velocity. The time histories at small time intervals for one, three, and five nodes were used as inputs to the NN. Depending on the run, approximately 70 inputs were chosen. The authors claimed to be able to predict stiffness changes to 4% accuracy. The authors found that the time history from a single, carefully selected node produced the best predictions.

Ceravolo and De Stefano (1995) used a BP NN to identify damage in a truss structure simulated by FEMs. The authors use the NN to identify the map from modal frequencies to the coordinates corresponding to the location of damage. Damage was modeled by removing elements of the truss. The network architecture was chosen to be 10 input nodes corresponding to 10 modal frequencies, a hidden layer with 10 nodes, and two output nodes corresponding to the x and y positions. Only single-damage scenarios were considered. The network was trained on 18 sample runs and cross-validated on 5 sample runs. The NN located the damage well. The authors do not discuss how noisy measurements or multiple damage would affect the results.

Kirkegaard et al. (1995) used recurrent NNs to predict solutions for general nonlinear ordinary differential equations (ODEs). This work is not a damage-identification paper as such because no structures were studied. The implication of the paper is that structural damage will change the equations of motion, which will lead to different time histories. The time history of the ODE was predicted using parameters called *delay coordinates* and a technique known as the *innovations approach.* The NN was used to identify the map from past sample data values to the next sample data value. The authors demonstrated both approaches to modeling a hysteretic oscillator system. The number of hidden nodes was chosen to be nine in one hidden layer. The training took approximately 2000 and 650 data points, respectively, with a single epoch. Recurrent NNs were determined that were able to predict nonlinear ODEs adequately. The authors do not comment on how the network training sample size or structure would change with the dimension of the ODE.

Klenke and Paez (1996) used two probabilistic techniques to detect damage in an aerospace housing component. The first technique used a probabilistic NN (PNN). The second technique used a probabilistic pattern classifier (PPC) of the authors' design. Both methods attempted to ascertain the existence of damage, but neither attempted to quantify the extent or location of the damage. The PNN applies a Bayesian decision criterion to determine set membership. In the problem studied, the two sets were either "damaged" or "undamaged." Data from each class were presented to the PNN, and probability density functions for each class were estimated. The idea was that for a piece of new data, an estimated likelihood was computed for membership into both classes, damaged and undamaged. The greater of these two likelihoods was taken as the guess for class membership. To apply the method, a probability density function needed to be estimated from the data. A Gaussian kernel-density estimator approximated the probability density function. The PPC method attempted to determine membership in a single class. Given class data such as instances of undamaged structures, the Gaussian kernel-density estimator was computed. The space of measurements was then mathematically transformed into normal and independent random variables. A new data instance could then be transformed into this same space, and a statistical chi-square test could be made to predict if the new instance lay in the same class. In both techniques, the necessary computational effort went up dramatically with the dimensionality of both the input and output spaces. It was therefore necessary to dimensionally reduce the input data to a manageable number of DoF, as well as to ensure that these DoF contained predictive information. For the aerospace housing component, the number of measurements was reduced to five static flexibilities estimated from experimental FRFs. The raw data were vibrational spectra. Taking small sequential subsets of the data created multiple data sets. Damage was modeled by five progressively worse cuts made in the housing. Both methods were perfect detectors of damage because in all cases damage was clearly identified.

Schwarz et al. (1996) used a BP NN to identify linear damage in spring–mass systems. The authors used a commercial package to implement the NN and were unspecific about the details of the network except to say that it was a three-layer BP network. The NN was used to identify the changes in spring constants as a result of changes in modal frequencies. The function mapping frequency shifts to changes in stiffness was multivalued in the sense that more than one stiffness change could cause the same frequency shift. The authors circumvented this problem by eliminating inconsistent data sets; only the smallest stiffness change that produced the desired frequency shift was retained. The NN was applied to a trial system consisting of two springs and two masses. One output was assigned to the stiffness change of each spring. The NN was trained with 1000 data points corresponding to changes in stiffness of up to approximately 100%. The authors found they could identify changes in stiffness to within 10%. One problem with this technique is that as the complexity increases to more realistic levels, the degeneracy becomes more of an issue. In addition, noise was not simulated, which would serve to further confound the algorithm. It is also known that regardless of the modeling technique, NNs included, modal frequency shifts are rather insensitive damage indicators.

Barai and Pandey (1997) carried out comparative studies on traditional neural networks (TNNs) (commonly used BP algorithm) and time-delay neural networks (TDNNs) architectures with BP learning algorithm for vibration signature analysis of a typical bridge truss with simulated damaged states. The training patterns were generated using a standard FEM program. The workings of both the networks were demonstrated by comparing the output with algorithmically generated performance parameters not considered in training. The performance issues related to TNN and TDNN were examined. For the cases illustrated, the performance of TDNN was found to be generally better as compared to that of the TNN. This was an important engineering significance as the measurements coming from various nodes involved simultaneously in the same network were good enough for identification purposes. It was concluded that the TDNN using a BP learning paradigm has great potential in damage identification.

Nakamura et al. (1998) carried out study using NNs to detect structural damage in an existing building damaged due to sever earthquake which occurred in Kobe, Japan, in 1995. NNs trained for undamaged state were used to furnish restoring force, which was compared to the corresponding time history of the system at a later stage during the monitoring process. The exercise was demonstrated for actual data obtained from ambient vibration measurements of a seven-storey steel building structure that was damaged under strong seismic motion. The network could differentiate between damaged storey and undamaged storey.

Marwala (2000) carried out studies on damage identification using a committee of NNs. Three distinct networks, as parts of the committee, were used for FRFs, modal properties (natural frequencies and mode shapes), and wavelet transforms (WTs) to identify the damage. Author demonstrated this approach for two examples. In first example of 3-DoF, 1936 data were collected. Training and testing data were 1436 and 500, respectively. An FRF network had 12 input units, 6 hidden units, and 3 output units. Modal properties network had 12 input units, 6 hidden units, and 3 output units. WT network had nine input units, six hidden units, and three output units. The results showed that the network performance was not influenced in the presence of 0%, 5%, and 10% Gaussian random noise. In second example of cylindrical shell, 120 data were collected. Training and testing data were 110 and 10, respectively. An FRF network had 15 input units, 8 hidden units, and 3 output units. Modal properties network had 23 input units, 12 hidden units, and 3 output units. WT network had 18 input units, 9 hidden units, and 3 output units. It was found that the committee approach gave the lowest mean square error, followed by modal property network. The FRF network and WT network gave the same value of mean square error.

As shown in the reviews contained in this section, the identification of damage using NNs is still in its infancy. Most researchers use BP NNs, although not many papers compared the performance of two different NN types. Most of the papers attempt to identify damage from information related to static or dynamic response. Most papers introduce damage in the structure by changing member shapes and/or cross-sectional areas. None of these produced a nonlinear dynamic system, which is what might be expected in real structures. Most of the papers assumed detailed knowledge of the actual structure including mass and stiffness matrices. A very few performed the identification of system parameters based on measured data so that no detailed knowledge of the structure was assumed. Generalizations of these nonmodel-based methods would seem to be more useful for practical applications.

COMPOSITE MATERIALS

The use of composites for engineering structures (aerospace, wind energy, civil infra-structure, etc.) is increasing rapidly due to several advantages such as higher specific strength and modulus, fewer joints, improved fatigue life, and higher resistance to corrosion. However, composite structures have complex damage mechanisms such as delamination between plies, fiber–matrix debonding, fiber breakage, and matrix cracking. These damages often occur below the surface due to foreign object impact, fatigue, and so on, and may not be visible. SHM of composite structures is a very active area of research. For an effective SHM, various techniques of damage detection such as guided waves, strain, and vibration-based techniques may be combined.

Ramadas et al. (2008) reported use of NN for detection, location, and sizing of transverse cracks in a composite beam. They combined four damage features, namely time of flight (TOF) and amplitude ratio of Lamb wave reflected from crack and first and second natural frequencies from vibrations of the structure. Because the amplitude of the reflected wave increases with increasing crack depth, the ratio can be used for predicting crack depth. TOF is used for predicting crack location because the arrival time of reflected wave group is linearly proportional to the loca-tion of crack. A cross-ply composite beam with a lay up sequence of $[0_2/90]_s$ was simulated using the FEM with cantilevered boundary condition. Figure 8.14 shows a typical signal wherein the various wave groups such as forward traveling wave, reflection from crack, reflection from free edge, and reflection from fixed boundary are indicated.

The authors (Ramadas et al., 2008) selected feedforward BP NN with three hidden layers, each having 10 neurons, and an output layer with two neurons (Figure 8.15). Tan-sigmoid functions were used for hidden neurons and linear neurons were used for the output layer. The input vector comprised TOF, amplitude ratio, first and sec-ond natural frequencies, and location and depth of transverse crack were used as outputs of the NN. The mean square error between target and NN prediction was minimized during training. Twenty different damages cases comprising various crack locations and depths were simulated. The NN predicted damage location and depth with an accuracy of about 90%. Also, it was observed that damage could be identified more effectively when damage features of more than one technique were combined.

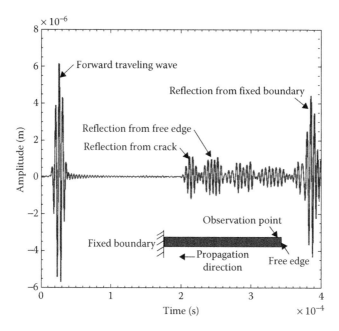

FIGURE 8.14 Typical signal obtained at observation point. (From Ramadas, C. et al., 2008, *Int. J. Smart Sensing Intell. Syst.*, 1, No. 4, 970–984. With permission.)

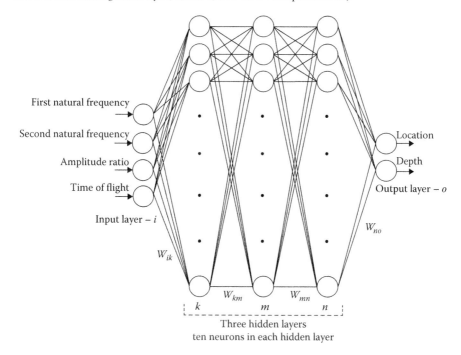

FIGURE 8.15 NN used for damage detection. (From Ramadas, C. et al., 2008, *Int. J. Smart Sensing Intell. Syst.*, 1, No. 4, 970–984. With permission.)

Moll et al. (2010) used a local linear NN to model the nonlinear dispersion curves in anisotropic plates. A challenging aspect of wave propagation in an anisotropic plate is that the group velocity of the fundamental modes is a function of the propagation angle in addition to the frequency–thickness product. The NN approach increased the angular resolution (even with sparse sensor network) and shortened the computational time for damage localization. For a local model network (Figure 8.16) with p inputs $\mathbf{u} = [u_1 \ u_2,\ldots,u_p]^T$, the output $\hat{y}$ can be calculated as the interpolation of M local model outputs $\hat{y}_i, i = 1,\ldots,M$ given by

$$\hat{y} = \sum_{i=1}^{M} \hat{y}_i(\mathbf{u})\Phi_i(\mathbf{u}), \quad \text{where} \sum_{i=1}^{M} \Phi_i(\mathbf{u}) = 1 \tag{8.30}$$

Here $\Phi_i(.)$ are called interpolation or validity or weighting functions. The validity functions describe the regions where the local models are valid and determine the contribution of each local model to the output. The validity functions are smooth functions between 0 and 1 to ensure smooth transition (no switching) between the local models. Also, the validity functions are such that the contributions of all local models sum up to 100%. An NN with six local models was used to model S_0-mode velocities for the experimental composite plate with nine piezoelectric patches (Figure 8.17). The model plot for group velocity (Figure 8.18) shows its variation with angle and frequency along with the experimental data in angular direction.

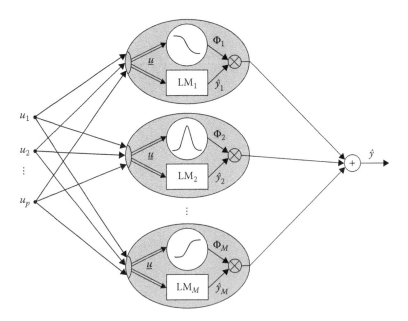

FIGURE 8.16 Local model network: the outputs of the local models are weighted with their validity function values and summed up. (From Moll, J. et al., 2010, *Smart Mater. Struct.*, 19, 045022–045038. With permission.)

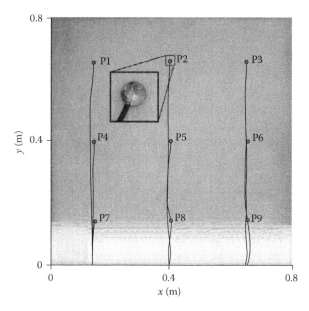

FIGURE 8.17   Glass-fiber-reinforced polymer experimental plate with nine piezoelectric patches. (From Moll, J. et al., 2010, *Smart Mater Struct.*, 19, 045022–045038. With permission.)

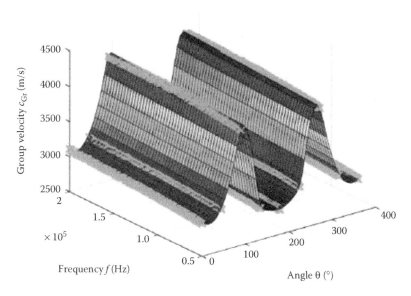

FIGURE 8.18   Model of experimental velocities for S_0-mode in comparison to the sparse measurement points shown in Figure 8.17. (From Moll, J. et al., 2010, *Smart Mater. Struct.*, 19, 045022–045038. With permission.)

The authors used this local linear NN approach for successful multisite damage localization wherein simulated signals were obtained by the spectral element method.

Aerospace Structures

Katsikeros and Labeas (2009) used a BP feedforward NN in the prediction of fatigue damage states of a typical aircraft cracked lap-joint structure (Figure 8.19). A lap-joint of eight columns and one row of rivet holes with cracks growing along the crack line was used in this study. The idea was to determine crack location and extent based on known strains at certain points of the structure. Strain data sets were obtained through finite element substructuring technique for an adequate number of different damage configurations. A large number of different crack patterns (more than 1100 in total) considered as possible damage scenarios were randomly generated, including cracks of several mm in length, located at any rivet hole in combinations from one up to four cracked holes.

The strain data were normalized using its far field value and the first 20 sine and cosine coefficients from its discrete Fourier transformation (DFT) were retained. A principal component analysis (PCA) of DFT coefficients was carried out and only the first six principal components (containing 98% of full information) were used as an NN input to achieve a significant reduction of the network size. An MLP NN with six input neurons and a single hidden layer of 12 neurons was trained for performing a nonlinear input–output mapping between the principal components of strain data and the crack pattern. The number of the output layer neurons was determined by the accumulated crack lengths along both sides of each rivet. To assess the performance

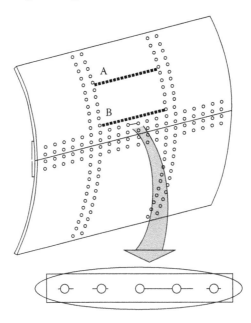

FIGURE 8.19 Multiple cracked panels of a typical aircraft fuselage butt-joint. (From Katsikeros, Ch. E., and Labeas, G. N., 2009, *Mech. Syst. Signal Process.*, 23, 372–383. With permission.)

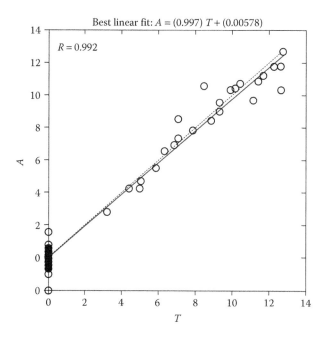

FIGURE 8.20   Network regression analysis over the validation set. (From Katsikeros, Ch. E., and Labeas, G. N., 2009, *Mech. Syst. Signal Process.*, 23, 372–383. With permission.)

of trained network, Figure 8.20 shows network outputs A versus target values T as open circles. An excellent fit is indicated because the dashed line (best linear fit) and the perfect fit (solid line—output equal to targets) are very close. Thus the trained network was successfully validated and proven capable for accurately predicting crack positions and lengths of a lap-joint structure.

Sundaram et al. (2009) investigated MLP NN-based SHM of co-cured and co-bonded carbon composite aircraft structures using Fiber Bragg Grating (FBG) sensors. This study includes prediction of load from observed strain pattern, detection of the presence and severity of skin–spar debonding, and determination of sensor malfunction (which is a particularly interesting aspect of this work). FBG sensor malfunction is typically indicated by its output reading zero due to a damaged sensor or its debonding from the structure. FEM of a skin/stiffener composite test box (Figure 8.21) was used for NN training that was verified with experimental data obtained for healthy conditions. The sensors were divided into three grids (comprising 16, 20, and 16 sensors) and a separate NN was created for each grid. Sensors were placed on the upper and lower surfaces of top and bottom skins at various locations along test box length.

The overall scheme for isolation of malfunctioning sensor comprised two networks, one each for *inverse solution* and *forward solution*, as indicated in Figure 8.22. The inverse solution network predicted load on the structure given real-time static strain pattern whereas forward solution network gave strain values for input load. For the grid with 16 sensors, the inverse solution network input layer comprised

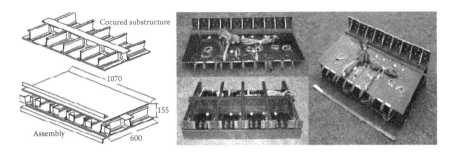

FIGURE 8.21   Schematic (left) and actual (right) composite test box. (From Sundaram, R. et al., 2009, *Proc. Int. Conf. on Adv. Smart Mater.*, Madurai, India. With permission.)

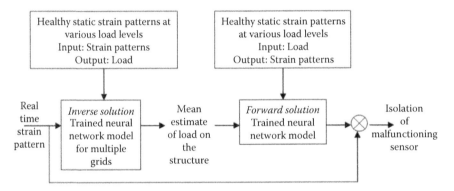

FIGURE 8.22   Architecture for isolation of malfunctioning sensor. (From Sundaram, R. et al., 2009, *Proc. Int. Conf. on Adv. Smart Mater.*, Madurai, India. With permission.)

16 neurons corresponding to 4 sensors in the grid at 4 different layers on the test box. The hidden layer consisted of 10 neurons and a single neuron was used for the output layer that predicted load value. The forward solution network comprised one input neuron, 4 hidden neurons, and 16 output neurons. A pure linear transfer function was used for both networks. The Levenberg–Marquardt BP algorithm was employed for training the inverse solution network and conjugate gradient optimization algorithm was used for the forward solution network. Figure 8.23 shows that NN predictions match very well with experimental strain values for the healthy case. When a particular sensor malfunctions, its strain value becomes zero. The distance between the (experimental) unhealthy strain value and the NN-predicted healthy strain is the largest leading to successful sensor malfunction detection in real time.

MECHANICAL SYSTEMS

Schlechtingen and Santos (2011) reported NN model-based approaches for wind turbine fault detection using online SCADA (supervisory control and data acquisition) data, which is usually available to wind turbine operators. Data from 10 different

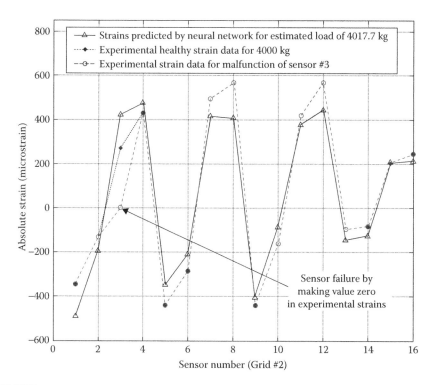

FIGURE 8.23   Detection of sensor malfunction for an applied load of 4000 kg. (From Sundaram, R. et al., 2009, *Proc. Int. Conf. on Adv. Smart Mater.*, Madurai, India. With permission.)

operating offshore turbines of 2 MW class were used for five real measured faults and anomalies, namely bearing temperature anomaly, two gearbox bearing damages, and two stator temperature anomalies. MLPs in combination with sigmoid transfer function were used for NN models. Figure 8.24 shows the architecture for autoregressive NN for predicting the generator bearing temperature. The input data included previous bearing temperature and time-lagged values of other correlated signals, namely electrical power output, nacelle temperature, generator stator temperature, and generator speed. Another network called an FSRC (full signal reconstruction) was also used to reconstruct the signal without previous bearing temperature (other inputs remained the same as autoregressive NN). The raw data were preprocessed for scaling/normalization, validity check, missing data, and lag removal (lag between changes of operational conditions and wind turbine signals) before input to the NNs. Three months of operational data (averaged over 10 minutes) were used for training the NNs. Figure 8.25 shows prediction error (averaged over one day) for bearing temperature using the FSRC model. It is observed that the error increases continuously after about first 100 days till catastrophic bearing damage. Results from autoregressive NN model are similar. Thus the NN approaches in Schlechtingen and Santos (2011) were able to correctly identify incipient faults in wind turbine bearing through online SCADA data.

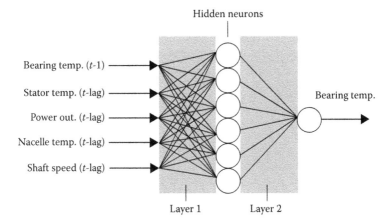

FIGURE 8.24 Autoregressive NN architecture for bearing temperature prediction. (From Schlechtingen, M., and Santos, I. F., 2011, *Mech. Syst. Signal Process.*, 25, 1849–1875. With permission.)

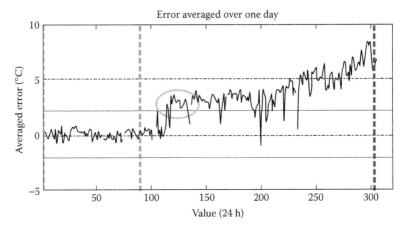

FIGURE 8.25 Averaged prediction error for bearing temperature using FSRC NN model. (From Schlechtingen, M., and Santos, I. F., 2011, *Mech. Syst. Signal Process.*, 25, 1849–1875. With permission.)

GA APPLICATIONS IN SHM

The use of GAs in SHM is not as extensive as the applications of NNs; however, a significant body of work is available in the published literature. This section presents only a sample of published results as illustrative examples. GAs have been applied for detecting minor damages in three-dimensional (3D) structures (Cha and Buyukozturk, 2015), condition monitoring of rotating mechanical systems (Saxena and Saad, 2007), identification of increasing levels of damage in airplane structures (Meruane and Heylen, 2010), and optimal sensor placement for damage detection (Yi et al., 2011).

Cha and Buyukozturk (2015) proposed a hybrid multiobjective genetic algorithms (GAs) to solve inverse problems for 3D structures in order to detect minor damage in multiple locations. Traditional damage detection methods based on modal properties have difficulty in detecting minor damages because they have little effect on the difference of the modal properties of the structure. Cha and Buyukozturk (2015) used modal strain energy (MSE) as a damage index and created various minor damage scenarios for the 3D structures to investigate the proposed multiobjective GAs. They showed that the method detects the exact locations and extents of the induced minor damages in the structure. The robustness of the proposed method was also investigated by adding 5% Gaussian random white noise as signal distortions in the computation of mode shapes used in the evaluation of MSE.

Saxena and Saad (2007) used GAs for optimizing NN design parameters for condition monitoring of mechanical systems. It was shown that GA-based selection of optimal NN features produced very good results in terms of training accuracy and classification success, even for vast search spaces. They performed experiments with constant crossover and mutation parameters to compare results between evolved NNs and fixed size NNs. Further, the number of hidden nodes, the number of input nodes, and the connection matrix can be evolved using GAs for adaptive NNs. The authors (Saxena and Saad, 2007) concluded that this technique may be applied to find out a desired set of good features for complex systems such as planetary gears where some faults cannot be easily distinguished.

Meruane and Heylen (2010) investigated model-based damage detection as a constrained nonlinear optimization problem. They used parallel GAs for a superior numerical performance in the evaluation of complicated or time-consuming objective functions. The GAs were used to explore the entire solution space and avoid local minima (generally encountered by conventional optimization approaches). Furthermore, GAs were not very sensitive to experimental noise or numerical errors. The objective function was based on operational modal data. The technique was verified with two experimental cases: aircraft structure subjected to three increasing levels of damage and a multiple cracked reinforced concrete beam subjected to a nonsymmetrical increasing static load. The parallel GA-based damage detection method was fast, detected experimental damage successfully, and avoided false damage detection, despite the presence of experimental noise or modeling errors.

Yi et al. (2011) used "generalized genetic algorithm" (GGA) to determine optimal sensor placement (OSP) during modal tests for SHM. The OSP plays an important role in SHM of large-scale structures. The dual-structure coding method (instead of binary coding) was implemented in this research, involving selection scheme, crossover strategy, and mutation mechanism. The tallest building in the north of China was used as the test structure. The sensor placements obtained by the GGA showed improved convergence and a better placement scheme compared to those by the exiting genetic algorithm.

CONCLUDING REMARKS

It is reflected from the review that the use of NNs for building models from data is growing steadily in the field of structural damage detection. Building such models requires intimate understanding of the data and knowledge of available NNs models and their applicability. A systematic approach that ensures those all-important aspects of the modeling can lead to building a good quality model. In this chapter, we discussed about commonly available NN models for classification and regression. Emphasis was given toward more frequently used BP-based NNs model. This chapter has raised some of the following important points in the context of their applicability to structural damage detection.

1. NN performance depends on the range of training samples and the scatter of training sample.
2. Usually larger amount of information (e.g., number of variables) in the input arrays does not necessarily improve the results of the network.
3. The configuration of NN architecture usually requires trial-and-error method, which is very time consuming. Although the processing of an NN encodes knowledge, the knowledge representation is vague and not easily understood. Unlike expert systems, NNs cannot explicitly explain their results. For an NN, it is very difficult to provide the end user with convincing explanations about how a decision was made.
4. As discussed in this chapter, an ANN is an excellent data-based prediction model. They are used owing to their demonstrated capabilities to create mappings from input to output data even if the data are noisy and when no model of the data exists. They can handle a variety of data types as well as time-variable data. Careful empirical evaluation of models created by the ANN is critical to selecting ANN architecture, setting their parameters, as well as to selecting between different ANN approaches. Unfortunately, from study it was found that there is little awareness to evaluation issues within the NN community (Reich and Barai, 1999). Proper evaluation of NNs model for structural identification is essential before putting it into practice.
5. Using NNs, it is possible to detect the location of damage and its extent if network is properly trained.
6. First, for real-world structures, the DoF of the system could be very large. It is believed that by performing several stages of model reduction, the DoF of the final design model might be relatively smaller in number. Second, the approach using the dynamic residual vector may be applied to such reduced models. Third, the role of noise in actual modal data measurement should be investigated; where minor damage is concerned, the noise could cause some false estimations. Finally, determination of the appropriate representation for damage information to reliably and precisely identify various states of damages requires further investigation. It is anticipated that an NN-based identification technique will be improved if a proper representation can be developed.
7. From the literature survey, it is obvious that the design of a reliable ANN is yet an unresolved issue in the field of damage assessment. A systematic study on various aspects of network simulation is needed to be carried out, here using the BP algorithm for supervised learning in MLP.
8. Most researchers have used BP NNs, although not many papers compared the performance of two different NN types. Most of these papers contained attempts to identify damage from information-related structural response and also, they have used change in member shapes and/or cross-sectional areas as damage. This is not usually expected in real structures. Also, many papers assume detailed knowledge of the mechanical structure including

mass and stiffness matrices. A few performed the identification of system parameters based on measured data so that no detailed knowledge of the structure was assumed. Generalizations of these nonmodel-based methods would seem to be more useful for practical applications.

9. The advantage of using NNs instead of traditional system identification methods is that once the network is trained, it requires virtually no computation time to identify the structural damage. This is a crucial advantage of the approach, especially for online damage detection problems.

10. Apart from BP algorithm-based network, many researchers have explored other algorithms for structural damage detection. It would be worthwhile, after one decade of NNs in this field, to take the stock of the present scenario and form certain guidelines to use them successfully in the field.

Today, there are neither standardized and reliable methods nor quantitative and objective guidelines for the in-service SHM of high safety and cost structures using NNs. Such structures are large in size, expensive to build, and difficult to maintain. Also they are susceptible to eventual and sudden damage. Typical examples of them are found among others in the construction, aerospace, off-shore, power generation, machine tool, and transport industries: highway bridges, dome-shaped roof trusses, spacecraft antennas, flight and railway vehicles, offshore structures, and so on. The ultimate objective of researchers in India must be to develop methods, procedures and guidelines of NNs for routine, nondestructive, in-service maintenance monitoring, and damage detection in these types of structures. In the context of India, following future problems can be taken up in the domain of structural identification.

1. Development of NNs system for online estimation of the cumulative damage at several critical points in structures. (e.g., bridges, dams, nuclear reactors, roads, etc.)

2. Drawbacks of conventional BP NNs have been overcome by TDNN, where patterns may vary over time and that require a period of time to be presented to network in the problems of vibration signature analysis. The work by Barai and Pandey (1997) was limited to simulated data of Indian railway steel bridge. It would be interesting to explore this kind of network for real-time applications to highway and railway bridges in India.

3. There is a need for carrying out experiments on incremental learning to compare performance of BP algorithm and other models such as cascade correlation, ART, and so on Structural identification systems should be capable of broadening their horizon and evolving to cope up with new information/situations. This is very important because the number of possible categories and new situations are unknown in several years of structural operation time. ANN can support the development of these systems.

4. Researchers have proposed that instead of using commonly used BP-based NNs; one can use hierarchy of NNs. This approach significantly reduces the computational requirements. Also, the approach is very well suited for parallel processing implementation in both hardware and software where each NN node can be an independent processor.

5. BP has been used as a supervised learning algorithm for structural identification. The convergence speed of this algorithm is often too slow. Several hours or even days of computing time is often required to train the NNs using the conventional serial workstations. Further, the total number of iterations for learning an example in NN is also in the order of thousands. The development of learning algorithms on general-purpose parallel computers such as PARAM 10000 with objective of reducing the overall computing time can be one of the approaches.

6. Currently, most of the NN applications are simulated in conventional computers. The next step will prospectively be the application of special VLSI (very large-scale integrated) neural chips to further accelerate the computational speed of NNs.

7. NN techniques can contribute to the further development of intelligent sensors for smart civil structures, where sensing, preprocessing, and decision making will be integrated into one unit.

8. In the context of fire hazards damage, NNs can be of great help and can be utilized for evaluating the same.

NNs are certainly the key technology for future intelligent engineering structures, provided applied appropriately.

REFERENCES

Adeli, H., and Yeh, C. (1989) "Perceptron learning in engineering design," *Computer-Aided Civil and Infrastructure Engineering*, 4, 247–256.

ASCE (2000a) "Artificial neural networks in hydrology. I: Preliminary concepts," *Journal of Hydrologic Engineering*, 5, No. 2, 115–123.

ASCE (2000b) "Artificial neural networks in hydrology. II: Hydrologic applications," *Journal of Hydrologic Engineering*, 5, No. 2, 124–137.

Barai, S. V. (1999) "The soft computing tool—Neural networks: An introduction and its applications to earthquake engineering," *Lecture notes for AICTE/ISTE—Short Term Course on Random Vibration and Applications to Earthquake Engineering*, December 21–31, 1999, IIT Kharagpur, Kharagpur, India, pp:1–31.

Barai, S. V. (2000) "Bibliography—Neural networks applications to damage detection." Available at: http://barai.sudhir.tripod.com/bibliography.html.

Barai, S. V., and Pandey, P. C. (1995a) "Performance of generalized delta rule in artificial neural networks for structural damage detection," *Engineering Applications in Artificial Intelligence*, 8, No. 2, 211–221.

Barai, S. V., and Pandey, P. C. (1995b) "Vibration signature analysis using artificial neural networks," *Journal of Computing in Civil Engineering*, ASCE, 9, No. 4, 259–265.

Barai, S. V., and Pandey, P. C. (1997) "Time-delay neural networks in damage detection of railway bridges," *Advances in Software Engineering*, 28, 1–10.

Berke, L., Patnaik, S. N., and Murthy, P. L. N. (1993) "Optimum design of aerospace structural components using neural networks," *Computers and Structures*, 48, No. 6, 1001–1010.

BrainMaker. (1989) Simulated Biological Intelligence User's Guide and Reference Manual, Third Edition, California Scientific, Sacramento, CA.

Brown, D. A., Murthy, P. L. N., and Berke, L. (1991) "Computational simulation of composite ply micromechanics using artificial neural networks," *Microcomputers in Civil Engineering*, 6, 87–97.

Carpenter, G., and Grossberg, S. (1987) "ART-2: Self-organization of stable category recognition codes for analog input patterns," *Applied Optics*, 26, No. 23, 4919–4930.

Ceravolo, R., and De Stefano, A. (1995) "Damage location in structures through a connectivistic use of FEM modal analyses," *International Journal of Analytical and Experimental Modal Analysis*, 10, No. 3, 176.

Cha, Y.-J., and Buyukozturk, O. (2015) "Structural damage detection using modal strain energy and hybrid multiobjective optimization," Computational Intelligence in Structural Engineering and Mechanics, *Computer-Aided Civil and Infrastructure Engineering*, 30, No. 5 (Special Issue), 347–358.

Cheng, M., and Popplewell, N. (1994) "Neural networks for earthquake selection in structural time history analysis," *Earthquake Engineering and Structural Dynamics*, 23, 303–319.

Dayoff, J. E. (1990) *Neural Networks Architecture: An Introduction*, Van Nostrand Reinhold, New York.

Doebling, S. W., Farrar, C. R., and Prime, M. B. (1998). "A summary review of vibration-based damage identification methods," *Shock and Vibration Digest*, 30, No. 2, 91–105.

Elkordy, M. F., Chang, K. C., and Lee, G. C. (1993) "Neural networks trained by analytically simulated damaged states," *ASCE Journal of Computing in Civil Engineering*, 7, 2, 130–145.

Elkordy, M. F., Chang, K. C., and Lee, G. C. (1994) "Applications of neural networks in vibration signature analysis," *ASCE Journal of Engineering Mechanics*, 120, No. 2, 251–265.

Flood, I., and Kartam, N. (1994a) "Neural networks in civil engineering I: Principles and understanding," *ASCE Journal of Computing in Civil Engineering*, 8, No. 2, 131–148.

Flood, I., and Kartam, N. (1994b) "Neural networks in civil engineering II: systems and applications," *ASCE Journal of Computing in Civil Engineering*, 8, No. 2, 149–162.

Flood, I., and Kartam, N. (1998) *Artificial Neural Networks for Civil Engineers: Advanced Features and Applications*, ASCE Publications, New York.

Fu, B., and Hajela, P. (1993) "Minimizing distortion in truss structures—A Hopfield network solution," *Computing Systems in Engineering*, 4, No. 1, 69–74.

Fwa, T. F., and Chan, W. T. (1993) "Priority rating of highway maintenance needs by neural networks," *ASCE Journal of Transportation Engineering*, 119, No. 3, 419–432.

Gagarin N., Flood, I., and Albrecht, P. (1994) "Computing truck attributes with artificial neural networks," *ASCE Journal of Computing in Civil Engineering*, 8, No. 2, 179–200.

Gan, M., and Liu, X. (1991) "Neural networks in structural preliminary design," in *Civil-Comp-91, Artificial Intelligence and Structural Engineering*, Civil-Comp Press, Oxford, pp. 285–293.

Ghaboussi, J., Garrett, J. H., and Wu, X. (1991) "Knowledge based modeling of material behaviour with neural networks," *ASCE Journal of Engineering Mechanics*, 117, No. 1, 132–153.

Giurgiutiu, V. (2008) *Structural Health Monitoring with Piezoelectric Wafer Active Sensors*, Academic Press, Waltham, MA.

Goh, A. T. C. (1994) "Seismic liquefaction potential assessed by neural networks," *ASCE Journal of Geotechnical Engineering*, 120, No. 9, 1467–1480.

Hajela, P., and Berke, L. (1991) "Neurobiological Computational models in structural analysis and design," *Computers and Structures*, 41, No. 4, 657–667.

Hajela, P., Fu, B., and Berke, L. (1991) "ART networks in automated conceptual design of structural systems," in *Civil-Comp-91, Artificial Intelligence and Structural Engineering*, Civil-Comp Press, Oxford, pp. 263–270.

Haykin, S. (1999), *Neural Networks: A Comprehensive Foundation*, 2nd edn., Prentice-Hall, Upper Saddle River, NJ.

Hecht-Nielsen, R. (1988) "Applications of counterpropagation networks," *Neural Networks*, 1, No. 2, 131–139.

Hegazy, T., and Moselhi, O. (1994) "Analogy based solution to markup estimation problem," *ASCE Journal of Computing in Civil Engineering*, 8, No. 1, 72–87.

Hinton, G. E., and Sejnowski, T. J. (1986) "Learning and relearning in Boltzmann machines," *Parallel Distributed Processing*, Vol. 1, MIT Press, Cambridge, MA, pp. 282–317.

Hoit, M., Stoker, D., and Cousolazio, G. (1994) "Neural networks for equation renumbering," *Computers and Structures*, 52, No. 5, 1011–1021.

Hopfield, J. J. (1982) "Neural networks and physical systems with emergent collective computational abilities," *Proceedings of the National Academy Sciences*, 79, 2554–2558.

Hopfield, J. J. (1984) "Neurons with graded response have collective computational properties like those of two-state neurons," *Proceedings of the National Academy Sciences*, 81, 3088–3092.

Hu, Y. H., and Hwang, J.-N. (Ed.) (2001), *Handbook of Neural Network Signal Processing*, CRC Press, Boca Raton, FL.

Jha, R., and He, C. (2004), "Adaptive neurocontrollers for vibration suppression of nonlinear and time varying structures," *Journal of Intelligent Material Systems and Structures*, 15, No. 9–10, 771–781.

Kamarthi, S. V., Sanvido, V. E., and Kumara, R. T. (1992) "NEUROFORM: Neural network systems for vertical formwork selection," *ASCE Journal of Computing in Civil Engineering*, 6, No. 2, 178–199.

Kamath, G. M., Sundaram, R., Gupta, N., and Rao, M. S. (2010) "Damage Studies in Composite Structures for Structural Health Monitoring using Strain Sensors," *Structural Health Monitoring*, 9, No. 6, 497–512.

Kartam, N., Flood, I., and Garrett, J. H. (1997) *Artificial Neural Networks for Civil Engineers: Fundamentals and Applications*, ASCE Publication, New York.

Karunanithi, N., Grenney, W. J., Whitley, D., and Bovee, K. (1994) "Neural networks for river Ow prediction," *ASCE Journal of Computing in Civil Engineering*, 8, No. 2, 201–220.

Katsikeros, Ch. E., and Labeas, G. N. (2009) "Development and validation of a strain-based Structural Health Monitoring system," *Mechanical Systems and Signal Processing*, 23, 372–383.

Kirkegaard, P., Nielsen, S., and Hansen, H. (1995) "Identification of non-linear structures using recurrent neural networks," in *Proceedings of the 13th International Modal Analysis Conference*, Nashville, TN, pp. 1128–1134.

Kirkegaard, P., and Rytter, A. (1994) "Use of neural networks for damage assessment in a steel mast," in *Proceedings of the 12th International Modal Analysis Conference*, Honolulu, Hawaii, pp. 1128–1134.

Klenke, S. E., and Paez, T. L. (1996) "Damage identification with probabilistic neural networks," in *Proceedings of the 14th International Modal Analysis Conference*, Dearborn, MI, pp. 99–104.

Kortesis, S., and Panagiotopoulos, P. D. (1993) "Neural networks for computing in structural analysis methods and prospects of applications," *International Journal for Numerical Methods in Engineering*, 36, 2305–2318.

Kosko, B. (1987) "Bi-directional associative memories," *IEEE Transactions on Systems, Man, and Cybernatics*, 18, No. 1, 49–60.

Kudva, J., Munir, N., and Tan, P. (1991) "Damage detection in smart structures using neural networks and finite element analysis," in *Proceedings of ADPA/AIAA/ASME/SPIE Conference on Active Materials and Adaptive Structures*, Alexandria, VA, pp. 559–562.

Leath, W. J., and Zimmerman, D. C. (1993) "Analysis of neural network supervised training with application to structural damage detection," in *Proceedings of the 9th VPI&SU Symposium on Damage and Control of Large Structures*, Blacksburg, VA, pp. 583–594.

Lippman, R. P. (1987) "An introduction to computing with neural nets," *IEEE ASSP Magazine*, April, pp. 6–22.

Manning, R. (1994) "Damage detection in adaptive structures using neural networks," in *Proceedings of the 35th AIAA/ASME/ASCE/AHS/ASC Structures, Structural Dynamics, and Materials Conference*, Hilton Head, SC, pp. 160–172.

Marwala, T. (2000) "Damage identification using committee neural networks," *ASCE Journal of Engineering Mechanics*, 126, No. 1, 43–50.

Masri, S. F., Chassiakos, A. G., and Caughey, T. K. (1993) "Identification of nonlinear dynamic systems using neural networks," *Journal of Applied Mechanics*, 60, No. 3, 123–133.

Meruane, V., and Heylen, W. (2010) "Damage detection with parallel genetic algorithms and operational modes," *Structural Health Monitoring*, 9, No. 6, 481–496.

Mitchell, M. (1996) *An Introduction to Genetic Algorithms*, MIT Press, Cambridge, MA.

Moll, J., Schulte, R. T., Hartmann, B., Fritzen, C.-P., and Nelles, O. (2010) "Multi-site damage localization in anisotropic plate-like structures using an active guided wave structural health monitoring system," *Smart Materials and Structures*, 19, 045022–045038.

Montalvao, D., Maia, N. M. M., and Ribeiro, A. M. R. (2006) "A review of vibration based structural health monitoring with special emphasis on composite materials," *Shock and Vibration Digest*, 38, No. 4, 295–324.

Moselhi, O., Hegazy, T., and Fazio, P. (1991) "Neural networks as tools in construction," *ASCE Journal of Construction Engineering and Management*, 117, No. 4, 606–625.

Murtaza, M. B., and Fisher, D. J. (1994) "Neuromodex—Neural network system for modular construction decision making," *ASCE Journal of Computing in Civil Engineering*, 8, No. 2, 221–233.

Nakamura, M., Masri, S. F., Chassiakos, A. G., and Caughey, T. K. (1998) "A method for non-parametric damage detection through the use of neural networks," *Earthquake Engineering and Structural Dynamics*, 27, 997–1010.

Narendra, K. S., and Parthasarathy, K. (1990) "Identification and control of dynamical systems using neural networks," *IEEE Transactions on Neural Networks*, 1, No. 1, 4–27.

Natke, H. G., and Yao, J. T. P. (1988). *Structural Safety Evaluation Based on System Identification Approaches*, Friedr Vieweg and Sohn, Braunschweig, Wiesbaden.

Nelson, M. M., and Illingworth, W. T. (1991) *A Practical Guide to Neural Nets*, Prentice Hall, Upper Saddle River, NJ.

Pandey, P. C., and Barai, S. V. (1994) "Nonlinear analysis of plates using artificial neural networks," *Journal of Structural Engineering* (India), 21, No. 1, 65–78.

Pandey, P. C., and Barai, S. V. (1995) "Multilayer perceptron in damage detection of bridge structures," *Computers and Structures*, 54, No. 4, 597–608.

Perry, J. L., and Baumgardt, D. R. (1991) "Seismic event identification using artificial neural networks," in *Proceedings of the IEEE 7th Conference on AI Applications*, Miami, FL, pp. 369–374.

Pidaparti, R. M. V., and Palakal, M. J. (1993) "Material model for composites using neural networks," *AIAA Journal*, 31, No. 8, 1533–1535.

Povich, C., and Lim, T. (1994) "An Artificial Neural Network Approach to Structural Damage Detection Using Frequency Response Functions," in *Proceedings of the 35th AIAA/ASME/ASCE/AHS/ASC Structures, Structural Dynamics, and Materials Conference*, Hilton Head, SC, pp. 151–159.

Raghavan, A., and Cesnik, C. E. S. (2007) "Review of guided-wave structural health monitoring," *Shock and Vibration Digest*, 39, 91–114.

Ramadas, C., Balasubramaniam, K., Joshi, M., and Krishnamurthy, C. V. (2008) "Detection of transverse cracks in a composite beam using combined features of Lamb wave and vibration techniques in ANN environment," *International Journal on Smart Sensing and Intelligent Systems*, 1, No. 4, 970–984.

Reed, S. (2009) "Artificial neural networks," in Boller, C., Chang, F.-K., and Fujino Y. (Eds.), *Encyclopedia of Structural Health Monitoring*, Vol. 2, John Wiley & Sons, Hoboken, NJ, pp. 611–624.

Reich, Y., and Barai, S. V. (1999) "Evaluating machine learning models for engineering problems," *Artificial Intelligence in Engineering*, 13, 257–272.

Rhim, J., and Lee, S. (1994) "A neural network approach for damage detection and identification of structures," in *Proceedings of the 35th AIAA/ASME/ASCE/AHS/ASC Structures, Structural Dynamics, and Materials Conference*, Hilton Head, SC, pp. 173–180.

Rogers, J. L. (1994) " Simulating structural analysis with neural networks," *ASCE Journal of Computing in Civil Engineering*, 8, No. 2, 252–265.

Rosenblatt, F. (1961) *Principles of Neurodynamics: Perceptrons and the Theory of Brain Mechanisms*, Spartan Books, Washington, DC.

Rumelhart, D. E., Hinton, G. E., and Williams, R. J. (1986) "Learning internal representations by error propagation," in Rumelhart D. E., and McCelland, J. L. (Eds.), *Parallel Distributed Processing: Exploration in the Microstructure of Cognitron, Vol. 1: Foundations*, MIT Press, Cambridge, Reading MA, pp. 318–362

Saxena, A., and Saad, A. (2007) "Evolving an artificial neural network classifier for condition monitoring of rotating mechanical systems," *Applied Soft Computing*, 7 (1), 441–454.

Schlechtingen, M., and Santos, I. F. (2011) "Comparative analysis of neural network and regression based condition monitoring approaches for wind turbine fault detection," *Mechanical Systems and Signal Processing*, 25, 1849–1875.

Schwarz, B. J., McHargue, P. L., and Richardson, M. H. (1996) "Using SDM to train neural networks for solving modal sensitivity problems," in *Proceedings of the 14th International Modal Analysis Conference*, Dearborn, MI, pp. 1285–1291.

Simpson, P. K. (1992) "Foundations of neural networks," in Sanchez-Sinencio, E. and Lau, C. (Eds.), *Artificial Neural Networks—Paradigms, Applications and Hardware Implementations*, IEEE Press, Piscataway, NJ.

Sohn, H., Farrar, C. R., Hemez, F. M., Shunk, D. D., Stinemates, D. W., and Nadler, B. R. (2003) "A review of structural health monitoring literature: 1996–2001," *Los Alamos National Laboratory Report*, LA-13976-MS. Los Alamos National Laboratory, Los Alamos, NM.

Spillman, W., Huston, D., Fuhr, P., and Lord, J. (1993) "Neural network damage detection in a bridge element," *SPIE Smart Sensing, Processing, and Instrumentation*, SPIE Vol. 1918, 288–295.

Stephens, J. E., and VanLuchene, R. D. (1994) "Integrated assessment of seismic damage in structures," *Microcomputers in Civil Engineering*, 9, 119–128.

Su, Z., and Ye, L. (2009) *Identification of Damage Using Lamb Waves: From Fundamentals to Applications*, Lecture Notes in Applied and Computational Mechanics, Volume 48, Springer, Berlin, Germany.

Szewczyk, Z. P., and Hajela, P. (1992) *Neural network based damage detection in structure*, Technical Report, RPI, Troy, NY.

Szewczyk, Z. P., and Hajela, P. (1994) "Damage detection in structures based on feature sensitive neural networks," *ASCE Journal of Computing in Civil Engineering*, 8, No. 2, 163–178.

Taylor, J. G., and Mannion, C. L. T. (1989) *New Developments in Neural Computing*, Taylor & Francis, New York, NY.

Topping, B. H. V., and Bahreininejad, A. (1997) *Neural Computing for Structural Mechanics*, Saxe-Coburg Publications, Edinburgh, UK.

Tsou, P., and Shen, M.-H. H. (1994) "Structural damage detection and identification using neural networks," *AIAA Journal*, 32, No. 1, 176–183.

Tung, A. T. Y., Wang, Y. Y., and Wong, F. S. (1993) "Assessment of liquefaction potential using neural networks," *Soil Dynamics and Earthquake Engineering*, 12, 325–335.

Vanluchene, R. D., and Sun, R. (1990) "Neural networks in structural engineering," *Microcomputers in Civil Engineering*, 5, 207–215.

Wasserman, P. D. (1989) *Neural Computing: Theory and Practice*, Van Nostrand Reinhold, New York.

Worden, K., Ball, A., and Tomlinson, G. (1993) "Neural networks for fault location," in *Proceedings of the 11th International Modal Analysis Conference*, Kissimmee, FL, pp. 47–54.

Wu, X., Garrett, J. H., and Ghaboussi, J. (1990) "Representation of material behaviour—Neural networks based models," *IEEE IJCNN-90*, Vol. I, pp. 229–234.

Wu, X., Ghaboussi, J., and Garrett, J. H. (1992) "Use of neural networks in detection of structural damage," *Computers and Structures*, 42, No. 4, 649–659.

Yeh, Y., Kuo, Y., and Hsu, D. S. (1993) "Building KBES for diagnosing PC piles with artificial neural network," *ASCE Journal of Computing in Civil Engineering*, 7, No. 1, 71–93.

Yen, G. G. (1994) "Identification and control of large structures using neural networks," *Computers and Structures*, 52, No. 5, 859–870.

Yen, G. G., and Kwak, M. K. (1993) "Neural network approach for the damage detection of structures," in *Proceedings of 34th AIAA/ASME/ASCE/AHS/ASC Structures, Structural Dynamics, and Materials Conference*, La Jolla, CA, pp. 1549–1555, AIAA-93-1485-CP.

Yi, T.-H., Li, H.-N., and Gu, M. (2011) "Optimal sensor placement for health monitoring of high-rise structure based on genetic algorithm," *Mathematical Problems in Engineering*, Article ID 395101, 12 pages, doi:10.1155/2011/395101.

Zeng, P. (1998) "Neural computing in mechanics," *Applied Mechanics Review*, 51, No. 2, 173–197.

9 Development of Fracture and Damage Modeling Concepts for Composite Materials

Ayad Arab Ghaidan Kakei, Jayantha Ananda Epaarachchi, M. Mainul Islam, and Jinsong Leng

CONTENTS

INTRODUCTION

The propagation and interaction of distributed microcracks are important conditions to lead to failure in brittle or quasi-brittle material. The interaction between microcracks that distributed in the brittle or quasi-brittle materials may cause the influence between microcracks; as a result, the mechanical properties of materials observably decrease. Although the influence of microcrack interaction has been extensively argued, it has not been resolved yet.

The microcracking in the elastic medium has been identified as the major damage-causing mechanism in brittle materials. Although the undamaged brittle material behaves isotropically, due to the result of microcracking, the brittle material under external loads in tension and compression responds differently (Su et al. 2007). The resulting anisotropic effective moduli of microcracked brittle material will differ depending on the magnitude and location and direction of the boundary loads, so the prediction of these changes will help in the understanding of the anisotropic damage.

EFFECT OF MICROCRACKS ON EFFECTIVE MODULI IN BRITTLE MATERIALS

Although various schemes have been proposed to describe the effective moduli of microcracked solid, the noninteracting microcracks are almost neglected in the simplest method such as Taylor's method or dilute concentration method. These methods could yield quite accurate result in some cases. However, using other methods such as self-consistent method (SCM), Mori–Tanaka method, differential method, and generalized SCM (GSCM), the effective moduli may be estimated for microcrack interaction and others. Feng and Yu (2000) pointed out that most of these methods' applications are limited to solids that are statistically homogeneous and subjected to uniform tractions or displacements. Those models are difficult to be incorporated into a model for estimating the effective moduli of materials with profuse microcracks. For coping the difficulty, the Mori–Tanaka method provided that the randomly microcracks' presence does not change the mean stress within the materials. Other methods such as SCM and GSCM have little advantage offer in more general cases with cumbersome derivations or numerical computations that are often encountered.

Disorder microcracks in brittle or quasi-disorder is the cause for many deformation and fracture mechanics. Over the past few decades, many theoretical models and methods are presented to solve this problem. Although many phenomenological and micromechanical damages have been established, still there are many unresolved problems. The major remaining problems were attributed as sensitivity of the materials to complex loading and mechanics of crack formation and development of secondary cracks into consideration of modeling (Xi-Qiao et al. 2004).

The growth and coalescence of microcracks are usually incorporated into a damage matrix of brittle and quasi-brittle solids. Because of the microcracks in brittle or quasi-brittle materials, the mechanical behavior shifts to the heterogeneous microstructures (Xi-Qiao et al. 2004).

Shen and Yi (2000) presented new equations to evaluate the effective bulk and shear moduli of solids with randomly dispersed cracks. They studied the effective moduli of linear elastic isotropic solids with randomly oriented microcracks as shown in Figure 9.1. The potential energy released by embedding a circular or spherical representative volume element (RVE) with microcracks into an infinite matrix is equal to that induced by introducing its effective medium into the identical infinite matrix.

$$\Delta f_{\text{effective}} = \Delta f_{\text{micro}} \tag{9.1}$$

where $\Delta f_{\text{effective}}$ and Δf_{micro} are the potential energies released by the effective medium and the microcracks embedded in the infinite matrix, respectively. $\Delta f_{\text{effective}}$ can be obtained by Eshelby's (1957) method. Two independent equations for the effective plane bulk modulus K and the effective shear modulus G of the microcracks under two remote loading conditions such as hydrostatic stress σ_k^0 and in-plane pure shear σ_G^0 were obtained from Equation 9.1 as

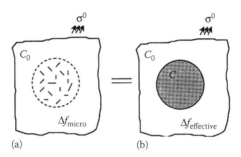

FIGURE 9.1 A new energy balance for the analysis of effective moduli of microcracked solids: (a) microcracks in infinite matrix and (b) microcracks' effective medium in the infinite matrix. (From Shen, L, and Yi, S, *International Journal of Solids and Structures*, 37, 3525–3534, 2000. With permission.)

$$\frac{1}{2K_0}\frac{K-K_0}{K+\xi(k-K_0)}=\frac{1}{\left(\sigma_k^0\right)^2}\frac{1}{A}\Delta f_{micro} \tag{9.2}$$

and

$$\frac{1}{2G_0}\frac{G-G_0}{G+\eta(G-G_0)}=\frac{1}{\left(\sigma_G^0\right)^2}\frac{1}{A}\Delta f_{micro} \tag{9.3}$$

where:

$K_0 = E_0/\left[2(1+v_0)(1-2v_0)\right]$ and $G_0 = E_0/\left[2(1+v_0)\right]$ denote the plane strain bulk and shear moduli of the matrix, respectively

E_0 and v_0 are Young's modulus and Poisson's ratio of the matrix, respectively

For two-dimensional (2D) plane strain solutions $\xi = 1/2(1 - v_0)$ and $\eta = (3 - 4v_0)/[4(1 - v_0)]$, and for three-dimensional (3D) plane strain solution, they replaced the circular RVE in two dimensions with the spherical RVE, $\xi = (1+v_0)/[3(1-v_0)]$ and $\eta = (8-10v_0/[15(1-v_0)]$, respectively. As solving microcrack interaction problems was difficult, the approximation of noninteracting crack solution was assumed to overcome the difficulty. Δf_{micro} for 2D and 3D isotropic problems, respectively, was written as (Kachanov 1992)

$$\Delta f_{micro} = -A\left(\frac{\pi}{2E_0'}\right)\rho\sigma_{ij}^0\sigma_{ij}^0 \tag{9.4}$$

and

$$\Delta f_{micro} = -V\frac{8(1-v_0^2)}{9(1-(v_0/2))E_0}\rho\left[\left(1-\frac{v_0}{5}\right)\sigma_{ij}^0\sigma_{ij}^0-\frac{v_0}{10}\left(\sigma_{kk}^0\right)^2\right] \tag{9.5}$$

where V is the volume of the spherical RVE and $E_0' = E_0/(1-v_0^2)$.

Substituting Equations 9.4 and 9.5 into the present energy balance equations (9.2 and 9.3) yielded a new noninteracting solution for 2D and 3D problems:

$$\frac{K}{K_0} = \frac{1}{\left[1+\left(1-v_0/1-2v_0\right)\pi\rho/\left(1-\left(\pi/2\right)\rho\right)\right]} \tag{9.6}$$

$$\frac{G}{G_0} = \frac{1}{\left[1+\left(1-v_0\right)\pi\rho/\left(1-\left(\pi/4\right)\rho\right)\right]} \tag{9.7}$$

and

$$\frac{k}{k_0} = \frac{1}{1+\left(16/9\right)\left(1-v_0^2/1-2v_0\right)\rho/\left[1-\left(32/27\right)\left(1+v_0\right)\rho\right]} \tag{9.8}$$

$$\frac{G}{G_0} = \frac{1}{1+\left(16/9\right)\left[1-v_0/\left(1-v_0/2\right)\right]\left(1-v_0/5\right)\rho/\left[1-\left(16/135\right)\left(7-5v_0/\left(1-v_0/2\right)\right)\left(1-v_0/5\right)\rho\right]} \tag{9.9}$$

The noninteracting solution completely neglects the interactions among cracks, but the SCM solutions take the interactions among cracks. The SCM solution is assumed that each crack is embedded into the circular or spherical RVE as the unknown effective media. The noninteracting solutions for the bulk and shear moduli are quite different from the conventional noninteracting solutions. A new SCM solution for 2D and 3D microcracked solids problems was

$$\bar{K} = 1-\frac{\pi\rho}{2}\left[1+\frac{1}{1-2v_0}\frac{\bar{K}}{\bar{G}}\right]\frac{\bar{K}}{1+\left(1/\left[2(1-v_0)\right]\right)\left(\bar{K}-1\right)} \tag{9.10}$$

$$\bar{G} = 1-\frac{\pi\rho}{2}\left[1+\frac{1}{1-2v_0}\frac{\bar{G}}{\bar{K}}\right]\frac{\bar{G}}{1+\left(3-4v_0\right)/\left[4(1-v_0)\right]\left(\bar{G}-1\right)} \tag{9.11}$$

and

$$\bar{K} = 1-\frac{16}{9}\rho\frac{1-v^2}{1-2v}\frac{\bar{K}}{1+\left(1+v_0\right)/\left[3(1-v_0)\right]\left(\bar{K}-1\right)} \tag{9.12}$$

$$\bar{G} = 1-\frac{16}{9}\rho\frac{1-v}{1-v/2}\left(1-v/5\right)\frac{\bar{G}}{1+\left(8-10v_0\right)/\left[15(1-v_0)\right]\left(\bar{G}-1\right)} \tag{9.13}$$

$$v = \frac{\left[\bar{K} - \left(1 - 2v_0/1 + v_0\right)\bar{G} \right]}{\left[2K + \left(1 - 2v_0/1 + v_0\right)\bar{G} \right]} \tag{9.14}$$

where:

$\bar{K} = k/K_0$ and $\bar{G} = G/G_0$ are the normalized effective plane strain bulk and shear moduli for 2D problems or the normalized effective bulk and shear moduli for 3D problems

v_0 and v are Poisson's ratios of the matrix and the effective medium, respectively

New noninteracting and SCM solutions were compared with various existing solutions (noninteracting, differential, and SCM) as shown in Figures 9.2 through 9.5. The result showed that at the high-density cracks both bulk and shear moduli are equal to zero because if crack density approaches infinity $(\rho \rightarrow \infty)$, the effective moduli of the RVE become zero. In addition, the result of shear moduli in the noninteracting solution is generally lower than that in the present SCM solution. However, both solutions cross each other in low crack density for shear moduli as shown in Figures 9.2 through 9.5.

Chen (2012) presented a novel numerical solution for the effective elastic moduli for 2D cracked medium by using the eigenfunction expansion variational method

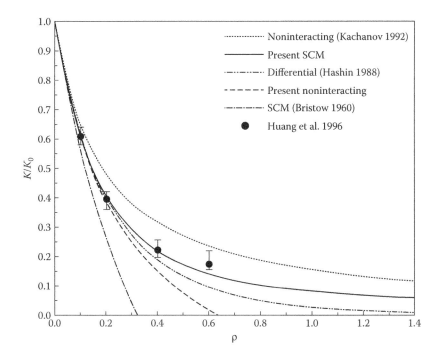

FIGURE 9.2 Effective plane strain bulk moduli for the 2D case. (From Shen, L, and Yi, S, *International Journal of Solids and Structures*, 37, 3525–3534, 2000. With permission.)

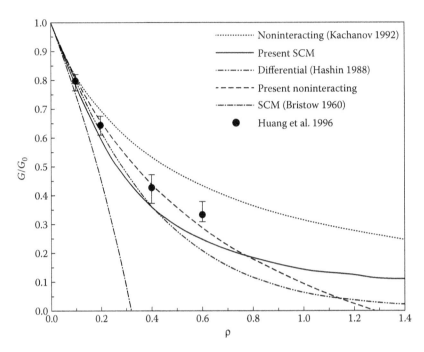

FIGURE 9.3   Effective shear moduli for the 2D case. (From Shen, L, and Yi, S, *International Journal of Solids and Structures*, 37, 3525–3534, 2000. With permission.)

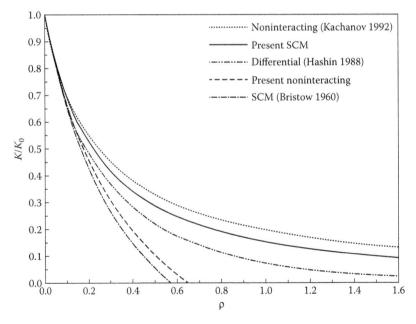

FIGURE 9.4   Effective bulk moduli for the 3D case. (From Shen, L, and Yi, S, *International Journal of Solids and Structures*, 37, 3525–3534, 2000. With permission.)

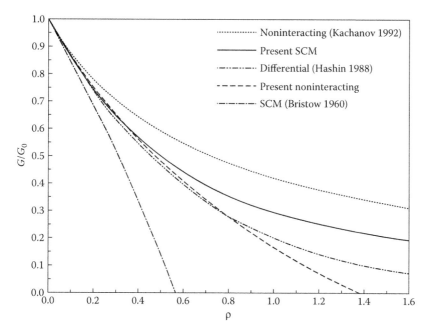

FIGURE 9.5 Effective shear moduli for the 3D case. (From Shen, L, and Yi, S, *International Journal of Solids and Structures*, 37, 3525–3534, 2000. With permission.)

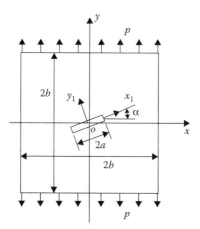

FIGURE 9.6 A cracked square element. (From Chen, YZ, *Engineering Fracture Mechanics*, 84, 123–131, 2012. With permission.)

(EEVM) as shown in Figure 9.6. The relevant strains, Young's modulus, and Poisson's ratio were evaluated for 2D cracked square plate under tension load in the y-direction as shown in Figure 9.7. The effective elastic modulus suggested for the randomly distributed crack was obtained from an average of many relevant values at different inclined angles of the cracks. A modified displacement field was proposed as follows:

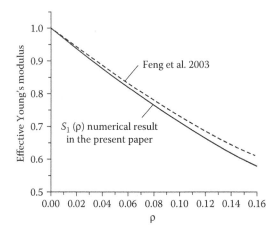

FIGURE 9.7 Comparison for the effective Young's modulus of elasticity for randomly distributed cracks. (From Chen, YZ, *Engineering Fracture Mechanics*, 84, 123–131, 2012. With permission.)

$$u(x, y) = u_a(x, y) + g_1 - \gamma y \tag{9.15}$$

$$v(x, y) = v_a(x, y) + g_2 - \gamma x \tag{9.16}$$

where g_1, g_2, and γ are three constants and represent the terms of translation and rotation of the body. These constants were obtained by applied boundary conditions of plate ends. $u_a(x, y)$ and $v_a(x, y)$ are displacement fields accordingly due to the presence of cracks. The strain components in the x- and y-directions were obtained as follows:

$$\varepsilon_x = \frac{\Delta u}{2b} \text{ with } \Delta u = \frac{1}{2b} \int_{-b}^{b} \left[u(b, y) - u(-b, y) \right] dy \tag{9.17}$$

$$\varepsilon_y = \frac{\Delta v}{2b} \text{ with } \Delta v = \frac{1}{2b} \int_{-b}^{b} \left[v(x, b) - v(x, -b) \right] dx \tag{9.18}$$

where Δu and Δv are the average elongation (or shortening) in the x- and y-directions. Young's modulus of elasticity and Poisson's ratio were obtained as follows:

$$E = \frac{P}{\varepsilon_y} \text{ and } v = \frac{-\varepsilon_x}{\varepsilon_y} \tag{9.19}$$

Because the crack was located in an inclined position, the above equation was rewritten as

$$E(\alpha) = \frac{P}{\varepsilon_y} \text{ and } v(\alpha) = \frac{-\varepsilon_x}{\varepsilon_y} \tag{9.20}$$

The crack was assumed in a random direction. Therefore, every individual inclined angle shares the same possibility. Therefore, the effective elastic constant was obtained by

$$E_{\text{eff}} = \frac{2}{\pi} \int_0^{\pi/2} E(\alpha)\,d\alpha \tag{9.21}$$

$$v_{\text{eff}} = \frac{2}{\pi} \int_0^{\pi/2} v(\alpha)\,d\alpha \tag{9.22}$$

The two integration constants were evaluated by using numerical solution. The methodology of the numerical work was a simple element which was isolated from the cracked medium. The traction force is not uniform on the top and bottom of the element. The average traction was found by integrating the traction along the length of cracks and divided by the length of crack (2a). The elastic properties for the cracked medium and all other parameters were evaluated by EEVM procedures. The elongation in the y-direction and the shortening in the x-direction were evaluated from the numerical solution and relevant strain. Young's modulus and Poisson's ratio were evaluated accordingly. The calculated Young's modulus was compared with Feng model (Feng et al. 2003), as shown in Figure 9.7, and has shown a good agreement.

Kushch et al. (2009) investigated a solid medium containing multiple cracks with prescribed orientation statistics as shown in Figure 9.8 by using a 2D unit cell

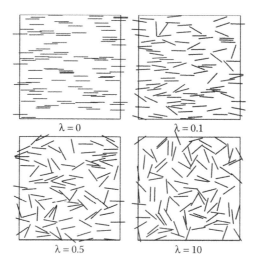

FIGURE 9.8 Cell model of a cracked solid. (From Kushch, VI et al., *International Journal of Solids and Structures*, 46, 1574–1588, 2009. With permission.)

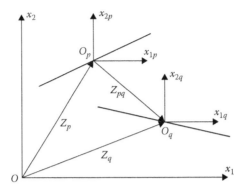

FIGURE 9.9 Global and local coordinate systems. (From Kushch, VI et al., *International Journal of Solids and Structures*, 46, 1574–1588, 2009. With permission.)

approach. The advantage of the unit cell approach is reducing a complex boundary value problem to an ordinary, well-posted set of linear algebraic equations. However, this approach assumed that the microgeometry of cracked solid with a unit cell containing multiple cracks. The medium was modeled by periodic structures as shown in Figure 9.9. The superposition principle technique of complex potentials and certain special functions was combined to calculate the effective elastic properties of a solid containing multiple cracks. Effective elastic properties were calculated by analytical averaging the strain and stress fields. This analytical solution was used to obtain the expression for the effective stiffness tensor by an exact finite element form. The local stress filed was integrated to obtain the exact closed-form expressions for the effective elastic moduli. The effective elastic moduli (C_{ijkl}^{*}) of the cracked solid was defined by the formula:

$$\sigma_{ij}^{*} = C_{ijkl}^{*}\varepsilon_{kl}^{*} \qquad (9.23)$$

where:
σ_{ij}^{*} is the volume-averaged of the macroscopic stress
ε_{kl}^{*} is the strain tensor

It was suggested that when the stress field corresponds to the uniaxial strain, strain may be equal to unity ($\varepsilon_{kl}^{*} = 1$), so Equation 9.23 was rewritten as $\sigma_{ij}^{*} = C_{ijkl}^{*}$. The result was compared with the existing approximation methods such as noninteraction approximation (Bristow 1960), differential scheme of the SCM (Zimmerman 1985), and its modification (Sayers and Kachanov 1991), as shown in Figure 9.10. The study showed that the effective elastic stiffness depends on the angular scattering of cracks.

There have been a number of studies that were investigated by the behavior of microstructural parameters in a cracked medium. However, the influence and contributions of microcracks to the elastic properties were not adequately addressed in the

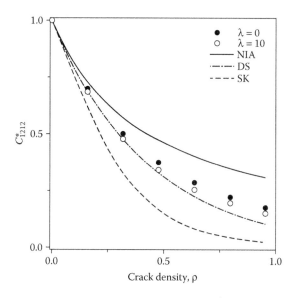

FIGURE 9.10 Comparison with the approximate theories: C^*_{1212} of a solid containing aligned and randomly oriented cracks. DS, differential scheme; NIA, noninteraction approximation; SK, Sayers and Kachanov. (From Kushch, VI et al., *International Journal of Solids and Structures*, 46, 1574–1588, 2009. With permission.)

modeling work. Bristow (1960) presented a formula to calculate the density of cracks ρ as a function of power β with crack diameters:

$$\rho = \frac{1}{A} \sum_i l_i^{\beta} \tag{9.24}$$

where:

l is the length of crack

A is the represented area

β is equal to 2 for the rectilinear cracks and 3 for the randomly oriented circular cracks

This formula was changed for nonrandom crack orientations by Kachanov (1980) from the scalar quantity to the second-rank tensor (α).

$$\alpha = \frac{1}{A} \sum \left(l^2 nn\right)^{(k)} \tag{9.25}$$

where:

n is a unit normal to crack

nn denotes a dyadic product with components $n_j n_j$

Kushch et al. (2009) presented another expression of crack density formula for a 2D unit cell:

$$\rho = \frac{l^2 N}{ab} \tag{9.26}$$

where:
 a and b are along the axes Ox_1 and Ox_2, respectively, as shown in Figure 9.9
 N is the number of cracks

These parameters account for individual crack contributions to the overall elastic properties, but they are affected by interactions between cracks.

Xi-Qiao et al. (2004) present the quasi-micromechanical model to calculate the complex damage associated with microcrack growth in the brittle and quasi-brittle materials containing many distributed microcracks as shown in Figure 9.11. The effective moduli of random heterogeneous elastic solid were obtained by the assembled volume averaging process within RVE. However, the relation between overall averages strain $\bar{\varepsilon}$ and uniform overall stress σ^∞, and the relation between overall average stress $\bar{\sigma}$ and uniform overall strain ε^∞ were written as follows:

$$\bar{\varepsilon} = S : \sigma^\infty, \bar{\sigma} = L\varepsilon^\infty \tag{9.27}$$

where:
 S is the effective compliance tensor
 L is the stiffness tensor

The average strain and stress were decomposed as

$$\bar{\varepsilon} = \bar{\varepsilon}^m - \bar{\varepsilon}^c, \bar{\sigma} = \bar{\sigma}^m - \bar{\sigma}^c \tag{9.28}$$

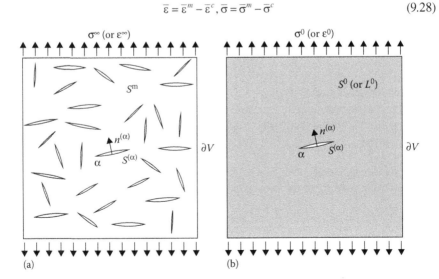

FIGURE 9.11 (a) An RVE of a microcracked solid and (b) the simplified model for calculating the opening displacement of a microcrack. (From Xi-Qiao, F et al., *Mechanics of Materials*, 36, 261–273, 2004. With permission.)

where:
   $\bar{\varepsilon}^m$ and $\bar{\sigma}^m$ denote the matrix strain and stress tensors averaged without cracks over the RVE, respectively
   $\bar{\varepsilon}^c$ and $\bar{\sigma}^c$ denote the microcrack-induced variations in the overall average strain and stress tensors, respectively

The variation of the volume-averaged stain due to the presence of microcrack was calculated as

$$\bar{\varepsilon}^c = \frac{1}{2V} \sum_{\alpha=1}^{N} S^{(\alpha)}(\bar{b}n + n\bar{b}) \tag{9.29}$$

where:
   the superscript α stands for a quantity of the αth microcrack
   $S^{(\alpha)}$, $\bar{b}$, and n denote the surface area, the average opening displacement discontinuity vector, and the unit vector normal to the crack faces, respectively

The normalized effective Young's modulus and shear modulus were obtained as

$$\frac{E}{E_m} = \left[1 + \frac{16\left(1-\left(v^m\right)^2\right)\left(10-3v^m\right)f}{45\left(2-v^m\right)\left(1-\xi f^{\eta}\right)} \right]^{-1}$$

$$\frac{G}{G_m} = \left[1 + \frac{32\left(1-v^m\right)\left(5-v^m\right)f}{45\left(2-v^m\right)\left(1-\xi f^{\eta}\right)} \right]^{-1} \tag{9.30}$$

where:
   E^m, G^m, and v^m are Young's modulus, the shear modulus, and Poisson's ratio of the pristine matrix, respectively
   f is the conventional scalar microcrack density parameter (Bristow 1960)
   ξ and η are the adjustable parameters that can be calculated by fitting experimental results or other theoretical results with a good accuracy

It should be noted that the conventional method was based on the concept of the effective medium or the effective field for estimating the effective moduli of microcracked solids, whereas the Xi-Qiao et al. (2004) model is established to simulate the constitutive response of microcrack-weakened brittle or quasi-brittle materials under complex loading. Furthermore, the suggested model assumed that all microcracks are distributed uniformly in the orientation space. As a consequence, Xi-Qiao et al. model calculates the microcrack interaction effects on the effective elastic moduli in a convenient manner. More detail about the modeling can be found in the work of Xi-Qiao et al. (2004).

   In an interesting work, Santare et al. (1995) presented the GSCM to calculate the anisotropic effective moduli of a medium containing microcracks with an arbitrary degree of alignment. The suggested numerical methods were used to account crack face contact and friction with the wave slowness for microcrack damage media. A finite element model (FEM) was used for GSCM to solve the analytical anisotropic crack inclusion boundary value problem by Su et al. (2007). It has been shown that the effective moduli were decreasing when the microcrack densities increased. Furthermore, the orientation of cracks has a significant effect on the effective moduli as shown in Figure 9.12. In addition, it has been shown that the response was quite different depending on whether the external loads are tension or compression as shown in Figure 9.13.

In a later study, Shen and Li (2004) presented an accurate numerical model to calculate the effective moduli of plates with various distributions of cracks. The crack line was divided into number of parts (M) to obtain the unknown traction on the crack line. There were six types of crack distributions and three kinds of cracks—four regular and two random distributions. More details are available in the work of Shen and Li (2004). An infinite matrix for strain energy was used in the FEM to calculate the effective moduli instead of a large square matrix. It was assumed that the strain energy of a circular sample due to the presence of microcracks had an expression as

$$\Delta f_{micro} = \frac{1}{E_0} \Delta f_{micro}^0 \tag{9.31}$$

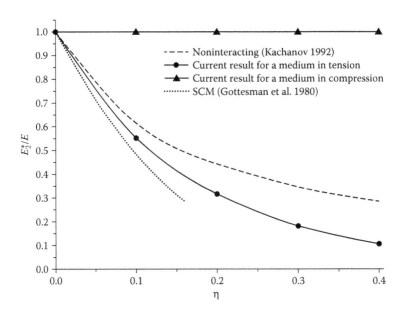

FIGURE 9.12 Normalized effective Young's modulus in the direction perpendicular to the cracks for aligned cracks versus crack density. (From Su, D et al., *Engineering Fracture Mechanics*, 74, 1436–1455, 2007. With permission.)

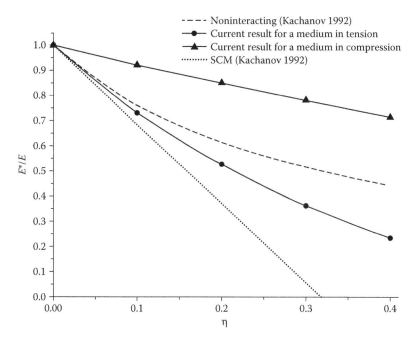

FIGURE 9.13 Normalized effective Young's modulus for randomly distributed cracks versus crack density. (From Su, D et al., *Engineering Fracture Mechanics*, 74, 1436–1455, 2007. With permission.)

where Δf^0_{micro} is the strain energy change associated with a plate matrix with unit Young's modulus, which was also independent of v_0. Therefore, E_1 and G_{12} were derived as

$$\frac{E_1}{E_0} = \frac{1}{\left(1 + \dfrac{2\Delta f^0_{micro}}{1 - \dfrac{3\Delta f^0_{micro}}{4}}\right)}$$

$$\frac{G_{12}}{G_0} = \frac{1}{\left(1 + \dfrac{1}{(1+v_0)} \dfrac{\Delta f^0_{micro}}{1 - \dfrac{1}{4}\Delta f^0_{micro}}\right)}$$

(9.32)

Comparison of the numerical results with some popular models is shown in Figure 9.14.

Ma et al. (2005) investigate the effect of nonuniform concentration distributions of microcracks in the brittle or quasi-brittle material and present a novel numerical method to evaluate the effect of microcrack growth and coalescence on the effective moduli and the tensile strength for brittle or quasi-brittle material. A plate of a unit thickness containing a large number of randomly distributed planer microcracks was considered

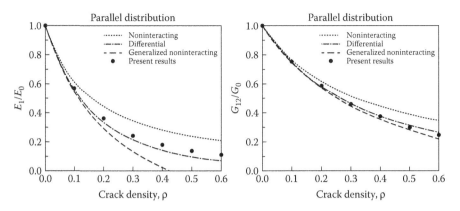

FIGURE 9.14 Comparison of some micromechanics models and the present numerical results associated with one-sized cracks. (From Shen, L, and Li, J, *International Journal of Solids and Structures*, 41, 7471–7492, 2004. With permission.)

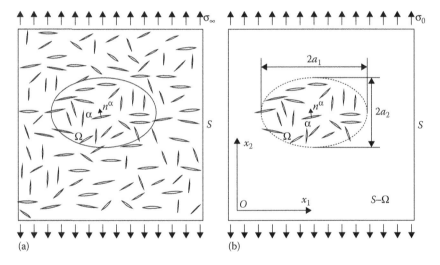

FIGURE 9.15 (a) A microcracked solid and (b) the simplified calculation model. (From Ma, L et al., *Engineering Fracture Mechanics*, 72, 1841–1865, 2005. With permission.)

in the study. The plate was subjected to a uniform stress σ_∞ in the far field, as shown in Figure 9.15a. The plane stress condition was assumed to be a simplified calculation. Some simplifications of the crack field were assumed and the subdomain Ω of the specimen surrounding the αth microcrack was defined, as shown in Figure 9.15b. To simulate anisotropic damage cases, an elliptical shape with the semiaxes a_1 and a_2 for the subdomain Ω was assumed. It was considered an αth microcrack in the subdomain Ω, whose length was denoted as $2l_\alpha$, as shown in Figure 9.16. According to Eshelby's inclusion theory, σ_∞ was replaced with an effective stress field σ_0 in order to correct the effect of removing cracks in the outside of the subdomain region. Therefore, the Ω

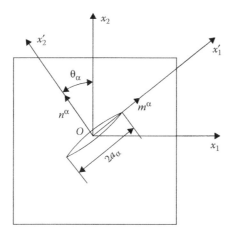

FIGURE 9.16   Global and local coordinate systems. (From Ma, L et al., *Engineering Fracture Mechanics*, 72, 1841–1865, 2005. With permission.)

region in the plate is a weaker zone (inclusion) with lower stiffness than the pristine matrix; thus, the average stress over Ω was expressed as

$$\sigma_\Omega = B : \sigma_\infty \tag{9.33}$$

where the fourth-order tensor B refers to the average stress concentration, which is given by

$$B = \left[I + P : (M - M_0) \right]^{-1} \tag{9.34}$$

$$P = M_0^{-1} : \left(I - S \right) \tag{9.35}$$

where:

M_0 and M denote the elastic compliance tensors of the pristine matrix and the subdomain Ω, respectively

I is the fourth-order identity tensor

S is the Eshelby tensor

Under the uniform traction boundary condition, the average stress σ_∞ over the domain Ω is equal to the far-field stress σ_∞. Ma et al. (2005) modified Equation 9.33 as σ_∞ was equal to σ_0. Interestingly, the proposed model was extended to simulate the coalescence process of microcracks that cause the fatal crack during the final rapture of specimen. The simulation of coalescence process was based on Feng et al.'s (2003) energy ratio (R) criteria. The parameter R represents the ratio between the released potential energy and the energy required to create two new surfaces along the broken ligament during the coalescence, which was written as

$$R = \frac{\Delta \Pi}{2c\gamma_s} \tag{9.36}$$

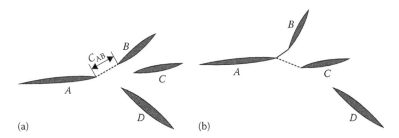

FIGURE 9.17   (a,b) Configurations of microcrack linkage. (From Ma, L et al., *Engineering Fracture Mechanics*, 72, 1841–1865, 2005. With permission.)

where:
  $\Delta\Pi$ denotes the release of potential energy because of the linkage of two neighboring microcracks
  c is the ligament size as shown in Figure 9.17
  γ_s is the surface energy per area of the brittle matrix

The coalescence process was simulated step by step and has depicted in Figure 9.18. The final rupture occurs when there are many successive coalescence steps. Thus, the strong interaction of microcracks and the linking of the failure path cause decrease in tensile strength and Young's modulus as shown in Figure 9.19.

The brief discussions to this point have clearly indicated that the mechanical properties of cracked materials depend on the distributions and interaction of microcracks in the material. The statistical distributions and interaction of microcracks lead to transport properties as the effective moduli and tensile strength. Such transport properties depend basically on the statistical distribution of orientations, sizes, and positions of microcracks. Therefore, the wealth of knowledge on microcrack modeling has undoubtedly established a strong foundation to investigate long-term behavior of brittle materials.

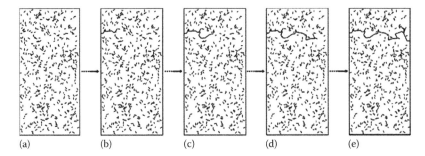

FIGURE 9.18   (a–e) Microcrack coalescence process of a specimen after 0, 7, 14, 16, 20, and 27 coalescence steps, respectively. (From Ma, L et al., *Engineering Fracture Mechanics*, 72, 1841–1865, 2005. With permission.)

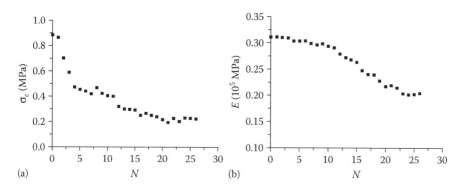

FIGURE 9.19 Changes of (a) the effective Young's modulus and (b) the tensile strength of the specimen during the microcrack coalescence process. (From Ma, L et al., *Engineering Fracture Mechanics*, 72, 1841–1865, 2005. With permission.)

LIMITATION OF NUMERICAL APPROACHES

The problems associated with modeling interaction of multiple microcracks increase the number of equations rapidly with the increase in the number of microcracks. It leads to limit the application of microcrack simulation method to a significantly smaller number of microcracks than microcracks existed at an actual failure of brittle material. Unfortunately, there is no method available yet for evaluating the interaction of microcracks of a large number of microcracks, as in most of actual materials. There are many simplified approaches to calculate the stress intensity factors (SIFs) of microcracks, but unfortunately the oversimplified approach is inappropriate for high concentration of cracks. The difficulty in obtaining the analytical solution of the SIFs of multiple interacting microcracks has opened the inroads to advanced numerical schemes such as the method of pseudo-tractions, the complex potential method, the double potential method, and the weight function method.

The SCM was used in many research projects to evaluate the effective elastic moduli for 2D solids with inhomogeneities, voids, or cracks. All previous papers were limited value for 2D crack densities lower than $1/\pi$ because the result will be zero stiffness of 2D solid. The zero stiffness occurs when the crack density is larger than $1/\pi$ in the case of a circular inclusion (Dan et al. 2007). However, an elliptical inclusion extends the crack density limits significantly for randomly oriented distribution of cracks.

The common numerical approximate theories such as self-consistent scheme (Budiansky and O'Connell 1976, Hoenig 1979) and the differential scheme (Hashin 1988) were proposed to account for interactions between the cracks. These extended models were based on one particle of an effective homogeneous medium with a single crack embedded, and therefore, the crack density is the only parameter associated in the models (Kushch et al. 2009). For this reason, the accuracy of those models for increased crack densities are highly uncertain.

EFFECT OF MICROCRACKS ON EFFECTIVE MODULI IN LAMINATED STRUCTURES

Most of the composite materials are used in the form of laminates consisting of a continuous matrix with different fiber orientations. This type of composite material is widely used as plate and shell structures in various engineering applications. Because fiber-reinforced composite materials (FRCs are generally an anisotropic material (or orthotropic material), the fracture and damage mechanics of FRCs remain as an unresolved problem. The cracks in FRC laminate may originate from microcracks or inherent flaws. The complete rupture of the lamina occurs when the crack propagations reach the interface under increased loading. The common cracks can be seen in the reinforced composite materials are transverse cracks, which appear along the fibers in constituent lamina and perpendicular to the interfaces. This type is not only seen in FRC laminate, but there are also common cracks to all layered structures, namely, the tunneling and channeling cracking, delamination, and interface decohesion.

The deformation and damage in laminated composite materials depend on the resin toughness and the state of stress, that is, mode I, mode II, mode III, or mixed mode of fractures as shown in Figure 9.20. Most brittle materials have much small deformation or damaged zone, but in ductile materials the damaged or deformation zone is bigger than that in brittle materials. In ductile materials, the damage and deformation system contains elastic and plastic deformation zones. However, in the brittle materials, the plastic damage is near to zero. Most of the resins in composite materials are brittle materials such as epoxy. However, Crasto et al. (1997) argued that when the load is increased, the microcracks appear in the zone in front of crack tips in the mode I stress state. These microcracks grow and coalescence, which results in crack advance. In mode II and mode III states, the microcracks grow until cracks reach the reinforced fibers. Sometimes, these cracks bound between plies due to fiber surfaces, which arrest these growing microcracks. These microcracks create macrocracks due to coalescence of cracks. This coalescence generally occurs at the fiber/matrix interface, as shown in Figure 9.21.

Matrix cracking grows parallel to the fibers within a ply block in composite materials. Tomohiro et al. (2005) pointed out that the mechanical properties of composite materials may be reduced due to matrix cracks. Matrix cracks may lead

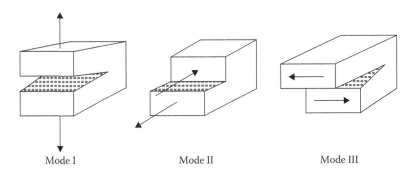

Mode I   Mode II   Mode III

FIGURE 9.20   Fracture modes.

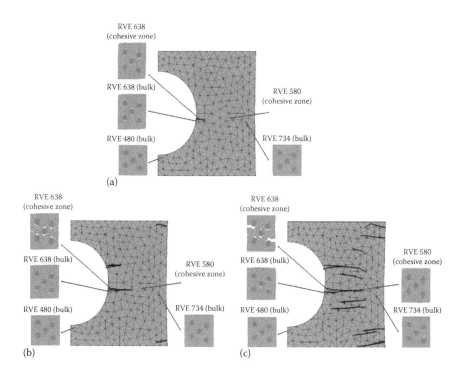

FIGURE 9.21 Formation and growth of ply interface cracks (delamination). Snapshots for selected times: (a) $t = 20.75$ s, (b) $t = 25.00$ s, and (c) $t = 27.50$ s. (From Souza, FV, and Allen, DH, *International Journal of Solids and Structures*, 48, 3160–3175, 2011. With permission.)

to other damage such as delamination and fiber breakage in laminated composite materials. Delamination can be regarded as interface crack between two adjacent plies. Tong et al. (1997) suggested that the initial stage of mechanical degradation in laminated composite materials is creation and progression of interlaminar matrix cracks. Therefore, matrix cracks can be considered as the first and foremost mechanism in the process of damage development.

There are three types of cracks that can be shown in the laminated composite materials. The cases of cracks depend on the cracks in the off-axis ply. The first type is known as off-axis ply cracks when laminated composite materials consist of cracks in plies inclined to an axis of symmetry. The second type is known as transverse cracks, in the case when the off-axis angle is 90°. The third type is known as tunneling cracks, in the case when the cracks are along ply directions.

Veer Singh and Talreja (2010) review that most of the previous studies mostly deal with the transverse cracks. Many of those studies are based on the energy-based approach to evaluate the damage in laminated composite materials (Nairn 2000). However, using the energy-based approach, data did not steadily agree with experimental observations. The energy-based approach combined with probabilistic considerations provides closer results for transverse cracks in cross-ply laminates.

The first type of damage mechanics approach for interlaminar or transverse cracks in laminated composite materials is micromechanics approach. This approach is related to the properties of ply level and ply thickness and used to understand the transverse cracks in cross-ply laminates. The transverse cracks cause the local separation of plies and fiber failures. McCartney (1997) analyzed the transverse cracks by using the microcracks approach for 90° plies in a symmetric laminate containing plies inclined to the axial loading direction. This type of approach generates the relationship between the measures of damage and the overall laminate properties. Although the continuum damage approach has this good feature, it has limitations of the certain material coefficients, which are usually determined experimentally for each laminate configuration. Recently, an approach has been presented to overcome this limitation. It is called synergistic damage mechanics as shown in Figure 9.22 (Veer Singh and Talreja 2009).

Transverse matrix cracks can be considered as a mode of failure. This type of cracks normally causes a small reduction in the overall stiffness of the structure. The big problem of transverse matrix cracks is difficult to detect during frequent inspections. McCartney (1997) pointed out that the transverse matrix cracks caused the leakage in the pressurized vessels and finally caused major delamination when the temperature increased.

Veer Singh and Talreja (2010) studied the off-axis ply cracks in multidirectional composite materials laminated under a quasistatic tensile in the longitudinal direction as shown in Figure 9.23. The study was based on the energy-based approach for off-axis ply cracks. The laminated energy parameters such as the energy release rate and arbitrary constants for crack tip were found from experimental methods. The 3D finite element analysis was used to calculate crack surface displacements (CSDs) because this model deals with off-axis cracks in multidirectional composite

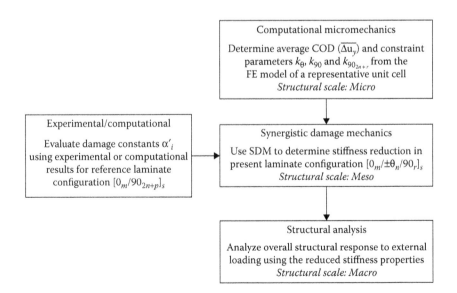

FIGURE 9.22 Multiscale synergistic methodology for analyzing damage behavior in a general symmetric laminate. (From Veer Singh, C, and Talreja, R, *Mechanics of Materials*, 41, 954–968, 2009. With permission.)

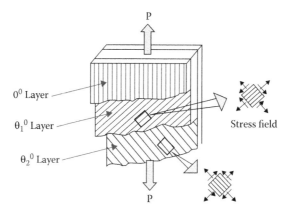

FIGURE 9.23 An off-axis laminate loaded in axial tension. (From Veer Singh, C, and Talreja, R, *International Journal of Solids and Structures*, 47, 1338–1349, 2010. With permission.)

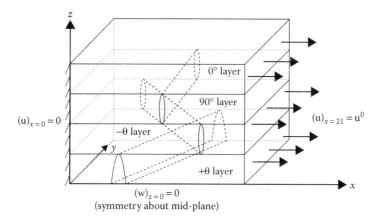

FIGURE 9.24 A representative unit cell for finite element analysis of $[0_m/90_p/\pm\theta_n]_S$ laminate. (From Veer Singh, C, and Talreja, R, *International Journal of Solids and Structures*, 47, 1338–1349, 2010. With permission.)

materials. An RVE that was a suitable 3D element to calculate the CSD is shown in Figure 9.24. It was assumed that a new crack appears between nearby cracks when maximum normal stress between them reaches to a critical strength value. The criterion for the formation new cracks was given as

$$W_{2N \to N} \geq NW_C \frac{1}{\sin\theta} t_0 \qquad (9.37)$$

where:
$W_{2N \to N}$ is the work to close N new cracks
t is the width of cracks
N is the number of cracks

θ is the angle of cracks

W_C is the material resistance (energy) per unit crack plane area per unit laminate width

According to the results, the ply cracking phenomenon comprises three stages as shown in Figure 9.25. However, the transverse cracks are initiated at the edge of width and progressively propagate along the width of the specimens depending on the ply orientation.

Varna et al. (1999) studied the effect of the ply orientations on the initiation of ply cracks, the propagation of cracks, and the change of elastic moduli due to damage accumulation. The initiation of cracks and the propagation of cracks happen in the off-axis plies normal to the loading direction. Talreja (1996) suggested that ply cracks cause the change of moduli; therefore, the synergistic damage mechanics approach may be applied to the investigation of ply cracks (Figure 9.26).

Yang et al. (2003) studied the crack propagation through the central layer of a symmetrical composite laminate materials based on the energy criteria. The study has shown that the critical stress for transverse crack propagation is less than that for tunneling crack propagation as shown in Figures 9.27 and 9.28. Therefore, the transverse cracking is a more common fracture mode in composite materials. The initial cracks may propagate first in the transverse direction until they reach the interface, but this propagation may be stopped if the outer layer is tougher than the inner layer. This may change the crack direction from the transverse crack to the tunneling crack under an increasing stress level. The delamination at the interfaces or damage in the outer layer may also occur when the crack tips touch the interfaces and the stress is increased in the FRC laminate. Because the embedded crack is comparatively smaller

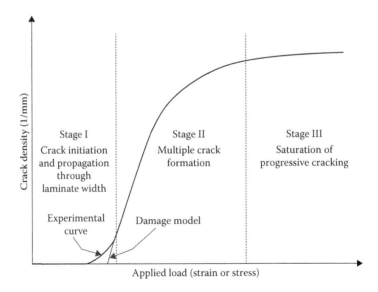

FIGURE 9.25 A typical damage evolution curve for transverse matrix cracking in composite materials. (From Veer Singh, C, and Talreja, R, *International Journal of Solids and Structures*, 47, 1338–1349, 2010. With permission.)

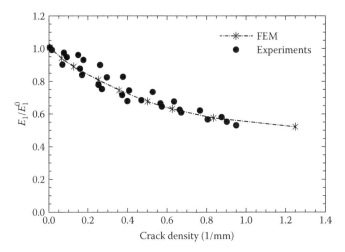

FIGURE 9.26 Finite element simulation of stiffness reduction for $[0/90_8/0_{1/2}]s$ laminate compared with experimental data. (From Varna, J et al., *Composites Science and Technology*, 59: 2139–2147, 1999; Veer Singh, C, and Talreja, R, *Mechanics of Materials*, 41, 954–968, 2009. With permission.)

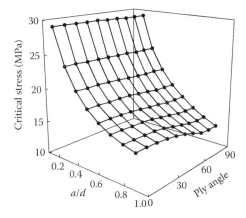

FIGURE 9.27 Variation of critical stress with a/d and ply angle *h* for tunneling crack. (From Yang, T et al., *Composite Structures*, 59, 473–479, 2003. With permission.)

than the length of the sample, the crack propagation problem will be reduced to 2D for 3D problems.

Crossman et al. (1980) showed that the higher crack density can be created experimentally in [±25/90] lamina in thinner 90° layers. As a consequence, the thinner plies and the inclination of the fibers with respect to the applied load assist in developing the crack density in the laminated composite materials.

There are a large number of methods that are used to predict the failure in composite materials. Unfortunately, the results from most theories differ significantly with the experimental observations. Most of numerical theories are based on continuum

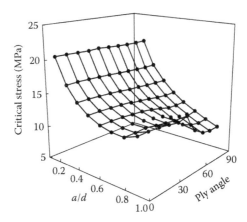

FIGURE 9.28 Variation of critical stress with a/d ($2d$ is the inner sublaminate thickness contains an elongated embedded crack width $2a$) and ply angle h for transverse crack. (From Yang, T et al., *Composite Structures*, 59, 473–479, 2003. With permission.)

approaches and the multiply crack condition of the continuum damage is not enough to predict the failure. Ochoa and Reddy (1992) suggested that it may be appropriate to look at the trends of strain energy release as a function of delamination size. Camanho et al. (2006) pointed out that over the past 40 years, the World-Wide Failure Exercise was done to check the status of currently available theoretical methods for predicting material failure in composite materials. The check result showed that most of the current theoretical methods do not close to experimental data. Most of the current theoretical methods need modification for simulating the real conditions.

CONCLUSION

The most common structural defects in brittle materials are microcracks. Small cracks significantly affect the stiffness and the strength of the material. However, the effective elastic moduli are sensitive to the density and orientation of the cracks. Therefore, a geometric model should be built which associates the crack interactions and the parameters of an actual microstructure of the cracked materials.

Fracture mechanics concepts have attended the energy based approaches. In brittle materials, the Griffith criterion is often used. However, using the energy-based approach, data did not steadily agree with experimental data. In addition, in laminated composite structures, it may be appropriate to look at the trends of strain energy release as a function of delamination size. Thus, most of the existing approaches need adjustment to simulate the actual problems.

REFERENCES

Bristow, JR (1960). Microcracks, and the static and dynamic elastic constants of annealed heavily cold-worked metals. *British Journal of Applied Physics*, 11: 81–85.
Budiansky, B, and O'Connell, RJ (1976). Elastic moduli of a cracked solid. *International Journal of Solids and Structures*, 12: 81–97.

Camanho, P, Dávila, C, Pinho, S, Iannucci, L, and Robinson, P (2006). Prediction of in stiu strengths and matrix cracking in composites under transverse tension and on-plane shear. *Composites: Part A*, 37: 165–176.

Chen, YZ (2012). A novel solution for effective elastic moduli of 2D cracked medium. *Engineering Fracture Mechanics*, 84: 123–131.

Crasto, A, and Kim, Y (1997), 'Hydrothermal Influence on the free-edge delamination of composites under compressive loading', Composite materials: Fatigue and Fractures, Vol. 6, ASTM STP 1285, E. A. Armanios, Ed, American Society for Testing and Materials, pp. 381–393.

Crossman, FW, Warren, WJ, Wang, ASD, and Law, GE, Jr (1980). Initiation and growth of transverse cracks and edge delamination in composite laminates. Part 2: Experimental results. *Journal of Composite Materials*, 14: 88–108.

Dan, S, Michael, H. S, and George, A. G, (2007), 'The effect of crack face contact on the anisotropic effective moduli of microcrack damaged media', Engineering Fracture Mechanics, Vol. 74, pp. 1436–1455.

Eshelby, JD (1957). The determination of the elastic field of an ellipsoidal inclusion and related problems. *Proceedings of the Royal Society of London. Series A*, 241: 376–396.

Feng, XQ, Li, J, and Yu, S (2003). A simple method for calculating interaction of numerous microcracks and its applications. *International Journal of Solids and Structures*, 40: 447–464.

Feng, XQ, and Yu, SW (2000). Estimate of effective elastic moduli with microcrack interaction effects. *Theoretical and Applied Fracture Mechanics*, 34: 225–233.

Hashin, Z (1988). The differential scheme and its application to cracked materials. *Journal of the Mechanics and Physics of Solids*, 36: 719–734.

Hoenig, A (1979). Elastic moduli of a non-randomly cracked body. *International Journal of Solids and Structures*, 15: 137–154.

Kachanov, M (1980). Continuum model of medium with cracks. *Journal of the Engineering Mechanics Division*, 106(EM5): 1039–1051.

Kachanov, M (1992). Effective elastic properties of cracked solids: Critical review of some basic concepts. *Applied Mechanics Reviews*, 45: 305–336.

Kushch, VI, Sevostianov, I, and Mishnaevsky, L, Jr (2009). Effect of crack orientation statistics on effective stiffness of microcracked solid. *International Journal of Solids and Structures*, 46: 1574–1588.

Ma, L, Xuyue, W, Xi-Qiao, F, and Shou-Wen, Y (2005). Numerical analysis of interaction and coalescence of numerous microcracks. *Engineering Fracture Mechanics*, 72: 1841–1865.

McCartney, LN (1997). Prediction of ply cracking in general symmetric laminates: Deformation and fracture of composites. *Proceedings of the 4th International Conference*. March 24–26, Institute of Materials, Manchester, pp. 101–110.

Nairn, JA (2000). Matrix microcracking in composites. In: R Talreja & JAE Manson (eds.), *Polymer Matrix Composites*. Vol. 2, Elsevier, Amsterdam, The Netherlands, pp. 403–432.

Ochoa, OO, and Reddy, JN (1992). *Finite Element Analysis of Composite Laminates*. Kluwer Academic Publishers, Dordrecht, The Netherlands.

Santare, MH, Crocombe, AD, and Anlas, G (1995). Anisotropic effective moduli of materials with microcracks. *Engineering Fracture Mechanics*, 52(5): 833–842.

Sayers, C, and Kachanov, M (1991). A simple technique for finding effective elastic constants of cracked solids for arbitrary crack orientation statistics. *International Journal of Solids and Structures*, 7: 671–680.

Shen, L, and Li, J (2004). A numerical simulation for effective elastic moduli of plates with various distributions and sizes of cracks. *International Journal of Solids and Structures*, 41: 7471–7492.

Shen, L, and Yi, S (2000). New solutions for effective elastic moduli of microcracked solids. *International Journal of Solids and Structures*, 37: 3525–3534.

Souza, FV, and Allen, DH (2011). Modeling the transition of microcracks into macrocracks in heterogeneous viscoelastic media using a two-way coupled multiscale model. *International Journal of Solids and Structures*, 48: 3160–3175.

Su, D, Santare, MH, and Gazonas, GA (2007). The effect of crack face contact on the anisotropic effective moduli of microcrack damaged media. *Engineering Fracture Mechanics*, 74: 1436–1455.

Talreja, R (1996), 'A synergistic damage mechanics approach to durability of composite material systems', In: Cardon A, Fukuda H, Reifsnider K, (edts). Progress in durability analysis of composite systems. Rotterdam: A.A. Balkema, pp. 117–129.

Tomohiro, Y, Takahira, A, Toshio, O, and Takashi, I (2005). Effects of layup angle and ply thickness on matrix crack interaction in contiguous plies of composite laminates. *Composites: Part A*, 36: 1229–1235.

Tong, J, Guild, F, Ogin, S, and Smith, P (1997). On matrix crack growth in quasi-isotropic laminates—I: Experimental investigation. *Composites Science and Technology*, 57: 1527–1535.

Varna, J, Joffe, R, Akshantala, NV, and Talreja, R (1999). Damage in composite laminates with off-axis plies. *Composites Science and Technology*, 59: 2139–2147.

Veer Singh, C, and Talreja, R (2009). A synergistic damage mechanics approach for composite laminates with matrix cracks in multiple orientations. *Mechanics of Materials*, 41: 954–968.

Veer Singh, C, and Talreja, R (2010). Evolution of ply cracks in multidirectional composite laminates. *International Journal of Solids and Structures*, 47: 1338–1349.

Xi-Qiao, F, Qing-Hua, Q, and Shou-Wen, Y (2004). Quasi-micromechanical damage model for brittle solids with interacting microcracks. *Mechanics of Materials*, 36: 261–273.

Yang, T, Liu, Y, and Wang, J (2003). A study of the propagation of an embedded crack in a composite laminate of finite thickness. *Composite Structures*, 59: 473–479.

Zimmerman, RW (1985). The effect of microcracks on the elastic moduli of brittle materials. *Journal of Materials Science*, 4: 1457–1460.

10 Life Prediction of Composite Rotor Blade

Ranjan Ganguli

CONTENTS

INTRODUCTION

Rotating wings seek to create an artificial airflow to generate power or thrust. Rotating structures are the important components of transport and energy industries. As such structures age, they are prone to health problems. Health monitoring of rotating structures has therefore received increasing attention in recent years [1–4]. Typically, existing rotor diagnostics rely on track and balance methods and do not isolate the type of damage [5–8]. Most research on rotor health monitoring has addressed metal blades [9–12]. However, the damage mechanics of a composite rotor are different [13–16]. Most wind turbine and helicopter rotor blades are made from composites, and predicting their damage state and life is important.

Several researchers have modeled a damaged rotor with isotropic material properties [9,11,17]. Some works considered crack models based on fracture mechanics [18], but these studies were limited to isotropic materials. A pioneering step addressing damage modeling in composite rotor blades was taken by Lakshmanan and Pines [19]. They modeled a transverse crack that extends across the entire width of the flexbeam of a bearingless rotor. However, there was a need for a detailed modeling of the damage modes in a composite rotor blade. Pawar and Ganguli [13] modeled matrix cracking in a helicopter rotor blade with cross-ply laminates and studied its behavior in forward flight. Pawar and Ganguli [14] further modeled matrix cracks in a generalized layup box and an airfoil section beam and studied their static behavior. They also studied [15] the effect of more severe damage modes such as debonding/delamination (DD) and fiber breakage (FB) due to matrix cracking for static condition [15] and for forward flight [16].

Typically, the first failure mode in the composite materials is dominated by matrix cracking. Matrix cracking triggers more severe damage modes such as DD and FB. When the matrix cracking effect saturates, a local delamination can emanate from the matrix crack tips. Delamination may lead to the dangerous damage mechanism of FB. Following matrix cracking and DD damage modes in composite materials, the matrix part of composite materials is debilitated and a proper transfer of the shear loads at the broken fibers is hindered.

Fatigue causes evolution of damage in the rotor blades. An analysis of the change in measurable system behavior of the composite rotor blade due to life consumption of the structure is needed to predict the life of composite rotor blades. This enables the use of the system measurements as "virtual sensors" for the blade life. Generally, the decay in structural properties of composite materials is estimated using phenomenological models [20]. Although the fatigue of metals is well understood, the analysis of fatigue in composites is complicated because the material properties of the constituents of composites are quite different and there are multiple failure modes [21].

Experiments have revealed that the variation in the modulus of elasticity due to damage growth in composite materials shows three distinct stages [22], as shown in Figure 10.1. Stage I shows a rapid stiffness reduction primarily caused by the development of transverse matrix cracks. Stage II shows damage growth caused by delamination and proceeds in an almost linear fashion with respect to cycles. Stage III shows rapid stiffness degradation due to initial fiber fractures leading to strand failures. A linear damage accumulation model was proposed by Nicholas and Russ [23]. Nonlinear damage accumulation functions were proposed by Subramanian et al. [24] and Halverson et al. [25]. These early accumulation models could not effectively capture all the three damage stages. For instance, the model proposed by Subramanian et al. [24] explains the fast damage growth during the early loading cycles in stage I but fails to describe the rapid damage growth in stage III. Halverson et al. [25] models stage III quite well but not stage I. Mao and Mahadevan [26] proposed a nonlinear model to capture damage evolution in composite materials subject to fatigue loading. Parameters of the proposed model were obtained from experimental data. Other life prediction modes based on micromechanics of composites [27]

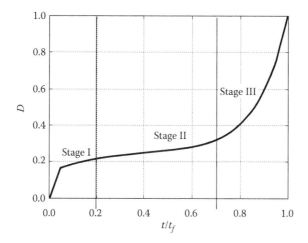

FIGURE 10.1 Fatigue damage growth curve for composite materials. Stage I: matrix cracking (MD); stage II: DD; and stage III: FB.

and finite element simulations [28] have been proposed. However, phenomenological models are attractive due to their simplicity.

It is possible to measure stiffness degradation directly on a test specimen. However, in real structures, it is easier to measure an auxiliary variable that is related to the stiffness. Such an auxiliary variable is often called a damage indicator [29]. Moon et al. [30] used natural frequencies of composite laminates as damage indicators. The fatigue damage state of the structure was estimated using the residual natural frequency, which was measured from vibration tests. Badewi and Kung [31] also linked fatigue of composite materials to changes in modal properties of graphite epoxy composites. They found that modal properties are a real-time indicator of damage growth in structures. It is possible to estimate the life of the composite rotor blade using the change in system behavior of the composite rotor blade. A structural health monitoring (SHM) system to predict the blade life can then be developed.

In general, there are two approaches to handle uncertainty in damage detection problems. The first approach is to prefilter the signals to remove the noise and outliers [32,33]. The second approach is to develop algorithms for detecting damage in the presence of noise. The SHM systems for the composite materials typically involve solving an inverse problem in which changes in some measurable properties of the structure are used to detect damage. The inverse problem is complicated because of uncertainty in the modeling, measurements, and signal processing. Therefore, the inverse problem for the development of an SHM system is often solved using soft computing methods such as neural network [34,35], genetic algorithm (GA) [36–38], and fuzzy logic [39–41] or by estimation methods such as Kalman filtering [42]. Soft computing methods are adept at the extraction of useful information from imprecise data. Recently, advanced soft computing methods that hybridize the best features of two or more soft computing methods have been developed. Such hybrid systems are often more accurate than the original ones. The genetic fuzzy system (GFS) is an advanced soft computing method that combines the approximate reasoning capabilities of the fuzzy systems with the learning capabilities of the GAs. The GFS for structural damage detection was pioneered by Pawar and Ganguli [43]. A GA was used for automating the process of rule generation for a fuzzy system for application to structural damage detection in beams. The GFS was further demonstrated for damage detection in a thin-walled composite beam [44].

In this chapter, physics-based damage models of the composite rotor blades are used, and the temporal stiffness degradation of the composite blade is linked to life consumption of the structure using curve fits based on a phenomenological fatigue damage model. Next, a simulation of the damaged rotor is used to develop a fuzzy logic system (FLS), which predicts the remaining life of the structure from measured response and loads data. Global and local damage in the composite rotor blade are addressed using two different GFSs. The first GFS predicts global life consumption using displacement-, force-, and moment-based measurement deltas. The measurement deltas are deviations between the undamaged and damaged blades. The second GFS predicts local life consumption using strain-based measurement deltas. The SHM system of a composite helicopter rotor blade is demonstrated using a two-cell airfoil section beam. Though the problem and numerical results are given for a helicopter rotor blade, the concepts are applicable to wind turbine rotor blades and other rotating structures under aerodynamic loading. This chapter is adapted from [4] and [45].

SIMULATION

Because it is difficult, if not impossible, to test the composite rotor blade for all damage situations, modeling and simulation are needed. The simulation of the damaged composite rotor blade is explained in three parts. The first part explains the mathematical model of a helicopter rotor system. The second part discusses the composite rotor blade cross-sectional analysis. The composite cross-sectional properties are the input to the mathematical model of the helicopter rotor system. The third part presents the modeling of the key damage modes in composite materials. These damage modes are incorporated in the mathematical model of the helicopter rotor system through the composite cross-sectional properties. Details of the forward flight simulation of a damaged composite rotor blade are available in the work of Pawar and Ganguli [16].

MATHEMATICAL MODEL OF A HELICOPTER ROTOR

The helicopter is represented by a nonlinear model of rotating elastic rotor blades dynamically coupled to a 6-degree-of-freedom rigid fuselage [12]. Each blade undergoes out-of-plane bending, in-plane bending, elastic twist, and axial displacement. Governing equations are derived using a generalized Hamilton's principle applicable to nonconservative systems:

$$\int_{\psi_1}^{\psi_2} \left(\delta U - \delta T - \delta W \right) d\psi = 0 \tag{10.1}$$

where δU and δT represent the virtual strain energy and the kinetic energy, respectively. The effect of the composite materials is included in the virtual strain energy through the elastic stiffness matrix. External aerodynamic forces on the rotor blade contribute to the virtual work variational, δW. The aerodynamic forces and moments are calculated using an inflow field from the Bagai–Leishman free wake model, and unsteady effects are incorporated using the Leishman–Beddoes model [12].

A finite element method is used to discretize the governing equation of motion. After finite element discretization using N elements, Hamilton's principle is written as

$$\int_{\psi_i}^{\psi_f} \sum_{i=1}^{N} \left(\delta U_i - \delta T_i - \delta W_i \right) d\psi = 0 \tag{10.2}$$

Each of the beam finite elements has 15 degrees of freedom. These degrees of freedom correspond to cubic variations in axial elastic and bending deflections, and quadratic variation in elastic torsion. There is interelement continuity of slope and displacement for bending deflections and continuity of displacements for elastic twist and axial deflections.

The numerical simulation is carried out at steady cruise flight. The finite element equations for each composite rotor blade are transformed to normal mode space for an efficient solution of the blade response. The nonlinear, periodic, and normal mode equations are then solved for steady response using a finite element in time method and a Newton–Raphson approach. Steady and vibratory components of the rotating frame blade loads are calculated by integrating the blade aerodynamic and inertia forces over the length of the rotor blade. Fixed frame hub loads are calculated by adding the individual blade loads. The equations for the blade response and vehicle equilibrium are solved simultaneously.

COMPOSITE ROTOR BLADE

The composite blade is modeled as a one-dimensional thin-walled beam undergoing extension, torsion, out-of-plane, and in-plane bending using the Chandra and Chopra model [46], which includes constrained warping torsion, and terms due to transverse shear. The stiffness matrix for the composite blade is given as [47]

$$\begin{bmatrix} N \\ M_x \\ -M_y \\ T_s \end{bmatrix} = \begin{bmatrix} K'_{11} & K'_{12} & K'_{13} & K'_{15} \\ & K'_{22} & K'_{23} & K'_{25} \\ sym. & & K'_{33} & K'_{35} \\ & & & K'_{55} \end{bmatrix} = \begin{bmatrix} W' \\ \varphi'_y \\ \varphi'_x \\ \varphi'_z \end{bmatrix} \tag{10.3}$$

where:
K'_{ij} are elements of the stiffness matrix after static condensation
N is the axial force
M_x is the torsion moment
M_y is the flap bending moment
M_z is the lag bending moment
W' is the derivative of the axial displacement

φ'_y is the curvature of flap bending
φ'_x is the curvature of lag bending
φ'_z is the derivative of torsion deflection

The terms in the stiffness matrix depend on the beam cross section and geometry, and are expressed in terms of A, B, and D matrices for composite laminates. The diagonal terms $EA = K'_{11}$, $EI_y = K'_{22}$, $EI_z = K'_{33}$, $GJ = K'_{55}$ are the axial, out-of-plane bending, in-plane bending, and torsion stiffness, respectively.

PROGRESSIVE DAMAGE ACCUMULATION

Matrix cracking, DD, and FB are the predominant damage modes in composite materials. These three damage modes are modeled at the lamina and the laminate level.

MATRIX CRACKING

The effect of matrix cracks in composite materials is included through the extension stiffness matrix A, the extension bending stiffness matrix B, and the bending stiffness matrix D. The stiffness matrices for the presence of matrix cracks $A^{(c)}$, $B^{(c)}$, and $D^{(c)}$ are obtained by subtracting the damage matrices ΔA, ΔB, and ΔD from the stiffness matrices A, B, and D of the undamaged laminates.

$$A^{(c)} = A - \Delta A \tag{10.4}$$

$$B^{(c)} = B - \Delta B \tag{10.5}$$

$$D^{(c)} = D - \Delta D \tag{10.6}$$

These stiffness matrices change with increasing crack density. The dimensionless crack density ρ^k for the ply k with thickness t^k and average crack spacing s^k is defined by

$$\rho^k = \frac{t^k}{s^k} \tag{10.7}$$

The stiffness matrices for the damaged laminate are obtained using the Adolfsson and Gudmundson matrix crack model [48], which links the strain increment due to an array of cracks to the total crack displacement.

DEBONDING/DELAMINATION

After a certain saturation crack density (ϕ_0), the stiffness gets saturated as further matrix cracks reduce the stiffness marginally. However, matrix cracks may induce more severe damage at the crack tip such as DD. The ply stiffness due to the presence of DD can be expressed as

$$Q_{xx}^M(\phi) = rE_{xx}^d(\phi) \tag{10.8}$$

$$Q_{yy}^M(\phi) = rE_{yy}^d(\phi) \tag{10.9}$$

$$Q_{yx}^{M}(\phi) = r\mu_{xy}^{d}(\phi)E_{xx}^{d}(\phi) \tag{10.10}$$

$$Q_{xy}^{M}(\phi) = r\mu_{yx}^{d}(\phi)E_{xx}^{d}(\phi) \tag{10.11}$$

$$Q_{ss}^{M}(\phi) = G_{xy}^{d}(\phi) \tag{10.12}$$

where:

$$r = \left[1 - \mu_{xy}(\phi)\mu_{yx}(\phi)\right]^{-1} \tag{10.13}$$

FIBER BREAKAGE

Fibers are the primary load-carrying elements of the fiber-reinforced composite material. Based on the fiber bundle theory, the effects of FB can be defined as

$$\begin{pmatrix} \bar{\sigma}_{xx} \\ \bar{\sigma}_{yy} \\ \bar{\sigma}_{xy} \end{pmatrix} = \begin{pmatrix} Q_{xx}^{M}(\phi) & Q_{xy}^{M}(\phi) & 0 \\ Q_{yx}^{M}(\phi) & Q_{yy}^{M}(\phi) & 0 \\ 0 & 0 & Q_{ss}^{M}(\phi) \end{pmatrix} \begin{pmatrix} d_f & 0 & 0 \\ 0 & d_f & 0 \\ 0 & 0 & d_f \end{pmatrix} \begin{pmatrix} \bar{\varepsilon}_{xx} \\ \bar{\varepsilon}_{yy} \\ \bar{\varepsilon}_{xy} \end{pmatrix} \tag{10.14}$$

where d_f is the degradation coefficient for FB:

$$d_f = e^{-(A_f/\delta^2)\beta} \tag{10.15}$$

As β is a constant for a given material, d_f varies with the area ratio A_f/δ^2.

LIFE PREDICTION

The stiffness degradation of the structure is related to the life of the structure. The stiffness degradation is expressed as a function of life of the structure using the Mao and Mahadevan model [26].

$$D = q\left(\frac{t}{t_f}\right)^{m_1} + (1-q)\left(\frac{t}{t_f}\right)^{m_2} \tag{10.16}$$

where:

$$D = \frac{E_0 - E}{E_0 - E_f}$$

E_0 is the initial stiffness ($t = 0$)
E_f is the stiffness at final failure time t_f
E is stiffness at any instant of time t

Equation 10.16 is used as a curve fit to link the physics-based damage with the remaining life of the structure. In contrast to other damage models for composite materials, the model in the equation captures the three phases of composite material degradation using one equation. Equation 10.16 is a mathematical model of curves of the type shown in Figure 10.1.

The values of the curve fitting parameters q, m_1, and m_2 need to be obtained. This is done by matching the stiffness reduction values obtained by physical damage modeling and by curve fitting (Equation 10.16) at the initial and the final life of the structure and at the transition points of matrix cracking to DD and DD to FB.

COMPOSITE ROTOR BLADE

A two-cell airfoil section blade with stiffness properties representing a stiff in-plane rotor is developed. Geometric properties and ply orientation of the two-cell airfoil section are illustrated in Figure 10.2. Ply elastic stiffness properties are $E_L = 206$ GPa, $E_T = 20.7$ GPa, $G_{LT} = 8.3$ GPa, and $\mu_{LT} = 0.3$. The length (l) of the blade is 5.08 m. All the laminates used in this chapter are selected from the family of $(0/\pm45/90)_s$ composites. This rotor blade is use to predict life.

Effect on Cross-Sectional Stiffnesses

The stiffness reduction resulting from the three damage modes are linked to the blade life consumption. Figure 10.3 shows the reduction in the normalized bending and torsion cross-sectional stiffness of the two-cell airfoil section beam due to the three damage modes. The stiffness reduction is expressed as a function of the blade life consumption using Equation 10.16.

$$D_1 = 0.3\left(\frac{t}{t_f}\right)^{0.2} + 0.7\left(\frac{t}{t_f}\right)^{8} \tag{10.17}$$

where:

$$D_1 = \frac{EI_{y0} - EI_y}{EI_{y0} - EI_{yf}}$$

EI_{y0} is the initial out-of-plane bending stiffness ($t = 0$)
EI_{yf} is the out-of-plane bending stiffness at final failure time t_f
EI_y is the out-of-plane bending stiffness at any instant of time t

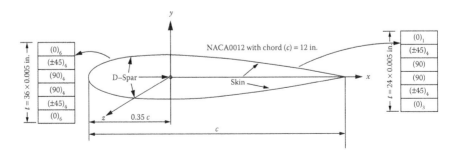

FIGURE 10.2   Details of a two-cell airfoil section blade.

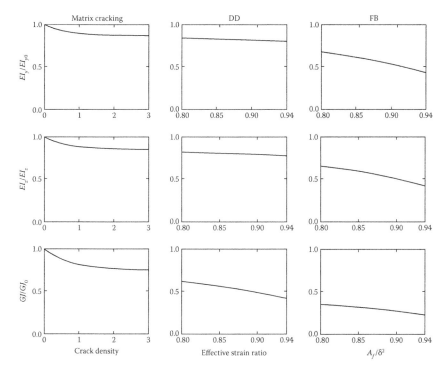

FIGURE 10.3 Change in stiffness of the composite rotor blade due to progressive damage accumulation.

The same model fits the out-of-plane and in-plane bending stiffness because degradation affects the normalized out-of-plane and in-plane stiffness identically.

For torsion stiffness,

$$D_2 = 0.6\left(\frac{t}{t_f}\right)^{0.3} + 0.4\left(\frac{t}{t_f}\right)^{8} \tag{10.18}$$

where:

$$D_2 = \frac{GJ_0 - GJ}{GJ_0 - GJ_f}$$

GJ_0 is the initial torsion stiffness ($t = 0$)
GJ_f is the torsion stiffness at final failure time t_f
GJ is the blade torsion stiffness at any instant of time t

Thus, D_1 and D_2 can be interpreted as continuum damage variables in bending and torsion, respectively.

The bending and torsion stiffness variation due to life consumption is shown in Figure 10.4. To understand the life consumption in the three damage modes, the

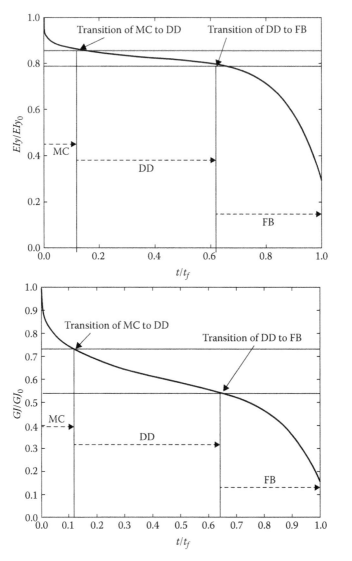

FIGURE 10.4 Decrease in bending and torsion stiffness with increase in life consumption of the structure. DD, debonding/delamination; FB, fiber breakage; MC, matrix cracking.

stiffness reduction plots are split into three zones of matrix cracking, DD, and FB. These correspond to stages I, II, and III of the phenomenological model shown in Figure 10.1. From Figure 10.4, it is observed that the life consumption due to matrix cracking is about 12%–15% of the total life. The life consumption due to DD is about 45%–55% of the total life. The remaining life of the structure is consumed by the final FB failure mode.

Effect on Static Response

Beam stiffness is a virtual indicator of damage growth in composites. A change in the beam stiffness can also change the parameters such as beam static response under a load that can be measured to assess the condition of the blade. The increase of the composite rotor blade response under unit static loads at the free end is given in Figure 10.5. About 12%–15% of the blade life is consumed by matrix cracking. The rate of change in static deflection in the DD zone is relatively slow. Almost 60%–65% of the blade life is consumed due to DD. Though FB is a catastrophic

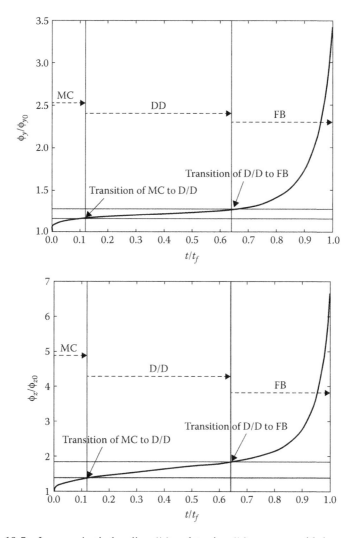

FIGURE 10.5 Increase in tip bending (ϕ_y) and torsion (ϕ_z) response with increase in life consumption of the structure, normalized to undamaged laminate values.

damage mode, the static response shows that the initial phases of FB do not manifest a sudden increase in the static response. We can postulate that the structure can be used up to the transition point of DD to FB zone.

ONLINE HEALTH MONITORING

Static deflection is suitable for nondestructive testing of composite blades. However, static deflection is not suitable for online health monitoring. To simulate online measurements, a numerical simulation of the composite rotor blade is performed. The composite damage models are incorporated into the simulation code. The rotor system response is further linked with life consumption through the relationship between the stiffness loss and the time. The measurement deltas between a damaged and an undamaged blade are used to create two GFSs for the prediction of life consumption due to global and local damage.

NUMERICAL SIMULATION OF MEASUREMENT DELTAS

The numerical simulation is used to study the effect of damage on the response of the structure. A four-bladed composite rotor undergoing progressive damage accumulation in the composite material is considered. Results are obtained in the forward flight at an advance ratio (nondimensional flight speed $\mu = V/\Omega R$) of $\mu = 0.3$, a moderate thrust condition $C_T/\sigma = 0.07$, lock number $\gamma = 6.34$, radius of gyration $mk_{m1}^2/m_0R^2 = 0.000174$, $mk_{m2}^2/m_0R^2 = 0.00061$, and $m/m_0 = 1$. The peak-to-peak values between the undamaged and damaged rotor blades are considered as measurement deltas. The measurement deltas based on deflections and forces are used to develop the global SHM, and those based on shear strains are used to develop the local SHM.

Figure 10.6 shows the displacement deltas due to matrix cracking and DD. Figures 10.7 and 10.8 show the loads during matrix cracking and DD damage mode.

To study the effects of local damage, the peak-to-peak changes in shear strains due to damage are calculated at five locations on the blade ranging from the root to the tip. These peak-to-peak changes in shear strains are given in Figures 10.9 and 10.10 for matrix cracking and DD, respectively. All the strains are calculated on the top of the beam and along the line passing through the point $0.35c$ ($c =$ chord length) on a two-cell airfoil section. The maximum change in shear strain occurs at the damage location along the blade length. Therefore, the shear strain is an useful indicator to locate damage.

FUZZY LOGIC SYSTEM

It is possible to observe the simulated data and use it to estimate damage and structural life. It is easier to develop an automated decision theoretic system to perform damage detection and life prediction. A GFS is developed in this chapter to address the issue of detecting and locating damage and residual life of the structure. The advantage of this approach over straightforward strain monitoring using thresholds is the ability to handle uncertainty and to provide linguistic guidelines for maintenance. A brief description of an FLS is given to ease the understanding of the GFS. Fuzzy logic is a soft computing method that processes numerical data and linguistic knowledge. An FLS maps an input feature vector into a scalar output [49]. It is expressed

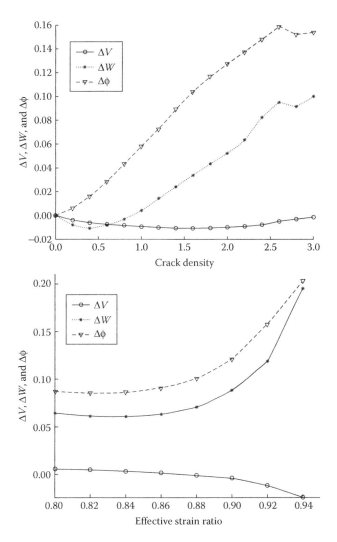

FIGURE 10.6   Peak-to-peak change in the tip in-plane bending (m), out-of-plane bending (m) and torsion (rad) deflection for increasing matrix crack density-and-effective strain ratio.

as a linear combination of basis functions and is an universal function approximator. A schematic diagram of an FLS is shown in Figure 10.11 for $\Delta\omega$ as crisp input and damage and damage location as crisp output. Here, $\Delta\omega$ is a measurement delta.

An FLS maps inputs to outputs using four basic components: rules, fuzzifier, inference engine, and defuzzifier. Rules are obtained from experts or from numerical data. The rules are expressed as a collection of IF-THEN statements such as "IF u_1 is HIGH, and u_2 is LOW, THEN v is LOW."

The fuzzifier converts input numbers into fuzzy sets. It activates rules that are expressed in terms of linguistic variables. An inference engine maps fuzzy sets to fuzzy sets and allows fuzzy sets to be combined. In some applications, numbers

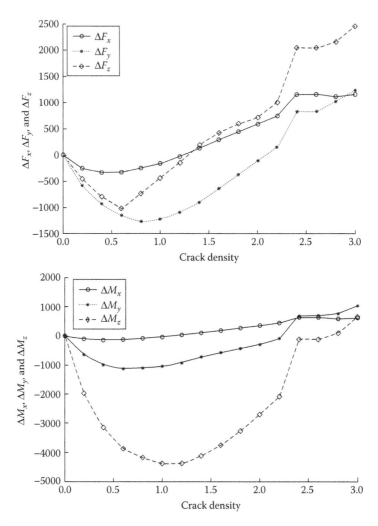

FIGURE 10.7   Peak-to-peak change in blade root forces (N) and moments (N-m) for increasing the matrix crack density.

are needed as an output of the FLS. In such cases, a defuzzifier is used to calculate numbers from linguistic values.

The design of the fuzzy system is difficult due to ambiguity about (1) the best rule set and (2) the need to tune the membership functions. The fuzzy rules and the membership functions must accurately estimate the mapping between the independent and dependent variables. Unfortunately, tuning of membership functions and rule generation are coupled. An appropriate number of fuzzy sets need to be selected. Most investigations use experience to come up with the number of fuzzy sets. The design of fuzzy systems can be automated by using an evolutionary algorithm such as the GA. This leads to the GFS whose design for the rotor life prediction problem is discussed in the "Genetic Fuzzy System" section.

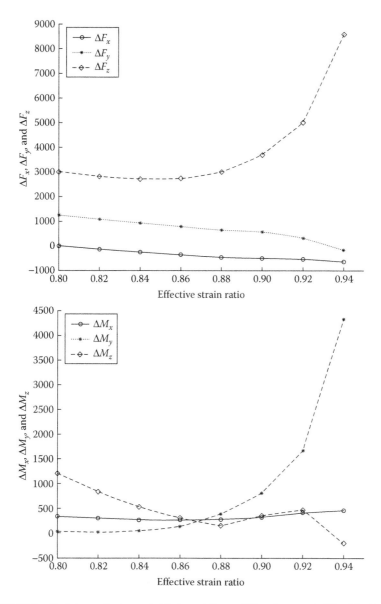

FIGURE 10.8 Peak-to-peak change in blade root forces (N) and moments (N-m) for increasing the effective strain ratio due to DD.

GENETIC FUZZY SYSTEM

There are two types of damage in composite rotor blades. The first type of damage is uniformly distributed along the whole blade and can occur due to the vibratory fatigue loads. The second possibility is that the damage is localized. This may occur due to a sudden impact by a foreign object. In this section, two GFSs are formulated. The global GFS (GGFS) predicts the physical damage and life consumption along

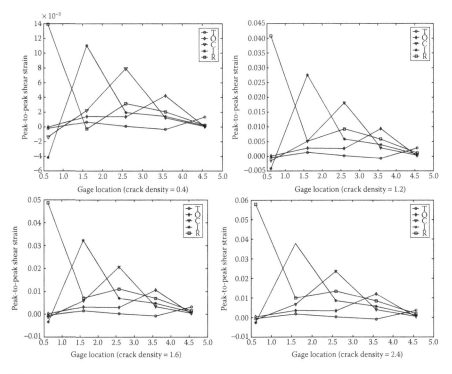

FIGURE 10.9 Peak-to-peak change in shear strains measured at various locations for various crack densities and cracks at various locations (root = 0 m and tip = 5.08 m). Crack locations: T = Tip, O = Outboard, C = Center, I = Inboard, and R = Root.

the whole blade. The local GFS (LGFS) predicts the physical damage and life consumption in various parts of the blade. The FB damage mode in composite materials is considered as a catastrophic damage mode. Therefore, FB is not considered while designing the GFS.

The SHM system development is shown in Figure 10.12. The global and local SHM differ in the choice of sensor measurements z in the figure.

Global Damage Detection

The GGFS is developed to predict the life consumption along the whole blade. Inputs to the GGFS are measurement deltas based on displacement, force, and moment. The outputs of the GGFS are various damage levels.

Physical damage and life consumption parameters are numbers. They are split into several zones based on the physical damage level and split into linguistic variables, as shown in Table 10.1. These classifications are based on the numerical simulations obtained earlier in this chapter. This table also enumerates the relation between the physical damage parameters and the life consumption parameters of the composite rotor blade. The linguistic classifications allow the damage parameters and the life consumption parameters to be grouped into small intervals, which are more robust to the presence of uncertainty. The different levels of damage are used to trigger

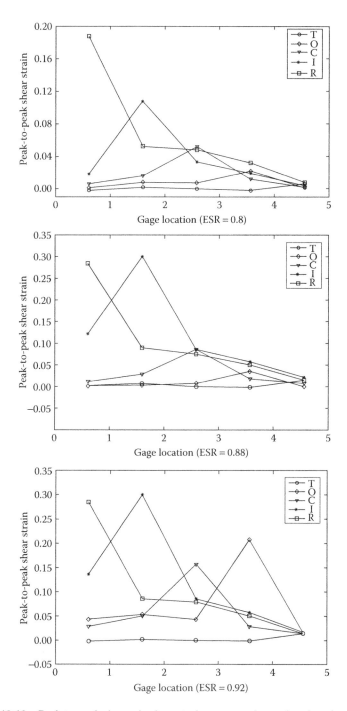

FIGURE 10.10 Peak-to-peak change in shear strains measured at various locations for various effective strain ratios and DD at various locations (root = 0 m and tip = 5.08 m). Crack locations: T = Tip, O = Outboard, C = Center, I = Inboard, and R = Root.

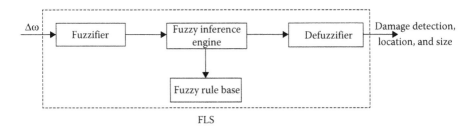

FIGURE 10.11 Schematic representation of an FLS for damage detection with measurement deltas ($\Delta\omega$) as inputs.

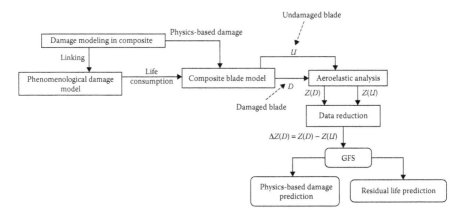

FIGURE 10.12 Schematic representation of development of structural health monitoring system for life prediction.

different alarm levels for prognostic action. The rotor blade is removed at the end of the DD stage when over 60% of the life is consumed.

The measurement deltas are treated as fuzzy variables. Gaussian fuzzy sets are used to define these input variables. The Gaussian membership function can be written as

$$\mu(x) = e^{-0.5\left((x-m)/\sigma\right)^2} \tag{10.19}$$

where:
    m is the midpoint
    σ is the standard deviation

Gaussian fuzzy sets provide smooth transition between the different sets [49]. Moreover, their nonzero value ensures that every rule in the fuzzy system fires. Changes in the measurement deltas are calculated using the simulation for different levels of damages along the whole blade. The midpoints of the Gaussian function are obtained by normalizing the changes in measurement deltas with their maximum values. The GFS is tested using the normalized noisy measurement delta (x), which is given by

$$x = m + u\alpha \tag{10.20}$$

TABLE 10.1
Linguistic Classification of Damage for the GFS for Global Damage Detection

Number	Damage Name	Damage Level	Life Consumption (Residual Life)	Prognostic Action
1	Undamaged	Nil crack density	Nil (100%)	OK
2	Very small crack density	Crack density 0–0.8	0%–2.5% (97.5%)	OK
3	Small crack density	Crack density 0.8–1.2	2.5%–5% (95%)	OK
4	Considerable crack density	Crack density 1.2–1.6	5%–7% (93%)	OK
5	High crack density	Crack density 1.6–2.0	7%–8.5% (91.5%)	OK
6	Very high crack density	Crack density 2.0–2.4	8.5%–10% (90%)	OK
7	Saturation crack density	Crack density 2.4–3.0	10%–12 % (88%)	OK
8	Transition of MC to DD	Crack density 3.0[a] to ESR 0.8[b]	12%–20% (80%)	WATCH
9	Slight DD	ESR 0.8–0.88	20%–43% (67%)	WATCH
10	Moderate DD	ESR 0.88–0.9	43%–50%(50%)	WATCH
11	Severe DD	ESR 0.9–0.92	50–56% (44%)	WATCH
12	Extreme DD	ESR 0.92–0.94	56%–62% (38%)	REMOVE

Note: Prognostic actions—OK: Blade is Okay, no action is required; WATCH: Put blade under watch; and REMOVE: Remove blade. Take for thorough inspection.

[a] Saturation crack density.

[b] The effective strain ratio (ESR) from where the effects of D/D become considerable.

Noise is added to the simulations to increase realism and develop a robust model-based diagnostic system. The noise-level parameter α defines the maximum variance between the computed value of m (normalized) and the simulated measured value x (normalized).

Rule Generation

Rules are obtained by fuzzification of the numerical values obtained from the simulation of the composite rotor blade. The fuzzy sets corresponding to ΔW, $\Delta\phi$, ΔF_x, ΔF_y, ΔF_z, ΔM_x, ΔM_y, and ΔM_z are generated by taking the change in measurements obtained from the simulation as midpoints of membership functions corresponding to a damage level. The degree of membership in the fuzzy set is calculated for each measurement delta corresponding to a given damage level. Each measurement delta is assigned to the fuzzy set with the maximum degree of membership. One rule is obtained for each damage level by relating the measurement deltas. The standard deviation is obtained by maximizing the success rate.

The fuzzy rules provide a knowledge base and represent how a maintenance engineer would interpret data to isolate a damage level using measurement deltas. The fuzzy rules are applied using product implication. Degrees of membership for each of the damage levels are obtained by applying fuzzy rules for a given measurement.

Damage-level isolation requires the most likely damage level to be found. The damage level with the highest degree of membership is chosen as the most likely damage level.

Tuning of the Rules

Consider that N is the total number of classifications and N_c is the number of correct classifications by the GFS. The success rate is defined as

$$S_R = \frac{N_c}{N} \times 100 \tag{10.21}$$

The success rate of the GFS is computed using N noisy training samples. The midpoints of the fuzzy sets are tuned using the simulations. The success rate is a function of the standard deviations (σ_{ij}) of the Gaussian functions for the fuzzy system, that is,

$$S_R = S_R\left(\sigma_{ij}\right) \tag{10.22}$$

The standard deviation of the Gaussian membership functions is calculated using GA for optimization of the success rate.

$$\text{Maximize } S_R\left(\sigma_{ij}\right) \tag{10.23}$$

$$\text{For } \sigma^{min} \leq \sigma_{ij} \leq \sigma^{max}, i = 1, 2, \ldots, M \text{ and } j = 1, 2, \ldots, P$$

where:
   M is the number of rules
   P is the number of measurements

Therefore, the success rate of the GFS is the objective function and the standard deviations σ corresponding to each rule are the design variables for the GA. The details of GA can be found in [50–52]. The use of optimization to design the fuzzy system leads to an optimal diagnostic system, which provides the best results for a given structure, measurement suite, and noise level in the data.

Local Damage Detection

The LGFS is developed to predict the damage at various locations along the blade. Inputs to the LGFS are strain-based measurement deltas at five locations, and outputs are physical damage parameters and life consumption parameter at different locations. The different locations considered are as follows: "Tip" ranges from 0% to 20% of the blade from the free end, "Outboard" from 20% to 40%, "Center" from 40% to 60%, "Inboard" from 60% to 80%, and "Root" from 80% to 100%.

Physical damage and life consumption parameters at each location are split into linguistic variables. Fuzzy rules are defined based on the shear strains obtained for a few key physical damage parameters. These shear strains are first obtained for physical damage parameters and then linked with the life of the blade. The linguistic

TABLE 10.2
Linguistic Classification of Damage for Local Damage Detection

Damage Name	Damage Level	Life Consumption (Residual Life)	Prognostic Action
Undamaged	Matrix crack density 0	Nil (100%)	OK
Small crack density	Matrix crack density 0.4	About 2% (98%)	OK
Moderate crack density	Matrix crack density 1.2	About 5% (95%)	OK
High crack density	Matrix crack density 1.6	About 7% (93%)	OK
Very high crack density	Matrix crack density 2.4	About 10% (90%)	OK
Slight DD	ESR 0.8	About 20% (80%)	WATCH
Moderate DD	ESR 0.88	About 43% (67%)	WATCH
Severe DD	ESR 0.92	About 56% (44%)	REMOVE

Note: Prognostic actions—OK: Blade is Okay, no action is required; WATCH: Put blade under watch; and REMOVE: Remove blade. Take for thorough inspection.

relations are shown in Table 10.2. Again, the blade is removed at the end of the DD zone when almost 60% of the blade life is consumed.

Strain-based measurement deltas $\Delta\varepsilon_{Tip}$, $\Delta\varepsilon_{Outboard}$, $\Delta\varepsilon_{Center}$, $\Delta\varepsilon_{Inboard}$, and $\Delta\varepsilon_{Root}$ are treated as fuzzy variables. Change in strains is calculated using the simulation for a combination of five different locations and seven different levels of damages, and are shown in Figures 10.9 and 10.10. Formulation of the LGFS system and calculation of the success rate are done using the algorithm discussed in Section Global Damage Detection for the GGFS.

Testing of GFS

The GGFS and LGFS are tested at various noise levels. All the measurements are normalized with their maximum value. Success rates of the GGFS are calculated for noise levels of 0.05, 0.10, 0.15, and 0.20. The GFS is tested using all measurements and the results are shown in Table 10.3. The success rates for the LGFS are tested at noise levels of 0.03, 0.05, and 0.10. Table 10.4 shows the success rate for all the rules with various noise levels.

Implementation of the SHM System

The SHM system can be implemented on the composite rotor blade to predict physical damage and residual life of the blade. As shown in Figure 10.13, for GGFS, tip deflection and root forces, and for LGFS, the strains measured at five locations can be compared with the database of measurements from the undamaged blade at a given flight condition. Further, the measurement deltas can be calculated and input to the GFS for prediction of the physical damage and residual life of the blade. As shown in Table 10.1 for global SHM and in Table 10.2 for

TABLE 10.3

Success Rate of Various Testing Noise Levels and Training Noise Level of 0.15 for Global Damage Detection Using All Measurements

Rule Number	Physics-Based Rule	All Measurements			
		$S_{R0.05}$	$S_{R0.1}$	$S_{R0.15}$	$S_{R0.20}$
1	Undamaged	100.00	100.00	100.00	100.00
2	Very small crack density	100.00	100.00	100.00	99.90
3	Small crack density	100.00	100.00	100.00	99.30
4	Considerable crack density	100.00	100.00	98.30	90.20
5	High crack density	100.00	100.00	100.00	100.00
6	Very high crack density	100.00	100.00	100.00	100.00
7	Saturation crack density	100.00	100.00	100.00	100.00
8	Transition of MC to DD	100.00	100.00	100.00	99.50
9	Slight DD	100.00	100.00	95.30	86.30
10	Moderate DD	100.00	100.00	98.90	94.40
11	Severe DD	100.00	100.00	98.00	94.60
12	Extreme DD	100.00	100.00	100.00	100.00
Average		100.00	100.00	99.21	97.02
Minimum		100.00	100.00	95.30	86.30

local SHM, the software can give direct instructions to maintenance engineers based on predictions of life consumption.

CLOSURE

Life consumption of the composite rotor blade is predicted by linking the physics-based damage models with phenomenological models. An automated online SHM approach is developed by using the simulated measurement deltas obtained from a numerical simulation. The GFS for prediction of global physical damage and life consumption is developed by using blade response and loads-based measurements. The GFS for prediction local physical damage and life consumption is developed using the strain-based measurement deltas.

It is found that the total life of the composite rotor blade can be divided into three stages based on the physics-based damage modes. The first stage is dominated by matrix cracking and shows rapid stiffness reduction, which consumes 12%–15% of the total blade life. The second stage is dominated by DD. In this stage, life consumption is about 45%–55% of the total life. The third stage is dominated by FB, which leads to the final failure of the composite rotor blade.

The success rate of the GGFS depends on the number of measurements, the type of measurements, and the noise level in the measurement data. The GGFS and LGFS give good results up to a noise level of 0.2 and 0.05, respectively.

TABLE 10.4

Success Rate of Various Rules for Local Damage Detection Using Strain-Based Measurement Deltas

Rule Number	Rule	$S_{R0.05}$	$S_{R0.10}$
1	Undamaged	100.00	100.00
2	Small CD at Tip	100.00	100.00
3	Small CD at Outboard	100.00	100.00
4	Small CD at Center	100.00	100.00
5	Small CD at Inboard	100.00	100.00
6	Small CD at Root	100.00	100.00
7	Moderate CD at Tip	98.60	81.10
8	Moderate CD at Outboard	99.60	81.50
9	Moderate CD at Center	99.90	86.00
10	Moderate CD at Inboard	100.00	95.50
11	Moderate CD at Root	100.00	99.90
12	High CD at Tip	99.90	47.20
13	High CD at Outboard	100.00	84.20
14	High CD at Center	99.80	68.40
15	High CD at Inboard	100.00	96.10
16	High CD at Root	100.00	83.80
17	Very high CD at Tip	99.30	84.10
18	Very high CD at Outboard	97.70	69.20
19	Very high CD at Center	100.00	89.90
20	Very high CD at Inboard	100.00	93.50
21	Very high CD at Root	100.00	99.40
22	Slight DD at Tip	100.00	100.00
23	Slight DD at Outboard	100.00	100.00
24	Slight DD at Center	100.00	100.00
25	Slight DD at Inboard	100.00	100.00
26	Slight DD at Root	100.00	100.00
27	Moderate DD at Tip	100.00	100.00
28	Moderate DD at Outboard	100.00	100.00
29	Moderate DD at Center	100.00	100.00
30	Moderate DD at Inboard	100.00	100.00
31	Moderate DD at Root	92.90	74.90
32	Severe DD at Tip	100.00	86.00
33	Severe DD at Outboard	100.00	100.00
34	Severe DD at Center	100.00	100.00
35	Severe DD at Inboard	100.00	100.00
36	Severe DD at Root	91.80	73.90
Average		99.43	91.52
Minimum		91.80	47.20

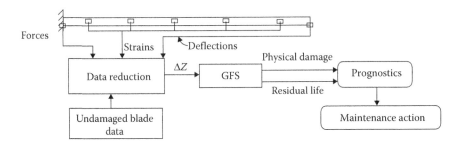

FIGURE 10.13 Schematic representation of implementation of SHM of a composite rotor blade for life prediction.

It should be noted that approximations are made in this chapter due to the use of phenomenological models and only three types of damage are considered. However, such assumptions are needed to obtain estimates of life of real structures.

ACKNOWLEDGMENT

The author acknowledges his former PhD student Prashant Pawar for his helping in conducting this research.

REFERENCES

1. Taylor, S.G., Park, G., Farinholt, K.M., and Todd, M.D., Diagnostics for piezoelectric transducers under cyclic loads deployed for structural health monitoring applications. *Smart Materials and Structures*, 22(2), 2013, 025024.
2. Arsenault, T.J., Achuthan, A., Marzocca, P., Grappasonni, C., and Coppotelli, G., Development of a FBG based distributed strain sensor system for wind turbine structural health monitoring. *Smart Materials and Structures*, 22(7), 2013, 075027.
3. Dos Santos, F.L.M., Peeters, B., Van der Auweraer, H., and Sandoval Góes, L.C., Modal strain energy based damage detection applied to a full scale composite helicopter blade. *Key Engineering Materials*, 569, 2013, 457–464.
4. Pawar, P.M., and Ganguli, R., Genetic fuzzy system for online structural health monitoring of composite helicopter rotor blades. *Mechanical Systems and Signal Processing*, 21(5), 2007, 2212–2236.
5. Wang, S., Danai, K., and Wilson, M., Adaptive method of helicopter track and balance. *Journal of Dynamic Systems Measurements and Control—Transactions of the ASME*, 127(2), 2005, 275–282.
6. Wang, S., Danai, K., and Wilson, M., A Probability-based approach to helicopter rotor tuning. *Journal of the American Helicopter Society*, 50(1), 2005, 56–64.
7. Rosen, A., and Ben-Ari, R., Mathematical modelling of a helicopter rotor track and balance: Theory. *Journal of Sound and Vibration*, 200(5), 1997, 589–603.
8. Ben-Ari, R., and Rosen, A., Mathematical modelling of a helicopter rotor track and balance: Results. *Journal of Sound and Vibration*, 200(5), 1997, 605–620.
9. Azzam, H., and Andrew, M.J., The use of math-dynamic model to aid the development of integrated health and usage monitoring. *Proceedings of the Institution of Mechanical Engineers, Part G: Journal of Aerospace Engineering*, 206(G1), 1992, 71–76.
10. Ganguli, R., Chopra, I., and Haas, D.J., Formulation of a helicopter rotor-system damage detection methodology. *Journal of the American Helicopter Society*, 41(44), 1996, 302–312.

11. Ganguli, R., Chopra, I., and Haas, D.J., Simulation of helicopter rotor-system structural damage, blade mistracking, friction and freeplay. *Journal of Aircraft*, 35(4), 1998, 591–597.

12. Ganguli, R., Health monitoring of helicopter rotor in forward flight using fuzzy logic. *AIAA Journal*, 40(12), 2002, 2773–2781.

13. Pawar, P.M., and Ganguli, R., On the effect of matrix cracks in composite helicopter rotor blade. *Composites Science and Technology*, 65(3/4), 2005, 581–594.

14. Pawar, P.M., and Ganguli, R., Modeling multi-layer matrix cracking in thin-walled composite rotor blade. *Journal of the American Helicopter Society*, 50(4), 2005, 354–366.

15. Pawar, P.M., and Ganguli, R., Modeling progressive damage accumulation in thin walled composite beams for rotor blade applications. *Composites Science and Technology*, 66(13), 2006, 2337–2349.

16. Pawar, P.M., and Ganguli, R., On the effect of progressive damage on composite helicopter rotor system behavior. *Composite Structures*, 78, 2007, 410–423.

17. Yang, M., Chopra, I., and Haas, D.J., Sensitivity of rotor-fault-induced vibrations to operational and design parameters. *Journal of the American Helicopter Society*, 49(3), 2004, 328–339.

18. Stevens, P.L., Active interrogation of helicopter main rotor faults using trailing edge flap actuation. PhD Dissertation, The Pennsylvania State University, Philadelphia, PA, 2001.

19. Lakshmanan, K.A., and Pines, D.J., Damage identification of chordwise crack size and location in uncoupled composite rotorcraft flexbeams. *Journal of Intelligent Material Systems and Structures*, 9(2), 1998, 146–155.

20. Fatemi, A., and Yang, L., Cumulative fatigue damage and life prediction theories: A survey of the state of the art for homogeneous materials. *International Journal of Fatigue*, 20(1), 1998, 9–34.

21. Cárdenas, D., Elizalde, H., Marzocca, P., Abdi, F., Minnetyan, L., and Probst, O., Progressive failure analysis of thin-walled composite structures. *Composite Structures*, 95, 2013, 53–62.

22. Degrieck, J., and Van Paepegem, W., Fatigue damage modeling of fibre-reinforced composite materials review. *Applied Mechanics Reviews*, 54(4), 2001, 279–300.

23. Nicholas, T., and Russ, S.M., Elevated temperature fatigue behavior of SCS-6/Ti 24Al 11Nb. *Material Science Engineering-A*, A153(1/2), 1992, 514–519.

24. Subramanian, S., Reifsnider, K.L., and Stinchcomb, W.W., A cumulative damage model to predict the fatigue life of composite laminates including the effect of a fibre-matrix interface. *International Journal of Fatigue*, 17(5), 1995, 343–351.

25. Halverson, H.G., Curtin, W.A., and Reifsnider, K.L., Fatigue life of individual composite specimens based on intrinsic fatigue behavior. *International Journal of Fatigue*, 19(5), 1997, 369–377.

26. Mao, H., and Mahadevan, S., Fatigue damage modelling of composite materials. *Composite Structures*, 58(4), 2002, 405–410.

27. Adibnazari, S., Farsadi, M., Koochi, A., and Khorashadizadeh, S.N., New Approach for fatigue life prediction of composite plates using micromechanical bridging model. *Journal of Composite Materials*, 49(3), 2014, 309–319.

28. Naderi, M., and Maligno, A.R., Finite element simulation of fatigue life prediction in carbon/epoxy laminates. *Journal of Composite Materials*, 47(4), 2013, 475–484.

29. Umesh, K., and Ganguli, R., Composite material and piezoelectric coefficient uncertainty effects on structural health monitoring using feedback control gains as damage indicators. *Structural Health Monitoring*, 10(2), 2011, 115–129.

30. Moon, T.C, Kim, H.Y., and Hwang, W., Natural frequency reduction model for matrix-dominated fatigue damage of composite laminates. *Composite Structures*, 62(1), 2003, 19–26.

31. Bedewi, N.E., and Kung, D.N., Effect of fatigue loading on the modal properties of composite structures and its utilization for prediction of residual life. *Composite Structures*, 37(3/4), 1997, 357–371.

32. Roy, N., and Ganguli, R., Filter design using radial basis function neural network and genetic algorithm for improved operational health monitoring. *Applied Soft Computing*, 6(2), 2006, 154–169.

33. Roy, N., and Ganguli, R., Helicopter rotor blade frequency evolution with damage growth and signal processing. *Journal of Sound and Vibration*, 283(3–5), 2005, 821–851.

34. Ganguli, R., Chopra, I., and Haas, D.J., Detection of helicopter rotor system simulated faults using neural networks. *Journal of the American Helicopter Society*, 42(2), 1997, 161–171.

35. Reddy, R.R.K., and Ganguli, R., Structural damage detection in a helicopter rotor blade using radial basis function neural networks. *Smart Structures and Materials*, 12(2), 2003, 232–241.

36. Ramanujam, N., Nakamura, T., and Urago, M., Identification of embedded interlaminar flaw using inverse analysis. *International Journal of Fracture*, 132(2), 2005, 153–173.

37. Iwasaki, A., and Todoroki, A., Delamination identification of CFRP structure by discriminant analysis using mahalanobis distance. *Key Engineering Materials*, 270–273(1–3), 2004, 1859–1865.

38. Meruane, V., and Heylen, W., An hybrid real genetic algorithm to detect structural damage using modal properties. *Mechanical Systems and Signal Processing*, 25(5), 2011, 1559–1573.

39. Chandrashekhar, M., and Ganguli, R., Uncertainty handling in structural damage detection using fuzzy logic and probabilistic simulation. *Mechanical Systems and Signal Processing*, 23(2), 2009, 384–404.

40. Chandrashekhar, M., and Ganguli, R., Damage assessment of structures with uncertainty by using mode-shape curvatures and fuzzy logic. *Journal of Sound and Vibration*, 326(3), 2009, 939–957.

41. Beena, P., and Ganguli, R., Structural damage detection using fuzzy cognitive maps and Hebbian learning. *Applied Soft Computing*, 11(1), 2011, 1014–1020.

42. Lee, J., Wu, F., Zhao, W., Ghaffari, M., Liao, L., and Siegel, D., Prognostics and health management design for rotary machinery systems—Reviews, methodology and applications. *Mechanical Systems and Signal Processing*, 42(1), 2014, 314–334.

43. Pawar, P.M., and Ganguli, R., Genetic fuzzy system for damage detection in beams and helicopter rotor blades. *Computer Methods in Applied Mechanics and Engineering*, 192(16–18), 2003, 2031–2057.

44. Pawar, P.M., and Ganguli, R., Matrix cracking detection in thin-walled composite beam using genetic fuzzy system. *Journal of Intelligent Material Systems and Structures*, 16(5), 2005, 381–468.

45. Pawar, P.M., and Ganguli, R., Fuzzy-logic-based health monitoring and residual-life prediction for composite helicopter rotor. *Journal of Aircraft*, 44(3), 2007, 981–995.

46. Chandra, R., and Chopra, I., Structural response of composite beams and blades with elastic couplings. *Composites Engineering*, 2(5–6), 1992, 347–374.

47. Ganguli, R., and Chopra, I., Aeroelastic optimization of a helicopter rotor with two-cell composite blades. *AIAA Journal*, 34(4), 1996, 835–841.

48. Adolfsson, E., and Gudmundson, P., Thermoelastic properties in combined bending and extension of thin composite laminates with transverse matrix cracks. *International Journal of Solids Structures*, 34(16), 1997, 2035–2060.

49. Kosko, B. *Fuzzy Engineering*. Prentice Hall, Upper Saddle River, NJ, 1997.

50. Ganguli, R., *Engineering Optimization: A Modern Approach*. CRC Press, Boca Raton, FL, 2012.

51. Deb, K., *Optimization for Engineering Design: Algorithms and Examples*. PHI Learning, New Delhi, India, 2004.

52. Mitchell, M., *An Introduction to Genetic Algorithms (Complex Adaptive Systems)*, A Bradford Book. MIT Press, Cambridge, 1998.

11 In Situ Structural Health Monitoring Systems for Aerospace Structures
Recent Developments

Ratneshwar (Ratan) Jha, Rani Warsi
Sullivan, and Ramadas Chennamsetti

CONTENTS

INTRODUCTION

Aerospace structures are subjected to high-design loads under complex operational and environmental conditions as well as discrete damaging loads such as impacts. Although maintenance constitutes a significant part of aircraft life cycle cost, the existing paradigm for ensuring structural safety may allow an undetected damage to grow to a critical size between major inspections (schedule-based maintenance) and cause catastrophic failure. In addition, the use of composites for aircraft structures is increasing rapidly due to several advantages such as lighter weight, fewer joints, improved fatigue life, and higher resistance to corrosion. However, composite structures have several forms of damage such as delamination between plies, fiber–matrix debonding, fiber breakage, and matrix cracking. These damages often occur below the surface due to fatigue, foreign object impact, and so on, and may not be visible.

As noted in a recent National Aeronautics and Space Administration (NASA) publication [1], visual inspection is the primary (approximately 80%) nondestructive inspection method used for maintaining vehicle safety between major inspections. Once a damage is detected during visual inspection, nondestructive evaluation (NDE) methods are used for damage quantification. There are several existing NDE techniques, including ultrasonic, acoustic emission (AE), radiography, shearography, and thermography [2,3]. Heida and Platenkamp [4] evaluated several NDE methods for their capability in detection, sizing, and depth estimation of defects present in composite aerospace structures. They concluded that all methods have their specific advantages and limitations making them suitable for a particular defect type. Bossi and Giurgiutiu [5] have reported NDE methods that are recommended (or not recommended) for detection of various types of damages in composites. Visual inspections have serious limitations, and the current NDE methods often necessitate the removal of individual components for evaluation and require point scanning; in general, they are time consuming and expensive.

A structural health monitoring (SHM) system capable of performing diagnostics efficiently (considering both time and cost) between major scheduled inspections and/or on demand during flight would lead to a paradigm shift for maintaining aerospace vehicle safety. The goal of SHM is to determine and classify damages (presence, location, and severity) for a structure exposed to varying environmental and operational conditions as well as instrumentation noise (i.e., "real-world" conditions) while eliminating false indications. SHM has tremendous potential to improve safety and reduce life cycle cost by detecting damages early and allowing a much more efficient maintenance schedule (condition-based maintenance). In addition, an onboard SHM system can provide on-demand health state of a vehicle and enable reconfiguration of onboard flight control laws in real time for best possible use of the residual control effectiveness following any damage.

SHM using Lamb waves [6] has shown a lot of promise for composite structures. Lamb waves are elastic waves resulting from superposition of guided longitudinal and transverse (shear) waves. Lamb waves travel in thin plates with unconstrained boundaries and have the capability of traveling long distances with little attenuation. Local stiffness change, crack, or delamination causes reflection, dispersion, attenuation, and mode conversion. Based on the measured time history of the propagated wave, traveling time, speed reduction, wave attenuation, mode conversion, and so on, parameters are extracted and used as damage identification variables. Development of Lamb wave-based SHM algorithms is a very active area of research [2–3,7–10]. Lamb waves (usually fundamental symmetric and asymmetric modes) can be used to detect both superficial and internal flaws. Signal processing plays a vital role in extracting damage information from wave propagation data. Due to the dispersive nature of the Lamb waves, accurate identification of localized events is required to determine the location of the damage in a structure [3].

SHM is essentially an inverse problem in which the measured data are used to predict the current condition of a structure. SHM algorithms may be classified as (1) model-based technique and (2) model-free monitoring data interpretation. The model-based SHM techniques require a validated mathematical model of the monitored structure (generally using the finite element [FE] method [FEM]) whose predictions are compared with the current response for damage detection. However, detection of small damages requires high-frequency Lamb waves, and FEM mesh size should be sufficiently refined (usually 10–20 nodes per wavelength) to obtain an accurate solution. This leads to very high-computational cost making FEM simulations problematic for SHM. Spectral FEMs (SFEMs), essentially FE in the frequency domain, are well suited for high-frequency wave motion analysis and SHM [11–13]. However, current SFEM modeling capability is limited to relatively simple structures (two-dimensional plates and their assemblies).

The model-free SHM methods involve interpretation of monitoring data by comparing the current responses to those of the baseline (healthy) structure. However, it is very difficult in practice to generate a large database for baseline structure. Without such a database available for computing a damage index (DI), variations in the operational and/or environmental conditions may mask the changes in the response caused by structural damage. The "time-reversal" method (TRM) is a novel approach to perform Lamb wave-based SHM without the need for baseline data. Applications of TRM using the pitch-catch method have received some attention [9,14–16].

Uncertainties in SHM arise from multiple sources, such as manufacturing, data acquisition, signal processing, computation of DI, and model parameters. Deterministic damage identification provides limited information for decision making and may introduce false alarms and unnecessary maintenance actions. Therefore, uncertainties in damage diagnostics must be scientifically included and accurately quantified. Probabilistic modeling and analysis, including traditional uncertainty propagation analysis and sensitivity analysis, and advanced Bayesian/entropy-based analysis with novel numerical techniques are key components for reliable diagnostics, effective condition-based maintenance, and life cycle cost reduction. Lopez and Sarigul-Klijn [17] have presented a comprehensive review of uncertainties involved in flight vehicle structural damage diagnosis.

The sensors constitute a vital part of any SHM system. The *Encyclopedia of Structural Health Monitoring* [2] presents a comprehensive view of the various SHM sensors, including electrical strain gauges, optical fibers, piezoelectric (lead zirconate titanate [PZT], polyvinylidene fluoride [PVDF]) sensors, electromagnetic sensors, and capacitive sensors. This chapter focuses on SHM systems using piezoelectric and optical fiber sensors only because they seem to be the most suitable sensing options. It is noted that the airbus approach to SHM development and deployment (Table 11.1) consists of four generations in which the SHM system is given step by step more complexity, features, and responsibility [18]. The comparative vacuum method is an example of generation 1 SHM system, whereas SHM systems using piezoelectric arrays and optical fiber sensors are classified as higher generation. Airbus is currently testing the impact damage detection using acoustic–ultrasonic system consisting of piezoelectric arrays for two kinds of fuselage ground validators as well as in-flight validators using A350/A340 aircraft [18].

TABLE 11.1
Airbus Approach to SHM Development and Deployment

Generation 0	For the monitoring of structural tests—already deployed in the course of the A380 and A350 XWB component and full-scale testing
Generation 1	For in-service aircraft with benefits in maintenance—has achieved technology readiness for selected technologies and given applications
Generation 2	For in-service aircraft with benefits in maintenance and weight savings at the component level—under development
Generation 3	For in-service aircraft with benefits in maintenance, weight savings at the aircraft level, and intrinsic quality assessment in manufacturing and assembly—under development

Source: Blumenfeld, L., ed. (2014). *Airbus Technical Magazine #54—Flight Airworthiness Support Technology.* http://www.airbus.com/support/publications/ (accessed May 15, 2015). With permission.

PIEZOELECTRIC ACTUATION/SENSING-BASED SHM OF AEROSPACE STRUCTURES

ELASTIC WAVE PROPAGATION: LAMB WAVES AND DAMAGE INTERACTIONS

Lamb waves are ultrasonic waves, which propagate in plate-like structures [19]. Because these waves are guided by top and bottom surfaces of the plate, they are also known as guided waves. When a guided wave propagates through a plate, the whole thickness of the plate undergoes deformation. Therefore, Lamb waves are sensitive to subsurface defects in addition to surface defects. Because Lamb waves are sensitive to defects such as delamination, disbond, matrix cracking, and fiber breakage [20]. Many researchers have explored the use of these waves for NDE and SHM applications [21].

In general, the features of Lamb waves, which are used for damage detection, are mode conversion, time of flight, change in amplitude, change in wave velocity, dispersion, reflection and transmission characteristics, confinement of particular modes at defective regions, and so on. However, before implementing SHM methodology using Lamb waves, it is necessary to understand *a priori* the interaction phenomena between the Lamb mode(s) employed for SHM and the defects to be monitored. The interaction phenomena depend on various factors such as the type of Lamb mode (such as symmetric or antisymmetric), the geometry and orientation of the defect, and the frequency of excitation.

When the fundamental antisymmetric Lamb mode (A_0) interacts with the front edge of a delamination/disbond in a composite laminate, it splits into two wave groups, and one wave group propagates in each sub-laminate. During this interaction, the fundamental symmetric Lamb mode (S_0) is generated, and it propagates in the sub-laminates [22–24]. At the rear edge of the delamination, the A_0 and S_0 modes interact with the edge and start propagating from one sub-laminate to the other. These modes are termed as "Turning Lamb modes" [25]. Propagation of high-amplitude Lamb modes from the rear edge of a delamination were observed in the pulse-echo mode [26]. The genesis of these waves is the Turning Lamb modes and reflections at the rear edge. In a symmetric laminate, if the delamination is located symmetrically, the generated S_o mode is confined only to the sub-laminates. Otherwise, the S_o mode propagates in the main laminate also.

Interaction of Lamb waves with through-hole damage in a composite laminate is presented in Reference [27]. From experiments, it is found that the amplitude of the transmitted wave across the hole is significantly less compared to that captured in the healthy regions. It is also shown that the stacking sequence of the laminate influences the scattering of A_o mode at a through hole [28]. From the studies carried out on uni-directional, bidirectional, and quasi-isotropic laminates, it is observed that the scattering directivity pattern of the A_o mode is influenced by a hole diameter-to-wavelength ratio and ply stacking sequence.

Benmeddour et al. [29] experimentally investigated the interaction of the fundamental Lamb modes with symmetrical notches in aluminum plates. The specific Lamb mode was excited by using two identical thin piezoceramic transducers placed at the opposite sides of the plate. The reflected and transmitted wave groups at the notch were captured using conventional transducers located on the plate surface in the front of and behind the defect. The power reflection and transmission coefficients were estimated for various depths of symmetrical notches. It is found that the A_o mode is more sensitive to the symmetrical notches than the S_o mode considering the power reflection coefficients. Ke et al. [30] developed a three-dimensional FE-based model to simulate the propagation of ultrasonic-guided Lamb wave that was generated and detected by an air-coupled ultrasonic scanning system along the structures made of anisotropic, viscoelastic materials, and their scattering by defects of complex shapes. Experimental validation of the FE model was carried out on an aluminum plate with a through-thickness hole, a glass–polyester composite plate with an impact damage, and a composite tank containing a Teflon film between its titanium liner and carbon–epoxy winding. Shen et al. [31] proposed a hybrid boundary element method approach to study the interaction of Lamb waves with transverse internal defects. Variations in transmission and reflection coefficients with respect to the defect height, the width, and the product of frequency and thickness were estimated. The above studies help in providing guidelines for internal defect detection using Lamb waves. It is clear that a thorough understanding of interactions of interrogating Lamb waves with defects is key for successful implementation of SHM schemes.

PIEZOELECTRIC TRANSDUCERS FOR LAMB WAVE ACTUATION AND SENSING

Selection of sensors for creating a network in SHM applications is vitally important. Lamb waves can be transmitted and received using both contact and noncontact transducers. In contact-based techniques, there are two categories. In the first category, an actuator/transducer is placed over the specimen, and in general, a thin layer of couplant (film) is present between the transducer and the specimen. The coupling of energy between the transducer and the specimen depends on the thickness of the film and the pressure applied on the transducer. In the second category, the actuator is permanently attached (bonded) to the specimen. Because the actuator is bonded to the specimen, good coupling of energy between the actuator and the specimen is ensured.

In general, the actuators/transducers in the first category are bulky and not suitable for *in situ* SHM applications. Transducers in the second category, active piezoelectric patches (wafers) that can act as both the transmitter and the receiver, are lightweight and can be an integral part of the host structure through embedding in composite laminates or surface bonding [32]. Because piezoelectric wafers are small in size, they require less power to excite Lamb waves. Therefore, this further reduces the size and weight of pulser–receiver equipment. Once the sensor network is integrated to a host structure, it is possible to carry out online and offline health monitoring of the structure. Because Lamb wave exhibits multimodal characteristics, several modes may propagate at a given frequency. Therefore, sometimes it is difficult to excite a desired Lamb mode using a network of piezoelectric patches. Piezoelectric actuators/sensors are available in various shapes and sizes for SHM applications [33].

PZT Actuators/Sensors for SHM—Implementation Examples

In SHM, efforts are always made toward the assessment of health and integrity of a structure. When structures are deployed in various operating conditions, it is possible that the stiffness and strength of the structure diminish, in general, with time. As a part of SHM, the effective mechanical elastic properties are evaluated using a network of sensors. Vishnuvardhan et al. [34] developed a single transmitter and multiple receiver (STMR) array patches, as shown in Figure 11.1a, using two flexible printed circuit board-based patches for (1) the reconstruction of the material state (material characterization) using a single-quadrant, double-ring STMR array, and (2) the SHM of anisotropic plate-like structures using a full-ring STMR array. Experimental demonstration was performed using a 3.15 mm-thick graphite–epoxy composite plate, as shown in Figure 11.1b. Genetic algorithm-based inversion algorithm using the velocity measurements of the fundamental symmetric Lamb mode (S_o) was used for prediction of elastic moduli of the plate. Data from the full-ring STMR array was used

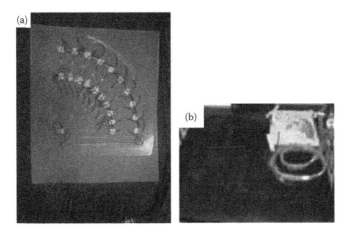

FIGURE 11.1   (a) Single-quadrant, double-ring STMR array. (b) The array positioned over graphite–epoxy laminate. (From Vishnuvardhan, J. et al., *NDT&E International*, 42, 193–198, 2009. With permission.)

in phase addition reconstruction algorithm for imaging reflections from edges and defects present in the anisotropic plates.

Lu et al. [27] investigated Lamb wave characteristics in a carbon fiber/epoxy composite panel of a stacking sequence $[0/\pm45_2/90/0_2]_s$ having five reinforced stiffeners (co-cured on its surface) and two holes as shown in Figure 11.2a. A total number of 18 patches were bonded in a 6×3 array, as shown in Figure 11.2b. Initially, studies were conducted on attenuation and dispersion of Lamb waves induced by stiffeners, which were perpendicular to the direction of propagation and dependent on the propagation velocity of Lamb waves.

A through hole of 11.9 mm diameter was drilled between S_9 and S_{15} sensors, as shown in Figure 11.2b. An inverse algorithm based on correlations between the digital damage fingerprints (DDFs) of Lamb wave signals, collected in different

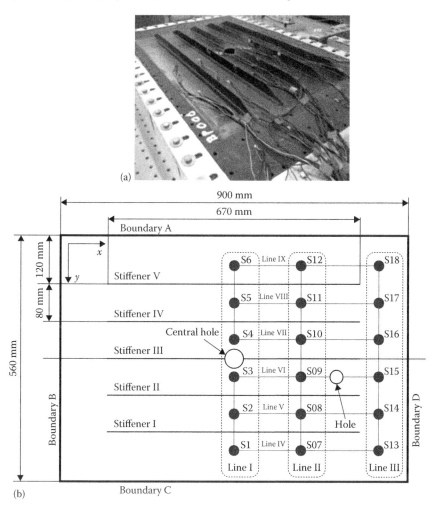

FIGURE 11.2 (a) CFRP plate with stiffeners. (b) Locations of piezo patches. (From Lu, Y. et al., *Composite Materials*, 43, 3211–3230, 2009. With permission.)

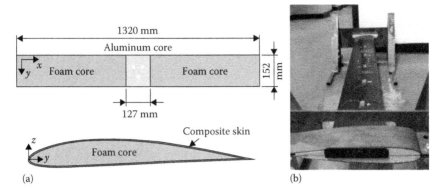

FIGURE 11.3   (a) Cross section and (b) composite wing fixed in the fixture for low-velocity impact test. (From Liu, Y. and Chattopadhyay, A., 2012, Journal of Intelligent Material Systems and Structures, 2074–2083. With permission.)

combinations of actuator–sensor paths, in the reference and damage structures, was developed for identification of the location of the hole. The proposed approach can be used for detection of delaminations, disbonds, impact damages, and so on.

Liu and Chattopadhyay [35] proposed a methodology for monitoring and quantifying the impact damage on a sandwich composite unmanned aerial vehicle (UAV) wing, with aluminum as the core and carbon-fiber-reinforced polymer (CFRP) as the skin, as shown in Figure 11.3. Macro-fiber composite (MFC) transducers were bonded on the surface of the wing to transmit and receive Lamb waves. A series of 10 low-velocity impact tests were carried out on the wing, and both the impact force and the velocity were recorded to estimate the impact energy absorbed by the wing. The whole impact area was divided into eight sections. In each test, the impact energy was 11 J. No visible damages were observed in the first eight tests, but after the ninth and tenth impacts, visible dents were observed on the composite skin. It was found that after the tenth impact, there were multiple types of damages, such as foam cracking, debonding between the foam core and the composite skin, fiber breakage, and delaminations in the composite skin. To predict the damages in the wing, nine statistical features of the signals were used. Compared to conventional principal component analysis (PCA) algorithms, it was shown that kernel PCA (KPCA) shows better damage quantification capability.

PZT-Based SHM—Algorithms for Damage Detection and Signal Processing

Signal processing and damage detection algorithms play an important role in successful implementation of SHM in any structure. Signal processing can be carried out in time and frequency domains, and also in joint time–frequency domain. Because Lamb waves are dispersive, the time–frequency domain is more convenient way for representation of the waves. Frequently used time–frequency representation techniques for Lamb waves are wavelet transform, Wigner–Ville distribution, and short-time Fourier transform (STFT).

Ng and Veidt [36] proposed a methodology that employs a network of sensors to sequentially scan the structure before and after damage by transmitting and receiving Lamb waves. Initially, a damage localization image is reconstructed by analyzing the cross-correlation of the scatter signal envelope with the excitation pulse envelope for each transducer pair. From this, a potential damage area is reconstructed by superimposing the image observed from each actuator and sensor signal path. The proposed methodology was verified through experiments and numerical studies. Lu et al. [27] developed an inverse algorithm based on correlations between the DDFs of wave signals in the reference and damaged structures for identification of damage. Cross-correlation coefficient, which varies from –1 to +1, can be used for the assessment of occurrence of damage [37]. Raw data from experiments can directly be used for the estimation of correlation coefficients and damage identification [38]. Lu et al. [27] extracted DDFs from experimental signals and used to achieve data compression for the estimation of correlation coefficients. Various combinations of actuator–sensor paths were used to predict the location of damage. Compared to the results from raw data, the proposed technique demonstrated an improved accuracy in damage detection.

Liu et al. [35] used nine statistical features in time and frequency domains on signals captured using MFCs. The time domain techniques employed for extracting the first five statistical features are root mean square (RMS), variance, peak value, peak-to-peak value, and kurtosis. Techniques employed for converting the signals to the frequency domain are Fourier transform and Hilbert transform. A DI based on the Mahalanobis distance was defined to quantify the impact damages. It was demonstrated that KPCA has better damage quantification capability than PCA. In the KPCA algorithm, the input features are transformed into a high-dimensional feature space using nonlinear mapping. It was shown that the KPCA can extract higher order and nonlinear information, and more robust damage quantification is possible.

Samaratunga et al. [13] formulated wavelet spectral finite elements (WSFEs) based on the first-order shear deformation theory for accurate and efficient wave propagation analysis at high frequencies. The WSFE method yields up to 2-order-of-magnitude reduction in the computational time as well as a direct relationship between the system input and output in the frequency domain (frequency-dependent dynamic stiffness matrix), which is well suited for SHM applications. Samaratunga et al. [39] demonstrated delamination detection in a composite laminate by combining the dynamic stiffness matrix of healthy structure (WSFE model) and the current measured response of the structure to compute the force necessary to create damage (Figure 11.4). The resulting DI provides quantitative information regarding the location and severity of damages. Also, this method does not need any baseline data (beyond those required for model validation), the knowledge of input is not necessary, and it is computationally very fast.

PZT Sensors for SHM—Phased Arrays for SHM and Time Reversal

Phased array contains multiple transduction elements, generally piezoelectric wafer active sensor (PWAS), which can be controlled electronically in order to have required time delays. Using the phased array, beam forming is possible and a strong Lamb

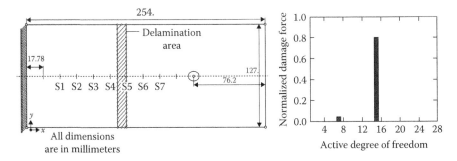

FIGURE 11.4 Cross-ply eight-layer composite plate with delamination (left) and DI values using experimental response measurements (right).

mode can be made to propagate in a desired direction. Giurgiutiu [40] demonstrated the detection of a crack in an aluminum plate using phased array of PWAS in the pulse-echo mode. It was also shown that good coupling is achieved when the length of PWAS is equal to half of the wavelength of a particular Lamb mode. Purekar and Pines [41] employed Lamb waves generated and received using a piezoelectric sensor array for detection of delamination location in the pulse-echo mode in a cross-ply laminate.

Time reversibility of Lamb waves is based on the fact that if there is no damage, then the excited signal can be reconstructed by sending back the received signal. Reconstruction of the excited signal fails if there is damage in the path of propagation of Lamb wave. Poddar et al. [42] proposed a baseline-free damage identification technique using the time reversibility of a Lamb mode. Initial studies were performed on a healthy metallic plate. The similarity between the waveforms (original and reconstructed) was quantified using the RMS deviation. A block mass, a notch, and an area were considered as representatives of various damages in an aluminum plate. They demonstrated that Lamb wave time reversal technique can detect the above representative damages in the baseline-free environment. Zhu [43] et al. proposed a time-reversal technique in the frequency–wavenumber domain for fast detection of plate damage using scattered flexural plate waves. Using the cross-correlation imaging condition in the frequency domain, damage images can be quantitatively obtained. The proposed technique was 2 orders of magnitude faster than the time-reversal technique in the time domain. Experimental validation on an aluminum plate with two artificial damages proved that the technique is a useful tool for faster damage identification in a real-time SHM scheme.

Watkins and Jha [16] developed a modified TRM (MTRM) which can reduce the hardware (PZT power amplifier, wiring, etc.) requirements significantly by using a single actuator and multiple sensors. For any signal path, the MTRM requires only one transducer to actuate signals and the other transducer to sense signals. This is in contrast with conventional TRM in which both transducers need to work as actuators and sensors. Furthermore, the developed MTRM (Figure 11.5) can be used with noncontact sensors such as a laser vibrometer. The MTRM was used for diagnosis of damage severity in a composite plate impacted incrementally to cause three levels of damage. It was observed that the damage indicator increased consistently (only for the signal path through the damage area) as the damage level was increased. The MTRM clearly identified the path of damage location and its severity.

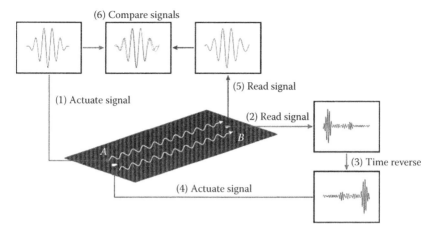

FIGURE 11.5 MTRM using a pitch-catch arrangement of transducers. (From Watkins, R. and Jha, R., *Mechanical Systems and Signal Processing*, 31, 345–354, 2012. With permission.)

OPTICAL FIBER SENSING–BASED SHM OF AEROSPACE STRUCTURES

Technological advances have enabled the development of a number of optical fiber sensing methods over the past few years. The most prevalent optical technique involves the use of fiber Bragg grating (FBG) sensors. These small, lightweight sensors have many attributes that enable their use for a number of measurement applications, particularly for SHM [44]. This section gives an overview of the implementation of FBG sensors for large-scale aerospace structures and applications.

FBG FUNDAMENTALS

Optical fibers consist of a light-transmitting glass filament that is surrounded by a cladding, as shown in Figure 11.6. These fibers are usually single-mode telecommunication-grade fibers that have a pure silica cladding and a germanium-doped silica core [45,46]. This enables the cladding to have a lower refractive index relative to the glass fiber, thus enabling light to propagate only within the core (4–9 μm diameter) of the optical fiber. FBGs, first demonstrated by Hill [47], are created by "writing" small segments in the photosensitive core of an optical fiber using an intense ultraviolet (UV) source such as a UV laser. Depending on the intensity of the light used, the refractive index of the fiber is permanently altered, and the resulting periodic variation is called a FBG. As shown in Figure 11.6, when a broadband light is transmitted into the optical fiber, a narrow band of light is reflected based on the spacing of the periodic variation and the refractive index variation that is present in the core of the fiber. The FBG scatters the light propagating inside the fiber core, and the out-of-phase components form destructive interference and cancel each other. The in-phase components of light sum constructively, forming a reflected spectrum with a center wavelength known as the Bragg wavelength λ_B, which is defined as [46,47]

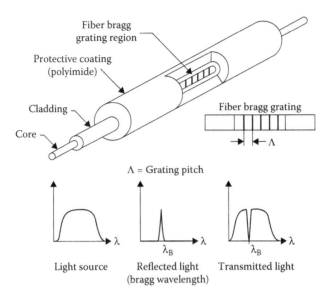

FIGURE 11.6 Schematic diagram of an FBG sensor with period Λ inside a single-mode optical fiber. (From Keulen, C.J. et al., *Journal of Reinforced Plastics and Composites*, 30, 1055–1064, 2011. With permission.)

$$\lambda_{\mathrm{B}} = 2n_{\mathrm{eff}}\Lambda \tag{11.1}$$

where:
    n_{eff} is the effective refractive index
    Λ is the grating period (spacing between gratings—see Figure 11.6)

The change in spacing of the periodic refractive index is a function of strain and temperature. Under loading (mechanical or thermal), the fiber stretches due to strain and thermal expansion, causing a change in the Bragg grating period and the refractive index, resulting in a shift of the Bragg wavelength. Therefore, in response to the strain ε and the change in temperature ΔT, the Bragg wavelength λ_{B} shifts by an amount $\Delta\lambda_{\mathrm{B}}$ as

$$\frac{\Delta\lambda_{\mathrm{B}}}{\lambda_{\mathrm{B}}} = \left(1 - P_e\right)\varepsilon + \left(\alpha_S + \alpha_f\right)\Delta T \tag{11.2}$$

where:
    P_e is the strain-optic coefficient
    α_S is the thermal expansion coefficient of the fiber
    α_f is the thermal-optic coefficient

Because both the temperature and the strain influence the reflected wavelength of an FBG, temperature compensation must be introduced for accurate strain measurements.

FBG Attributes

The small size and mass of FBGs make these sensors prime candidates for SHM applications in the aerospace industry. A primary benefit of using the FBG technology is the multiplexability of the optical sensors that enables the monitoring of a high-density strain distribution using a single fiber [48,49]. In applications that require a large number of sensors, significant cost savings per sensor is realized compared to the installation cost of conventional foil strain gauges. Unlike conventional sensors, FBGs are nonconductive, electrically passive, and immune to electromagnetic interference, enabling their use in environments subject to noise, corrosion, or high voltage [50]. These diminutive sensors can be directly integrated into polymer matrix composite (PMC) structures [46] or surface mounted [44] on the test object. FBGs, which have the capability to measure very large strains (>10,000 μm/m), can be used to monitor advanced PMCs that are often used in critical load-bearing structures. FBGs are not distance dependent, and have long-term stability and good corrosion resistance [51].

FBG Issues

There are also challenges when using FBGs. Foremost among the issues, these sensors show high temperature dependence. Kreuzer [51] states that the ratio of the change in wavelength to the reference Bragg wavelength $\Delta\lambda_B/\lambda_B$ caused by 1°C is proportional to $\Delta\lambda_B/\lambda_B$ caused by 8 μm/m mechanical strain. Therefore, temperature compensation must be included. Also, at high temperatures, there can be distortion of the optical fibers due to the mismatch in the thermal expansion coefficients of the cladding, core, and any coating. Because Bragg gratings are located in the center of the optical fiber, they are displaced from the surface of the test article and the sensor location must be taken into consideration. There are also issues concerning the mounting or attachment of these sensors. Betz et al. [52] investigated attaching methods for optical sensors and concluded that surface mounting is preferable to other methods. Particularly in the case of embedding FBGs within test structures, the relationship between the strain distribution and the sensor signal is complex and requires clear understanding to obtain accurate results. Also, the issue of ingress and egress of the optical fiber are major concerns due to the fragile nature of the fibers [53,54]. These fragile sensors are not as rugged as strain gauges and, therefore, must be handled more carefully. Typical issues in the installation of optical fiber sensors include delamination of sensing probes and a high risk of fiber breakage [54]. Additionally, the cost of optical interrogators, used for measurement of the FBGs, can be prohibitive for use on small aircraft. As such, research is ongoing for development of low-cost systems [55].

FBG Applications in Aerospace Structures

SHM refers to the process of observing the structural and/or mechanical responses of a system over time for the purposes of damage detection and characterization. An effective SHM method can increase efficiency, reduce maintenance and inspection costs, and extend the service life of mechanical systems and structures. As discussed above, the many attributes of FBG sensors make them ideal for SHM applications. An overview of literature in the field of optical methods indicates that FBG sensor technology is considered a promising and widely used method for evaluating and monitoring structural system response and damage detection [56].

The monitoring of structural performance is increasingly gaining importance for many applications, particularly in the aerospace industry. Load monitoring and damage detection are two critical aspects of aircraft SHM. Typically, load monitoring is accomplished by measuring local strains, and damage detection involves the monitoring of acoustic signals [57]. Review papers in the area of SHM using optical fiber sensors of large-scale composite structures include References [49,56,58].

Strain Measurement

The widest use of FBGs is for measurement of strain, temperature, and pressure. Primarily, the shift in the wavelength or the refractive index is measured by the optical system, and the strain is obtained by using Equation 11.2. A number of studies deal with large aerospace vehicles and their components. Most work involves the monitoring of strain data. Alvarenga et al. [59] used FBGs on a lightweight UAV to determine the real-time deformation shape. The NASA Dryden Flight Research Center (DFRC) has developed an instrumentation system and analysis techniques that combine to make distributed structural measurements practical for lightweight vehicles. The DFRC's fiber-optic strain sensing (FOSS) technology enables a multitude of lightweight, distributed strain measurements [60,61]. A number of studies involve ground testing on aircraft to obtain measurements from FBGs mounted on the surface of the test structure. Nicholas et al. [62] used the FOSS system to determine the deflected wing shape and the out-of-plane loads of a large-scale carbon composite wing of an ultralight full-scale UAV. The composite wing (length 5.5 m), subjected to concentrated and distributed loads, was instrumented with an optical fiber (cladding diameter = 127 μm, strain optic coefficient $P_e = 0.1667$) on its top and bottom surfaces, resulting in approximately 780 strain sensors bonded to the wings. Compared to conventional foil strain gauges, the FOSS system provided a much higher density (every 12.5 mm) of strain measurements, thereby revealing the details of the structure under load. The strain response of the wing and the high density of sensors revealed not only the structural response but also the structural details. The circles in Figure 11.7 indicate the location of the whiffletree load points. Similarly, Childers et al. [48] performed static load tests on an advanced composite transport wing to obtain strain measurements from thousands of FBGs and compared these to measurements obtained from electrical foil gauges. The high density of measurements revealed the structural behavior that is typically obtained from numerical models. To develop a real-time UAV SHM system, Gupta et al. [63] investigated a number of issues regarding FBGs such as sensor integration, location, and embedment. They used actual flight test data to monitor the vibration and load signature during flight conditions. Kim et al. [55] used a low-speed, low-cost FBG interrogator to obtain temperature-compensated strain measurements of a full-scale small aircraft wing structure subjected to various temperatures and loading conditions. As a part of the European Smart Intelligent Aircraft Structures project, Ciminello

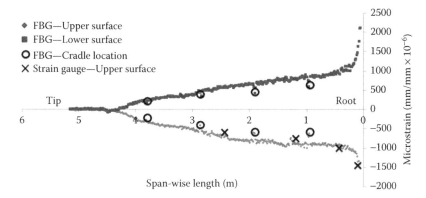

FIGURE 11.7 Strain distribution from FBGs on the composite wing of an ultralight aerial vehicle. (From Nicolas, M.J. et al., Fiber Bragg grating strains to obtain structural response of a carbon composite wing. In *ASME Conference on Smart Materials, Adaptive Structures and Intelligent Systems*, 2013 Snowbird, UT. Paper No. SMASIS2013-3265, V002T05A012 pp (8 pages). With permission.)

et al. [64] developed an in-flight shape monitoring system that uses the chord-wise strain distribution obtained from a network of FBG sensors for an adaptive trailing edge device.

FBG sensors can also be integrated into the test structure. Dvorak et al. [65] embedded FBG sensors into the carbon/epoxy spar caps and in the adhesive joints of a glass/epoxy composite wing to monitor strains during a static loads test. Amano et al. [66] embedded multiplexed FBG sensors into an aerospace carbon/epoxy advanced grid structure to measure mechanical strains of all ribs in order to evaluate damage under low-velocity impact loading. Ruzek et al. [67] embedded FBGs in CFRP fuselage stiffened panels and performed compression after impact tests. The FBG-based system successfully acquired all buckling modes of the CFRP panels.

FBG-based systems have been used in the structural testing of composite overwrapped pressure vessels (COPVs), which are used to contain high-pressure fluids in propulsion in aerospace vehicles [68]. Banks et al. [69] embedded FBGs to measure strain distributions in Kevlar COPVs during stress rupture tests. Mizutani et al. [70] conducted real-time strain measurements using FBG sensors on a composite liquid hydrogen tank mounted on a reusable rocket vehicle.

For space applications, McKenzie and Karafolas [71] presented a review of fiber optic sensing for satellites, launchers, atmospheric entry vehicles, the International Space Station, space structures, and solar sails. Physical phenomena such as strain and temperature distributions, acceleration, pressure, and spacecraft attitude were obtained using optical sensing methods.

Damage Detection

CFRP is used in many advanced aerospace applications, primarily due to its high specific strength and stiffness in the fiber direction. However, even with modest loading, CFRPs develop a complicated internal damage that can include transverse cracking, fiber fracture, and delamination [72]. AE is often used as a nondestructive method for early detection of damage. Mendoza et al. [73,74] report on the development of a wireless in-flight FBG-based AE monitoring system known as FAESense™. FBG sensors successfully captured the strain and AEs of a composite wing of a solar-powered aircraft that was subjected to a load-to-failure test in a solar simulation environment [75]. The strain variations and AE signals were used to verify the occurrence and location of the damage. Xiao et al. [57] demonstrated the use of FBGs to simultaneously monitor both strains and acoustic signals for monitoring load and damage in aircraft structures. Betz et al. [52] developed a damage identification system for aerospace structures that uses FBGs to sense ultrasound by obtaining the linear strain component produced by Lamb waves.

CONCLUDING REMARKS

Successful development and implementation of SHM technologies in aerospace vehicles involve several stakeholders such as manufacturers, academia, equipment suppliers, regulatory agencies, and airlines. SAE Aerospace Recommended Practice [76] provides guidelines for implementation of SHM on a fixed wing aircraft. Although diagnostics for simple systems has been successful, SHM technology development is required for more complex systems [77]. Further work is needed, including the correlation between loads, in-service experiences on components, and the actual damage [78]. Airbus considers the testing of SHM system's robustness in representative environments as the most important task to mature future service applications [18].

REFERENCES

1. Hunter, G.W., Ross, R.W., Berger, D.E., Lekki, J.D., Mah, R.W., Perey, D.F., Schuet, S.R., Simon, D.L. and Smith, S.W. (2013). A concept of operations for an integrated vehicle health assurance system. *NASA/TM-2013-217825*, January.
2. Boller, C., Chang, F.-K., Fujino, Y., eds. (2009). *Encyclopedia of Structural Health Monitoring*. John Wiley & Sons, Hoboken, NJ.
3. Diamanti, K. and Soutis, C. (2010). Structural health monitoring techniques for aircraft composite structures. *Progress in Aerospace Sciences* 46: 342–352.
4. Heida, J. H. and Platenkamp, D. J. (2011). *Evaluation of Non-Destructive Inspection Methods for Composite Aerospace Structures*. International Workshop of NDT Experts, Prague, Czech Republic, October 10–12, 2011.
5. Bossi, R.H. and Giurgiutiu, V. (2014). Nondestructive testing of damage in aerospace composites. In *Polymer Composites in Aerospace Industry*, P.E. Irving and C. Soutis (eds.)., Elsevier, Philadelphia, PA, pp. 413–448.
6. Lamb, H. (1917). On waves in an elastic plate. *Proceedings of the Royal Society of London*, 93(648): 114–118.
7. Raghavan, A. and Cesnik, C. (March 2007). Review of guided-wave structural health monitoring. *The Shock and Vibration Digest*, 39(2): 91–114.
8. Su, Z. and Ye, L. (2009) Identification of damage using lamb waves: From fundamentals to applications. *Lecture Notes in Applied and Computational Mechanics*, Vol. 48, Springer-Verlag, London, UK.
9. Giurgiutiu, V. (2008) *Structural Health Monitoring: With Piezoelectric Wafer Active Sensors*, Elsevier, Waltham, MA.

10. Giurgiutiu, V. (2014). Structural health monitoring (SHM) of aerospace composites. In *Polymer Composites in Aerospace Industry*, P.E. Irving and C. Soutis (eds.)., Elsevier, Waltham, MA, pp.449–507.

11. Doyle, J.F. (1997). *Wave Propagation in Structures: Spectral Analysis Using Fast Discrete Fourier Transforms*, 2nd ed., Springer-Verlag, New York.

12. Gopalakrishnan, S. and Mitra, M. (2010). *Wavelet Methods for Dynamical Problems: With Application to Metallic, Composite, and Nano-Composite Structures*. CRC Press, Boca Raton, FL.

13. Samaratunga, D., Jha, R. and Gopalakrishnan, S. (2014). Wavelet spectral finite element for wave propagation in shear deformable laminated composite plates. *Composite Structures*, 108: 341–353.

14. Park, H.W., Kim, S.B. and Sohn, H. (2009). Understanding a time reversal process in Lamb wave propagation. *Wave Motion*, 46(7): 451–467.

15. Poddar, B., Kumar, A., Mitra, M. and Mujumdar, P. M. (2011). Time reversibility of a Lamb wave for damage detection in a metallic plate. *Smart Materials and Structures* 20: 025001 (10pp).

16. Watkins, R. and Jha, R. (2012). A modified time reversal method for Lamb wave based diagnostics of composite structures. *Mechanical Systems and Signal Processing*, 31: 345–354.

17. Lopez, I. and Sarigul-Klijn, N. (2010). A review of uncertainty in flight vehicle structural damage monitoring, diagnosis and control: Challenges and opportunities. *Progress in Aerospace Sciences*, 46: 247–273.

18. Blumenfeld, L., ed. (2014). *Airbus Technical Magazine #54—Flight Airworthiness Support Technology*. http://www.airbus.com/support/publications/ (accessed May 15, 2015).

19. Rose, J.L. (1999). *Ultrasonic Waves in Solid Media*. Cambridge University Press, Cambridge, UK, pp. 101–128.

20. Ihn, J.B. and Chang, F.K. (2004). Detection and monitoring of hidden fatigue crack growth using a built-in piezoelectric sensor/actuator network: II. Validation using riveted joints and repair patches. *Smart Materials and Structures* 13: 621–630.

21. Raghavan, A. and Cesnik, C.E.S. (2007). Review of guided wave structural health monitoring. *The Shock and Vibration Digest*, 39(2): 91–114.

22. Hayashi, T. and Kawashima, K. (2002). Multiple reflections of Lamb waves at a delamination. *Ultrasonics*, 40: 193–197.

23. Ramadas, C., Balasubramaniam, K., Joshi, M. and Krishnamurthy, C. V. (2009). Interaction of primary anti-symmetric Lamb mode with symmetric delaminations: Numerical and experimental studies. *Smart Materials and Structures*, 18(8): 085011.

24. Ramadas, C., Balasubramaniam, K., Joshi, M. and Krishnamurthy, C.V. (2010). Interaction of guided Lamb waves with an asymmetrically located delamination in a laminated composite plate. *Smart Materials and Structures*, 19(6): 065009.

25. Ramadas, C., Balasubramaniam, K., Joshi, M. and Krishnamurthy, C. V. (2011). Numerical and experimental studies on propagation of Ao mode in a composite plate containing a semi-infinite delamination: Observation of turning modes. *Composite Structures*, 93(7): 1929–1938.

26. Ip, K. H. and Mai, Y.W. (2004). Delamination detection in smart composite beam using Lamb waves. *Smart Materials and Structures*, 13: 544–551.

27. Lu, Y., Ye, L., Wang, D. and Zhong, Z. (2009). Time-domain analyses and correlations of Lamb wave signals for damage detection in a composite panel of multiple stiffeners. *Composite Materials*, 43(26): 3211–3230.

28. Veidt, M. and Ng, C.-T (2011). Influence of stacking sequence on scattering characteristics of the fundamental anti-symmetric Lamb wave at through holes in composite laminates. *Journal of Acoustical Society of America*, 129(3): 1280–1287.

29. Farouk, B., Grondel, S., Assaad, J. and Moulin, E. (2009). Experimental study of the A_0 and S_0 Lamb waves interaction with symmetrical notches. *Ultrasonics*, 49: 202–205.

30. Ke, W., Castaings, M. and Bacona, C. (2009). 3D finite element simulations of an air-coupled ultrasonic NDT system. *NDT&E International*, 42: 524–533.

31. Wang. S., Huang, S. and Zhao, W. (2011). Simulation of Lamb waves interactions with transverse internal defects in an elastic plate. *Ultrasonics*, 51: 432–440.

32. http://www.piceramic.com/products/piezo-elements.html (accessed Dec 27, 2015).

33. Vishnuvardhan, J., Ajith, M., Krishnamurthy, C.V. and Balasubramaniam, K. (2009). Structural health monitoring of anisotropic plates using ultrasonic guided wave STMR array patches. *NDT&E International*, 42: 193–198.

34. Liu, Y. and Chattopadhyay, A. (2013). Low-velocity impact damage monitoring of a sandwich composite wing. Journal of Intelligent Material Systems and Structures, 24(17): 2074–2083.

35. Ng, C.T. and Veidt, M. (2009). A Lamb-wave-based technique for damage detection in composite laminates. *Smart Materials and Structures*, 18: 074006.

36. Hay, T.R., Royer, R.L., Gao, H.D., Zhao, X. and Rose, J.L. (2006). A comparison of embedded sensor Lamb wave ultrasonic tomography approaches for material loss detection. *Smart Materials and Structures*, 15(4): 946–951.

37. Wang, D., Ye, L., Lu, Y. and Su, Z. (2009). Probability of the presence of damage estimated from an active sensor network in a composite panel of multiple stiffeners. *Composites Science and Technology*, 69: 20542063.

38. Samaratunga, D., Kim, I., Jha, R. and Gopalakrishnan, S. (2011). Composite delamination detection using wavelet spectral finite element and damage force indicator method. In *AIAA2011-1953, 19th AIAA/ASME/AHS Adaptive Structures Conference*, April 4–11, Denver, CO.

39. Giurgiutiu, V. (2005). Tuned Lamb wave excitation and detection with piezoelectric wafer active sensors for structural health monitoring. *Intelligent Material Systems and Structures*, 16: 291–305.

40. Purekar, A.S. and Pines, D.J. (2010). Damage detection in thin composite laminates using piezoelectric phased sensor arrays and guided Lamb wave interrogation. *Intelligent Material Systems and Structures*, 21: 995–1010.

41. Poddar, B., Kumar, A., Mitra, M. and Mujumdar, P.M. (2011). Time reversibility of a Lamb wave for damage detection in a metallic plate. *Smart Materials and Structures*, 20(2): 025001.

42. Zhu, R., Huang, G.L. and Yuan, F.G. (2013). Fast damage imaging using the time-reversal technique in the frequency–wavenumber domain. *Smart Materials and Structures*, 22(7): 075028.

43. Nicolas, M.J., Sullivan, R.W. and Richards, W.L. (2013). Fiber Bragg grating strains to obtain structural response of a carbon composite wing. In *ASME Conference on Smart Materials, Adaptive Structures and Intelligent Systems*, Snowbird, UT, Paper No. SMASIS2013-3265, V002T05A012pp. (8 pages).

44. Kinet, D., Megret, P., Goossen, K., Qiu, L., Heider, D. and Caucheteur, C. (2014). Fiber Bragg grating sensors toward structural health monitoring in composite materials: Challenges and solutions. *Sensors*, 14: 7394–7419.

45. Keulen, C.J., Yildiz, M. and Suleman, A. (2011). Multiplexed FBG and etched fiber sensors for process and health monitoring of 2- and 3-D RTM components. *Journal of Reinforced Plastics and Composites*, 30: 1055–1064.

46. Hill, K.O. and Meltz, G. (1997). Fiber Bragg grating technology fundamentals and overview. *Journal of Lightwave Technology*, 15: 1263–1276.

47. Childers, B.A., Froggatt, M.E., Allison, S.G., Moore, T.C.S, Hare, D.A., Batten, C.F. et al. (2001). Use of 3000 Bragg grating strain sensors distributed on four eight-meter optical fibers during static load tests of a composite structure. NASA Langley Research Center, Hampton, VA, Document ID: 20040086084, 10p.

48. Kahandawa, G.C., Epaarachchi, J., Wang, H. and Lau, K.T. (2012). Use of FBG sensors for SHM in aerospace structures. *Photonic Sensors*, 2: 203–214.

49. Richards, W.L., Parker, J.A.R., Ko, W.L., Piazza, A. and Chan, P. (2012). *RTO AGARDograph 160: Application of Fiber Optic Instrumentation*. NASA Center for AeroSpace Information, Hanover, MD.

50. Kreuzer, M. (2006). *Strain Measurement with Fiber Bragg Grating Sensors*. HBM, Darmstadt, Germany, S2338-10e.

51. Betz, D.C., Staudigel, L., Trutzel, M.N. and Kehlenbach, M. (2003). Structural monitoring using fiber-optic bragg grating sensors. *Structural Health Monitoring*, 2: 145–152.

52. Kang, H.K., Park, J.W., Ryu, C.Y., Hong, C.S. and Kim, C.G. (2000). Development of fibre optic ingress/egress methods for smart composite structures. *Smart Materials and Structures*, 9: 149–156.

53. Kim, S.W., Kang, W.R., Jeong, M.S., Lee, I. and Kwon, I.B. (2013). Deflection estimation of a wind turbine blade using FBG sensors embedded in the blade bonding line. *Smart Materials and Structures*, 22 (12): 125004.

54. Kim, J.H., Lee, Y.G., Park, Y. and Kim, C.G. (2013). Temperature-compensated strain measurement of full-scale small aircraft wing structure using low-cost FBG interrogator. *Proceedings of the Sensors and Smart Structures Technologies for Civil, Mechanical, and Aerospace Systems*, SPIE 86922P. doi: 10.1117/12.2011720.

55. Guo, H., Xiao, G., Mrad, N. and Yao, J. (2011). Fiber optic sensors for structural health monitoring of air platforms. *Sensors*, 11: 3687–3705.

56. Xiao, G.G., Guo, H., Mrad, N., Rocha, B. and Sun, Z. (2012). Towards the simultaneous monitoring of load and damage in aircraft structures using fiber Bragg grating sensors. *Proceedings of the 22nd International Conference on Optical Fiber Sensors*, SPIE 8421BD; doi:10.1117/12.975070.

57. Minakuchi, S. and Takeda, N. (2013). Recent advancement in optical fiber sensing for aerospace composite structures. *Photonic Sensors*, 3: 345–354.

58. Alvarenga, J., Derkevorkian, A., Pena, F., Boussalis, H. and Masri, S. F. (2012). Fiber-optic strain senor-based structural health monitoring of an uninhabitated air vehicle, 6636–6641, *63rd International Astronautical Congress*, Naples, Italy, October 2012.

59. Bakalyar, J. and Jutte, C. (2012). Validation tests of fiber optic strain-based operational shape and load measurements, *53rd AIAA/ASME/ASCE/AHS/ASC Structures, Structural Dynamics and Materials Conferences*, AIAA 2012–1904, April 23–26, Honolulu, HI.

60. Richards, W.L., Parker, J.A.R., Ko, W.L., Piazza, A. and Chan, P. (2012). Application of Fiber Optic Instrumentation, RTO AGARDograph 160, *Flight Test Instrumentation Series*, Vol. 22, AC/323(SCI-228)TP/446.

61. Nicolas, M.J., Sullivan, R.W., Richards, W.L. and Bakalyar, J.A. *Structural Response of a Carbon Composite Wing Using Fiber Bragg Gratings, Proceedings of the American Society for Composites*, Arlington, TX, pp. 1642–1658.

62. Gupta, N., Augustin, M.J., Sathya, S., Sundaram, R., Prasad, M.H., Pillai et al. (2011). Flight data from an airworthy structural health monitoring system for an unmanned air vehicle using integrally embedded fiber optic sensors, *International Workshop on Structural Health Monitoring*, Stanford, CA. 479–486.

63. Ciminello, M., Concilio, A., Flauto, D. and Mennella, F. (2013). FBG sensor system for trailing edge chord-wise hinge rotation measurements. *Proceedings of the Sensors and Smart Structures Technologies for Civil, Mechanical, and Aerospace Systems*, SPIE 869221, April 19, doi:10.1117/12.2012017.

64. Dvorak, M., Had, J., Ruzicka, M. and Posvar, Z. (2011). Monitoring of 3D composite structures using fiber optic Bragg grating sensors *Proceedings of the 8th International Workshop on Structural Health Monitoring*, September 13–15, Stanford, CA. 1595–1602.

65. Amano, M., Okabe, Y., Takeda N. and Ozaki, T. (2007). Structural health monitoring of an advanced grid structure with embedded fiber bragg grating sensors. *Structural Health Monitoring*, 6: 309–324.

66. Ruzek, R., Kudrna, P., Kadlec, M., Karachalios, V. and Tserpes, K.I. (2013). Strain and damage monitoring in CFRP fuselage panels using fiber Bragg grating sensors. Part II: Mechanical testing and validation. *Composite Structures*, 107, January 2014, pp. 737–744..

67. McLaughlan, P.B. and Grimes-Ledesma, L.R. (2011). *Composite Overwrapped Pressure Vessels, A Primer* NASA/SP–2011–573, NASA Johnson Space Center, Houston, TX.

68. Banks, C.E., Grant, J., Russell, S. and Arnett, S. (2008). Strain measurement during stress rupture of composite over-wrapped pressure vessel with fiber Bragg gratings sensors, *Proceedings of the Smart Sensor Phenomena, Technology, Networks, and Systems*, SPIE 69330O, April 11, doi: 10.1117/12.776419.

69. Mizutani, T., Takeda, N. and Takeya, H. (2006). On-board strain measurement of a cryogenic composite tank mounted on a reusable rocket using FBG sensors. *Structural Health Monitoring*, 5: 205–214.

70. McKenzie, I. and Karafolas, N. (2005). Fiber optic sensing in space structures: The experience of the European space agency. *Proceedings of the SPIE*, Vol. 5855, SPIE, Bellingham, WA.

71. Beaumont, P.W.R. (1989). Failure of fibre composites: An overview. *Journal of Strain Analysis for Engineering Design*, 24: 189–205.

72. Mendoza, E., Prohaska, J., Kempen, C., Esterkin, Y. and Sun, S. (2013). In-flight fiber optic acoustic emission sensor (FAESense™) system for the real time detection, localization, and classification of damage in composite aircraft structures, *Proceedings of the Photonic Applications for Aerospace, Commercial, and Harsh Environments IV*, SPIE 87200K, May 31, doi:10.1117/12.2018155.

73. Mendoza, E., Prohaska, J., Kempen, C., Esterkin, Y., Sun, S. and Krishnaswamy, S. (2014). Fiber optic system for the real time detection, localization, and classification of damage in composite aircraft structures. *Proceedings of the 23rd International Conference on Optical Fibre Sensors*, SPIE 91577E, June 2, doi: 10.1117/12.2064895.

74. Kim, D.H., Lee, K.H., Ahn, B.J., Lee, J.H., Cheong, S.K. and Choi, I.H. (2013). Strain and damage monitoring in solar-powered aircraft composite wing using fiber Bragg grating sensors. *Proceedings of the Sensors and Smart Structures Technologies for Civil, Mechanical, and Aerospace Systems*, SPIE 869222, April 19, doi:10.1117/12.2009232.

75. SAE Aerospace Recommended Practice (2013). *Guidelines for Implementation of Structural Health Monitoring on Fixed Wing Aircraft*, ARP6461, Issued September 2013.

76. Esperon-Miguez, M., John, P. and Jennions, I. (2013). A review of integrated vehicle health management tools for legacy platforms: Challenges and opportunities. *Progress in Aerospace Sciences*, 56: 19–34.

77. Stolz, C. and Neumair, M. (2010). Structural health monitoring, in-service experience, benefit and way ahead. *Structural Health Monitoring*, 9(3): 209–217.

12 Metal Core Piezoelectric Fiber and Its Application

Hongli Ji, Jinhao Qiu, Hongyuan Wang, and Chao Zhang

CONTENTS

INTRODUCTION

In the recent decades, structural health monitoring (SHM) has attracted more and more interest in the engineering community. Its applications cover a broad bandwidth from civil engineering to aerospace applications, especially in aircraft industry. In fact, the idea of SHM is to allow nondestructive testing (NDT) methods become an integral part of the structures [1,2]. As a novel piezoelectric device, metal core piezoelectric fiber (MPF) has great potential to be a structurally integrated transducer for guided-wave SHM. Compared with common piezoelectric ceramics with in-plane isotropic characteristic [3–5], MPF exhibits unique directivity in Lamb wave sensing, which can locate the direction of the incoming wave without any information about the time of flight or the speed of propagation.

METAL CORE PIEZOELECTRIC FIBER

STRUCTURE OF MPF

To overcome the brittleness of piezoelectric ceramic and develop new device with easy-to-intergrate shapes, various options have been explored [6,7]. Among them, MPFs are fabricated using the extrusion method [8,9] and have been improved recently. As shown in Figure 12.1a, the geometry of an MPF consists of three parts: (1) metal (Pt) core, (2) surface electrode, and (3) piezoelectric ceramic fiber. The outer surface of an MPF is sputtered with a metal layer as a surface electrode, and the center is a metal core that is surrounded by a layer of piezoelectric ceramics. Thanks to the

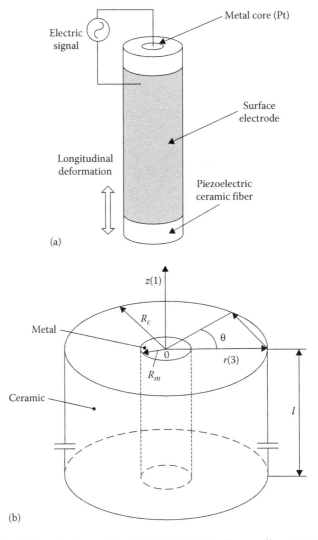

FIGURE 12.1 (a) The structure of single MPF. (b) MPF in the coordinate system.

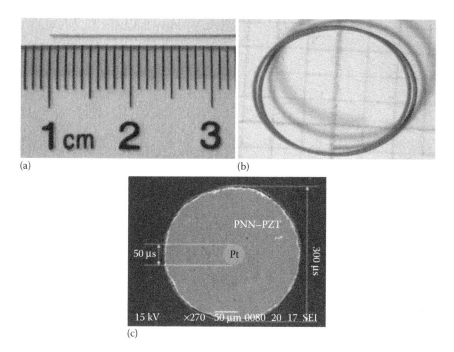

FIGURE 12.2   (a) Rectilinear-shaped MPF. (b) Curvilinear-shaped MPF. (c) SEM imaging.

metal core, MPFs overcome the brittleness of conventional piezoelectric fibers and can be used as a sensor or an actuator conveniently with two electrodes: the central metal core and the surface metal layer [10,11]. Moreover, due to the ductility of metal core, MPF can be processed into a variety of shapes according with reality structures. Figure 12.2 shows an MPF with rectilinear shape, an MPF with curvilinear shape, and a scanning electron microscope (SEM) imaging in cross section, respectively. The metal core of MPF is made of platinum (Pt), and that of fibroid piezoelectric ceramic is made of $Pb(Nb,Ni)O_3-Pb(Zr,Ti)O_3$ (PNN–PZT) slurry mixed with organic solvent. The length of a single MPF can range from several millimeters to centimeters.

STRAIN SENSITIVITY OF MPF

The geometry of MPF and the coordinate system are given in Figure 12.1b. Both the polarization and the electric field of MPF are radial. According to the convention of the local coordinate system in piezoelectric materials, subscript '3' denotes the radial direction of MPF and subscript '1' denotes the length direction of MPF. The radii of the piezoelectric fiber and the metal core are denoted by R_c and R_m, respectively, and the length of MPF is denoted by l.

Due to the slender shape of MPF, the strain component in the transverse direction can be ignored. Considering the electrical boundary condition of the MPF sensor to be an open circuit and ignoring the shear lag effect, the constitutive equations of the piezoelectric ceramic of MPF can be written by

$$S_{zz} = s_{11}^E T_{zz} + d_{31} E_r \tag{12.1}$$

$$D_r = d_{31} T_{zz} + \varepsilon_{33}^T E_r \tag{12.2}$$

where:

S_{zz} and T_{zz} are the strain and stress in the length direction
D_r is the electric displacement
E_r is the electric field in the radial direction
s_{11}^E is the elastic coefficient at a constant electric field
d_{31} is the piezoelectric coefficient
ε_{33}^T is the dielectric constant at a constant stress

After substitution of Equation 12.1 into Equation 12.2, the constitutive relation can be rewritten as

$$D_r = \frac{S_{zz} - d_{31} E_r}{s_{11}^E} \times d_{31} + \varepsilon_{33}^T E_r \tag{12.3}$$

Under the electrical boundary condition of an open circuit, the total charge on the electrode of an MPF sensor is zero; thus,

$$\int_0^l \int_0^{2\pi} D_r \cdot r \cdot d\theta dz = 0 \tag{12.4}$$

where the integration is performed over each electrode [12]. Because there is no free charge inside the piezoelectric ceramic fiber, the same expression is valid for any cylindrical area of radius r satisfying $R_m < r < R_c$. Because the electric field is in the radial direction, the following relationship can be obtained [13]:

$$E = \frac{V}{r \ln(R_c / R_m)} \tag{12.5}$$

Solving Equation 12.3 for E_r and integrating according to Equation 12.4, the following voltage expression is obtained:

$$V = \frac{(R_m + R_c) \ln(R_c / R_m)}{2l(d_{31} - \varepsilon_{33}^T s_{11}^E / d_{31})} \int_l S_{zz} dz \tag{12.6}$$

The Lamb wave field was excited by a circular actuator bonded to the surface of the aluminum plate. The expressions derived by Raghavan and Cesnik for Lamb waves excited by circular piezoceramics [14] are applicable here. The expressions of the displacement fields are as follows:

$$u_{r'}(r',z'=\pm b,t)=-\pi i\frac{T_0 a}{\mu}e^{i\omega t}\sum_{\xi^S}J_1(\xi^S a)\frac{N_S(\xi^S)}{D_S'(\xi^S)}H_1^{(2)}(\xi^S r')\ \text{(symmetric modes)}\ (12.7)$$

$$u_{r'}(r',z'=\pm b,t)=-\pi i\frac{T_0 a}{\mu}e^{i\omega t}$$

$$\sum_{\xi^A}J_1(\xi^A a)\frac{N_A(\xi^A)}{D_A'(\xi^A)}H_1^{(2)}(\xi^A r')\ \text{(antisymmetric modes)}$$

(12.8)

where:

$J_1()$ is the first class of first-order Bessel function
$H_1^{(2)}()$ is the second class of first-order complex Hankel function
The other parameters in the equations are defined as follows:

$$\alpha=\sqrt{\omega^2/c_L^2-\xi^2},\beta=\sqrt{\omega^2/c_T^2-\xi^2},c_L=\sqrt{(\lambda+2\mu)/\rho},c_T=\sqrt{\mu/\rho}$$

$$N_S=\xi\beta(\xi^2+\beta^2)\cos(\alpha b)\cos(\beta b)$$

$$D_S=(\xi^2-\beta^2)\cos(\alpha b)\sin(\beta b)+4\xi^2\alpha\beta\sin(\alpha b)\cos(\beta b)$$

$$N_A=\xi\beta(\xi^2+\beta^2)\sin(\alpha b)\sin(\beta b)$$

$$D_A=(\xi^2-\beta^2)\sin(\alpha b)\cos(\beta b)+4\xi^2\alpha\beta\cos(\alpha b)\sin(\beta b)$$

where:

λ and μ are Lame's constants for the plate material
ρ is the material density
The subscript S corresponds to the symmetric Lamb modes
The subscript A corresponds to the antisymmetric Lamb modes
The wave number ξ of a specific mode for a given ω is obtained from the solutions of the Rayleigh–Lamb equation for free waves in an isotropic plate, which is given as follows:

$$\frac{\tan\beta b}{\tan\alpha b}=\left(\frac{-4\alpha\beta\xi^2}{\xi^2-\beta^2}\right)^{\pm 1}$$

(12.9)

where the positive exponent is for symmetric Lamb modes and the negative one is for antisymmetric Lamb modes.

Assume that an MPF sensor bonded to the upper surface of an isotropic plate of thickness $2b$ and its response to the harmonic strain field excited by a circular crested actuator. The radius of the circular actuator is denoted by a; the radii of the piezo-electric fiber and the metal core are denoted by R_c and R_m, respectively; and its length is denoted by l. The origin of thickness coordinate, $z'=0$, is at the mid-plane of the

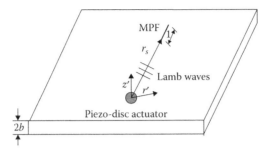

FIGURE 12.3   MPF response in the Lamb wave field due to circular piezo-disc actuator.

plate (Figure 12.3). The underlying assumption is that the longitudinal strain of the MPF sensor is the same as the strain component of the plate at the bonding location. Hence, the following relationship holds:

$$S_{zz} = S_{r'r'} \tag{12.10}$$

Substitution of Equations 12.7, 12.8, and 12.10 into Equation 12.6 gives

$$V^S = -\pi i \frac{T_0 a}{\mu} e^{i\omega t} \frac{(R_m + R_c)\ln(R_c/R_m)}{2l(d_{31} - \varepsilon_{33}^T s_{11}^E / d_{31})}$$

$$\sum_{\xi^S} \xi^S J_1(\xi^S a) \frac{N_S(\xi^S)}{D_S'(\xi^S)} \int_{r_S}^{r_S+l} H_1'^{(2)}(\xi^S r') dr' \ (\text{symmetric modes}) \tag{12.11}$$

$$V^A = -\pi i \frac{T_0 a}{\mu} e^{i\omega t} \frac{(R_m + R_c)\ln(R_c/R_m)}{2l(d_{31} - \varepsilon_{33}^T s_{11}^E / d_{31})}$$

$$\sum_{\xi^A} \xi^A J_1(\xi^A a) \frac{N_A(\xi^A)}{D_A'(\xi^A)} \int_{r_S}^{r_S+l} H_1'^{(2)}(\xi^A r') dr' \ (\text{antisymmetric modes}) \tag{12.12}$$

To examine the validity of the derived theoretical models, an experiment was performed on a $1200 \times 600 \times 2$ mm aluminum plate (Young's modulus $E = 70$ GPa, Poisson's ratio $\nu = 0.34$, density $\rho = 2700$ kgm^{-3}). A circular PZT of 8 mm in diameter and 0.48 mm in thickness, and an MPF sensor with a length of 20 mm long and a diameter of 300 μm were bonded to the surface of the plate. The PZT actuator was bonded on the extended axis of MPF sensor, and the distance between PZT and MPF was 20 cm (Figure 12.3). For the Lamb wave excitation, we have used a standard technique, which includes a function generator for generating a five-cycle 5 V Hanning windowed tone burst, with an amplifier to drive the PZT transducer. A National Instruments data acquisition board was used to acquire the sensor data. The theoretical and experimental signal amplitudes, normalized by peak amplitude over the tested frequency range, are compared over a range of frequencies for the S_0 and A_0 modes in Figure 12.4. The signals shown in the figure

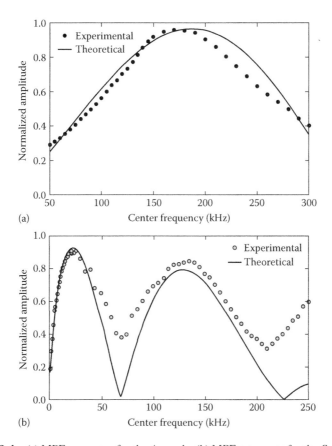

FIGURE 12.4   (a) MPF responses for the A_0 mode. (b) MPF responses for the S_0 mode.

have been averaged 128 times to reduce the noise levels. The experimental normalized amplitude curve in the figure matches well with the theoretical predictions, for predicting both the peak frequency of response and the overall trend. There is a slight error in the prediction of the peak frequency. This error can be attributed to the shear lag phenomenon, which is related to the assumption made by Raghavan and Cesnik [14] in the derivation pertaining to force transfer only along the free edges of the piezo.

RESPONSE OF MPF TO FLEXURAL LAMB WAVES

Assume that an MPF with an efficient length of l is mounted on the surface of an isotropic plate with a thickness of $2d$ in the Cartesian coordinate system (x, y, z) and the axial direction of MPF is parallel to x-axis. Lamb wave propagates in the plane (x', z) along the direction of x'. The origin of the reference system is the source of Lamb wave, which is at the middle of the plate (Figure 12.5). Considering the electrical boundary condition of MPF sensor to be an open circuit and ignoring

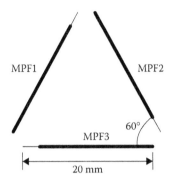

FIGURE 12.5 Configuration of the MPF rosette.

the shear lag effect, the MPF sensor voltage response to flexural Lamb waves can be written as

$$V = M \int_l S_{x'x'} \mathrm{d}x' \tag{12.13}$$

In the above formula, M expresses a constant, which is only related to the dimension and material parameters of MPF:

$$M = \frac{(R_m + R_c)\ln(R_c/R_m)}{2l(d_{31} - \varepsilon_{33}^T s_{11}^E / d_{31})} \tag{12.14}$$

where:

R_m and R_c are the radii of the metal core and piezoelectric fiber, respectively

d_{31}, ε_{33}^T, and s_{11}^E represent the piezoelectric coefficient, the dielectric coefficient at a constant stress, and the elastic coefficient at a constant electric field, respectively

$S_{x'x'}$ is the Lamb wave in-plane strain of the upper surface, which can be expressed as

$$S_{x'x'}\big|_{z=d} = i\xi^2 A(\tanh \alpha d - \frac{2\alpha\beta}{\xi^2 + \beta^2} \tanh \beta d) e^{i(\xi x' - \omega t - (\pi/2))} \tag{12.15}$$

where:

i is the imaginary unit

ξ is the wavenumber of Lamb wave at a specific angular frequency of ω

A is a constant related to Lamb wave magnitude

Parameters α and β can be defined as follows:

$$\alpha = \sqrt{\xi^2 - \omega^2/c_L^2}, \beta = \sqrt{\xi^2 - \omega^2/c_T^2} \tag{12.16}$$

where c_L and c_T are the bulk longitudinal and shear velocities, respectively.

As is mentioned before, the strain component in the transverse direction can be ignored due to the slender shape of MPF. Therefore, $S_{x'x'}$ can be projected only on the x-direction. The voltage response of an MPF to the Lamb wave can be given as

$$V = iHe^{-i(\omega t+(\pi/2))} \cos^2\theta \int_{x_1}^{x_2} e^{i\xi x\cos\theta} dx \tag{12.17}$$

where:
    θ is the angle between x- and x'-directions
    x_1 and x_2 are the x-coordinates of the starting and end points of MPF, respectively
    H expresses a constant, which is related to the incoming wave and the dimension
       and material parameters of MPF and can be defined as

$$H = M\xi^2 A\left(\tanh\alpha d - \frac{2\alpha\beta}{\xi^2+\beta^2}\tanh\beta d\right) \tag{12.18}$$

According to the sum-to-product formula and Euler equation, Equation 12.17 can be transformed into

$$V = \frac{2H}{\xi}\cos\theta\sin\frac{\xi\cos\theta\cdot l}{2}e^{-i(\omega t-\xi x_0\cos\theta)} \tag{12.19}$$

where $x_0 = (x_1 + x_2)/2$. The magnitude term of the above equation is

$$\tilde{V} = \left(\frac{2H}{\xi}\right)\cos\theta\sin\left(\frac{\xi l\cos\theta}{2}\right) \tag{12.20}$$

The above equation is derived from the hypothesis in pure harmonic waveform of a flexural mode Lamb wave. However, both theoretical derivation and experimental verification show that the MPF response to the flexural mode Lamb wave is larger than the responses to other modes at a low-frequency band (30 kHz). Thus, by tuning the central frequency in signal extraction process to a low frequency, the response can be regarded as a single-mode response for simplicity without the loss of generality.

STRUCTURE OF MPF ROSETTE

PACKAGING METHOD

As the same principle of the strain rosette based on the electrical strain gauges, MPF rosette combines three MPFs as shown in Figure 12.5. To meet the application requirements, there are many kinds of rosettes according to their different geometries. However, the delta-type rosette with angle 120° between the adjacent MPFs takes advantage of the symmetry and is very suitable for measuring the principal strain and angle.

Single MPF is very weak and brittle, which leads a tedious process in bonding the MPF to the structure. In addition, the electrodes on the MPF are very hard to connect with the wires. In order to improve the performance of the MPF rosette and simplify the process of placing three MPFs onto the structure, a method for packaging the MPF rosette is proposed. A schematic of the MPF rosette is shown in Figure 12.6. The fabrication process combines the flexible printed circuit technique

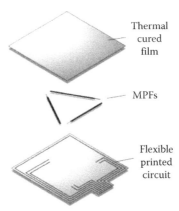

FIGURE 12.6 Packaging process of the MPF rosette.

used in the electronics industry with the modified composite manufacturing process. The four primary constituents of the MPF rosette are polyimide film, thermal cured film, MPFs, and flexible printed circuit. After integration and standardization, the packaged MPF rosette is shown in Figure 12.7. Then, three MPFs can be directly placed by pasting the packaged rosette onto the structure with epoxy adhesive. The electrodes of the MPF rosette can be easily used by a six-line port. Considering that the metal core electrodes of MPFs in a rosette share the common ground, the signals from an MPF rosette can be connected to the following voltage amplifier by a four-line port. It should be mentioned that because of the use of flexible printed circuit, thermal cured film, and polyimide film, the stiffness of the whole MPF rosette has been increased, which reduces the sensitivity of MPF. However, the experimental results in later discussion show the effectiveness of the packaged MPF rosette in sensing Lamb waves.

FIGURE 12.7 Actual picture of the MPF rosette.

DIRECTIVITY OF MPF ROSETTE

Due to the MPF characteristics, the sensitivity of flexural Lamb waves depends on the angle between the sensor lengthwise direction and the wave propagation direction. According to Equation 12.20, the response voltage is proportional to an angle-dependent factor $f(\theta)$, which can be written as

$$f(\theta) = \left| \cos\theta \sin\left(\frac{\xi l \cos\theta}{2}\right) \right| \qquad (12.21)$$

Considering a 2-mm-thick aluminum plate with Young's modulus $E = 70$ GPa, Poisson's coefficient $\nu = 0.3$, and material density $\rho = 2780$ kg/m³, the wavenumbers of the flexural Lamb wave at different frequencies are calculated by three-dimensional (3D) elasticity theory [15], as shown in Table 12.1. According to Equation 12.21, Figure 12.8 shows the directivity of MPF with different frequencies. When the wave

TABLE 12.1
Wavenumber of Flexural Lamb Waves

Frequency (kHz)	Wavenumber (m⁻¹)
15	168
20	196
25	221
30	244

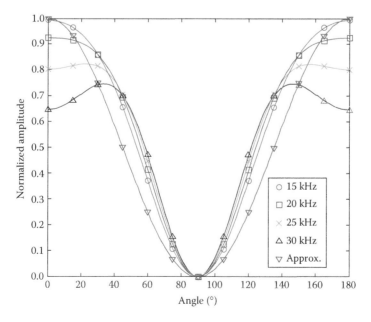

FIGURE 12.8 Directivity of MPF with different frequencies.

propagation direction is in the range of 40°–140°, there is no significant change in the normalized amplitude with different frequencies. The reason is that the projection length of the MPF in wave propagation direction is small when the angle is close to 90°. Compared with the wavelength, the small projection length of MPF can be ignored. Thus, the change in wavelength does not affect the normalized amplitude significantly.

According to the Taylor series of sine function, $\sin(\xi/\cos\theta/2)$ in Equation 12.21 can be simplified into $\xi/\cos\theta/2$, which is proportional to $\cos\theta$. Thus, when ξl is small, $f(\theta)$, which is only related to the angle θ, can be approximated as

$$f(\theta) \approx \cos^2 \theta \qquad (12.22)$$

The error in the above equation is related to the product ξl. Because ξ depends on the frequency of Lamb wave, choosing a suitable frequency is an important task in measuring the wave propagation direction. As presented above, in order to guarantee that the voltage response contains few components in other modes of Lamb waves except the flexural mode, a low-frequency range of 15–30 kHz is chosen for the analysis. As shown in Figure 12.8, the maximum error between Equaions 12.21 and 12.22 gets larger as the frequency increases. This phenomenon indicates that Equation 12.22 is more reasonable in the low-frequency range. Thus, this chapter chooses the frequency 15 kHz to determine the wave propagation direction. The maximum error between Equations 12.21 and 12.22 is 0.118 at 15 kHz. However, it should be mentioned that another way to improve the accuracy of approximation (Equation 12.22) is the use of smaller sensors. To locate higher frequency sources, smaller sensors also meet the requirement that the product ξl is small.

Using the approximate equation (12.22), three responses of an MPF rosette can be written as

$$V_{i\max} = \tilde{V}_{\max} \cos^2(\theta + \delta_j), \text{ for } \delta_{1,2,3} = 0°, 120°, 240° \qquad (12.23)$$

where $\tilde{V}_{\max}$ is the maximum response of an MPF when the incident Lamb wave is in the axial direction of the fiber.

It should be mentioned that Equation 12.23 has already ignored the difference in material parameters in each MPF. In practical applications, although the fabrication process achieves good consistency of MPFs, minor difference in the sensitivity of MPFs is unavoidable. To eliminate this influence, the gain coefficients of amplifiers are fine-tuned to guarantee that each two MPFs have the same voltage amplitude when the impact is located on their angle bisector. After adding Equation 12.23, $\tilde{V}_{\max}$ can be derived as

$$\tilde{V}_{\max} = \frac{2}{3}(V_{1\max} + V_{2\max} + V_{3\max}) \qquad (12.24)$$

In order to eliminate the effect caused by the different amplitudes of Lamb waves, the normalized amplitude can be given by

$$NA_i = \frac{V_{i\max}}{\tilde{V}_{\max}} = \frac{3}{2} \frac{V_{i\max}}{\displaystyle\sum_{i=1}^{3} V_{i\max}} \tag{12.25}$$

Substituting Equation 12.23 into Equation 12.25, the normalized amplitudes can be expressed as

$$NA_i = \cos^2(\theta + \alpha_i) = f(\theta + \alpha_i) \tag{12.26}$$

Because the normalized amplitude NA_i is only related to θ, the wave propagation direction can be calculated by the response voltages of an MPF rosette.

An experiment for testing the directivity of MPF rosette is shown in Figure 12.9a. The tested specimen is an aluminum plate with thickness 2 mm. A total of 13 different positions are used to impact the structure. The angle between adjacent impact positions is 15°. The impact has a distance of 100 mm away from the MPF rosette and is repeated 3 times in each impact position. The instruments include a voltage amplifier for four MPF rosettes, a NI PXIe-1082 chassis with a PXIe-8133 controller, and a digitizer PXI-5105. The signals from an MPF rosette are

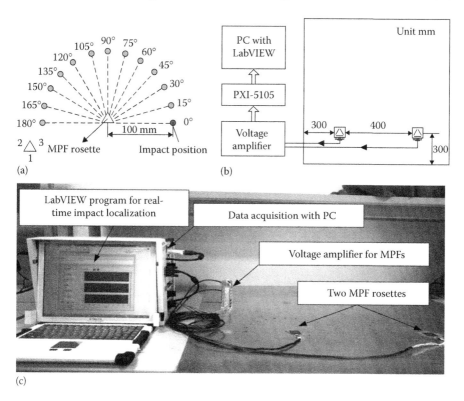

FIGURE 12.9 (a) Test for MPF directivity. (b) Experimental schematic. (c) Picture of an experimental equipment.

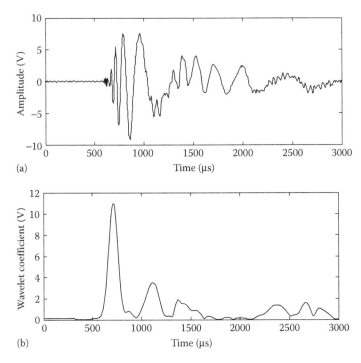

(a)

(b)

FIGURE 12.10 (a) Voltage signal in the time domain of MPF1. (b) CWT coefficient at 15 kHz of MPF1 response.

magnified 80 times with a passband of 500–200 kHz and are recorded at a sampling frequency of 1 MS/s, which will be triggered when one of the signals reaches a threshold level of 1 V and the previous history of each signal is also acquired by pre-trigger recording.

Figure 12.10a shows a typical response of MPF1 to an impact at the point 0° in Figure 12.9a. Because the phase velocity of Lamb waves varies with the frequency, the broadband signal exhibits a complex dispersive effect. In order to extract the voltage response to a particular mode of waves at 15 kHz, continuous wavelet transform (CWT) is used in this chapter. As one of the time–frequency analysis methods, CWT can be written as

$$\text{CWT}(a,b) = \frac{1}{\sqrt{a}} \int_R v(t) \psi^* \left(\frac{t-b}{a} \right) dt \qquad (12.27)$$

where:

$v(t)$ is the voltage response

$\psi^*(t)$ is the conjugate of the wavelet function $\psi(t)$ with the scale variable a and the position variable b

The wavelet function $\psi(t)$ that determines the analysis results is the most important factor in CWT. The complex Morlet wavelet is utilized here and can be expressed as

$$\psi(t) = \frac{1}{\sqrt{\pi\gamma}} \exp(-\frac{t^2}{\gamma}) \exp(j\omega_0 t)$$ (12.28)

where:

γ represents the width of the window

ω_0 represents the central angular frequency

By adding a Gaussian window with a narrowband, the complex Morlet wavelet can extract a relative narrowband signal [16]. According to the requirement for complex Morlet wavelet $\gamma\omega_0^2 \geq 1$, the parameters ω_0 and γ are equal to 6.28 and 1.5, respectively. The relation between the frequency of CWT and the scale variable a can be calculated as

$$f = \frac{f_s}{a}$$ (12.29)

where f_s is the sampling frequency. By choosing the central frequency $f = 15$ kHz, the coefficient of CWT is plotted in Figure 12.10b. In order to calculate the normalized amplitude of an MPF rosette, the maximum coefficient of CWT is used to represent the amplitude of each MPF response at 15 kHz. Then, the normalized amplitudes NA_i are calculated by Equation 12.25.

Table 12.2 shows the normalized amplitudes of three impacts at the angle 0°. Because the amplitudes of the impact energy vary, the maximum CWT coefficients of each MPF are different. However, the normalized amplitude eliminates the influence of the impact energy. In order to validate the directivity of the MPF rosette, the normalized amplitudes of MPF1 in different angles are plotted in Figure 12.11 with the mark "*." Within the margin of error, the experimental results are well matched with the theoretical analysis. When the wave propagation direction is vertical to MPF, the normalized amplitude reaches the minimum. As the wave propagation is close to the lengthwise direction of MPF, the normalized amplitude tends to 1.

TABLE 12.2

Experimental MPF Rosette Responses of Three Impact Tests at the Same Location (θ=0°, Figure 12.9a)

Impact Tests	Maximum CWT Coefficient			Normalized Amplitude			Theoretical Approximate Amplitude		
	MPF1	MPF2	MPF3	MPF1	MPF2	MPF3	MPF1	MPF2	MPF3
Impact 1	10.33	3.38	3.13	0.92	0.30	0.28	$\cos^2 0°$	$\cos^2 120°$	$\cos^2 240°$
Impact 2	8.83	2.56	2.85	0.93	0.27	0.30	$= 1$	$= 0.25$	$= 0.25$
Impact 3	8.58	2.83	2.74	0.91	0.30	0.29			

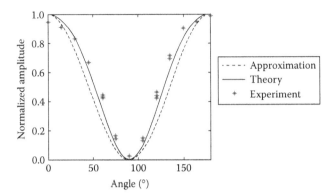

FIGURE 12.11   Directivity of MPF1 with the central frequency of 15 kHz.

PRINCIPLE OF IMPACT DAMAGE LOCALIZATION WITH MPF ROSETTES

IMAGING METHOD FOR IMPACT LOCALIZATION

Imaging Algorithm

Impact localization in a plane can be achieved by determining the wave propagation directions of two MPF rosettes. In order to determine the angle θ between the wave propagation direction and the MPF, the error curve between NA_i and $f(\hat{\theta}+\alpha_i)$ is defined in the form of standard deviation as

$$e(\hat{\theta}) = \sqrt{\frac{1}{3}\sum_{i=1}^{3}\left[N\tilde{A}_j - f(\hat{\theta}+\alpha_i)\right]^2}, \quad \text{for } \delta_{1,2,3} = 0°, 120°, 240° \qquad (12.30)$$

where $\hat{\theta}$ is the estimation of the wave propagation direction. According to Equation 12.26, the error curve $e(\hat{\theta})$ will be 0 when $\hat{\theta} = \theta$. In fact, it should be mentioned that $e(\hat{\theta})$ will hardly be 0 due to the unavoidable measurement error and the approximation of $f(\theta)$ in Equation 12.22. However, the error curve $e(\hat{\theta})$ can be used in the evaluation of the wave propagation direction. When the assumed $\hat{\theta}$ is close to θ, $e(\hat{\theta})$ tends to the minimum.

In an active SHM, the delay-and-sum algorithm [17,18] and tomography method [19] are usually used to create a digital image to highlight the area of the damage. The basic principle of the delay-and-sum algorithm is calculating the intensity of image, which equals to the probability of the damage in the position of the pixel. To show the impact location by an intensity map and avoid solving the equations, an imaging method based on the error curve $e(\hat{\theta})$ is proposed in this chapter. As shown in Figure 12.12, the wave propagation direction $\hat{\theta}_j$ at an arbitrary point O with the coordinate (x, y) can be calculated as

$$\hat{\theta}_j = \arctan\left(\frac{x - x_j}{y - y_j}\right) \qquad (12.31)$$

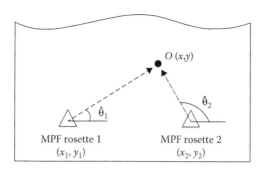

FIGURE 12.12   Imaging method for impact localization.

where (x_j, y_j) is the coordinate of the jth MPF rosette. The image intensity $I(x, y)$ at the point O is defined as

$$I(x, y) = \sum_{j=1}^{N_s} \frac{e(\hat{\theta}_j)}{N_s} \tag{12.32}$$

where N_s is the number of MPF rosettes.

Considering that the image intensity $I(x, y)$ is an average value of the error $e_j(\hat{\theta}_j)$, the actual impact location will be shown by the area with lower intensity in the image.

Impact Localization Test

Two MPF rosettes are boned to the tested specimen. As shown in Figure 12.9b, all equipment is the same as the experiment for testing the directivity of MPF rosette. The distance between two MPF rosettes is 400 mm. Four typical impact positions are tested to validate the proposed method. The impact is excited by dropping a steel ball. In each test, six signals from two MPF rosettes are recorded with a sampling frequency of 1 MS/s and a trigger threshold of 1 V. The signals are magnified 80 times and filtered with a passband of 500–200 kHz.

Using the proposed imaging method, the localization results are summarized in Figures 12.13 through 12.16. The actual impact position is shown by a cross symbol (X), and the MPF rosettes are shown by triangle symbols (Δ) in the image. To analyze the wave direction estimation based on MPF rosettes, the error curves $e_j(\hat{\theta}_j)$ are also listed along with the actual incident angle of each rosette.

In the images, the impact position is clearly shown by a bright focalized spot, which represents the lower error $e_j(\hat{\theta}_j)$. Taking impact 1 in Figure 12.13 as an example, the error curve of MPF rosette 1 reaches the minimum near the angle 62°, which indicates the actual wave propagation direction. MPF rosette 2 shows that the wave propagation direction is near 120°. Using the imaging method, the estimation of the impact position with the least error is (486, 667). The error between the estimation position and the actual position is 8.06 mm. The other results are summarized in Table 12.3. A significant error is seen in Figure 12.16, which is located along the line connecting two MPF rosettes. It is because when the impact is on or near the line between two MPF rosettes, the two propagation directions are close to parallel and

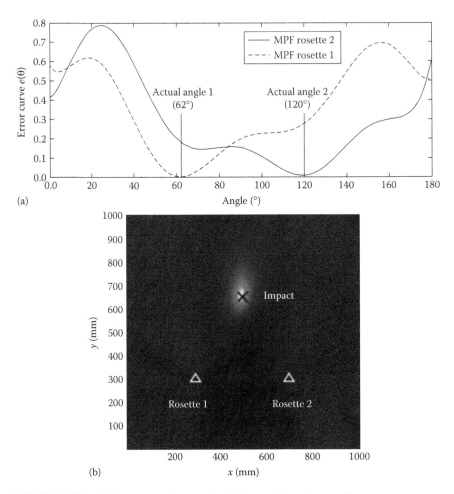

FIGURE 12.13 (a) Error curve of impact 1. (b) Impact 1 imaging result.

have an infinite number of intersections. Thus, it can be explained that the bright area in Figure 12.16b is dispersed along the line and fails to estimate the impact location. This "blind zone" can be corrected by using a third MPF rosette offering redundant information of other propagation direction.

Bayesian Inference Strategy for Impact Localization

Bayesian Model

Traditional localization strategy usually needs two rosette sensors and utilizes two wave propagation directions identified by those two rosettes, respectively, and then considers the intersection of two directions as the identified source [3,20–22]. The impact location cannot be detected successfully when two identified directions almost coincide with each other. Naturally, more rosettes should be used in location

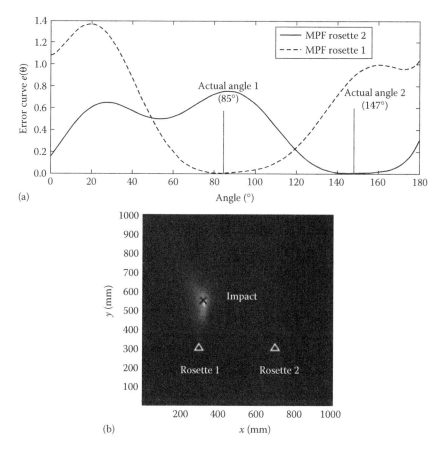

(a)

(b)

FIGURE 12.14 (a) Error curve of impact 2. (b) Impact 2 imaging result.

identification; however, due to unexpected structural uncertainties and unavoidable measurement errors, it is probable that there exists more than one intersection when more identified directions are used to implement whole-structure covered impact localization, especially in the circumstance that the impact is relatively far from the rosette locations.

Under this situation, a Bayesian inference strategy for impact identification emerges as a fusion of multiple intersections, and a probability density distribution imaging will be obtained to indicate the likelihood to be the actual impact point.

For the detection of impact location in plate-like structures, an MPF rosette network comprising N_s rosettes is shown in Figure 12.17. According to Equation 12.26, the calculated normalized magnitude can be defined as

$$NA_{ij}^c = \cos^2(\theta + \delta_i) = \cos^2\left(\arctan\left(\frac{y_p - y_j}{x_p - x_j}\right) + \delta_i\right) \qquad (12.33)$$

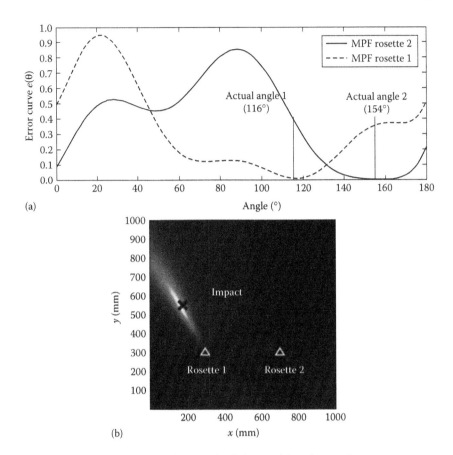

(a)

(b)

FIGURE 12.15   (a) Error curve of impact 3. (b) Impact 3 imaging result.

where:

(x_p, y_p), (x_j, y_j) are the coordinates of the center location of the impact and the jth rosette ($j = 1, ..., N_s$), respectively

The subscript i ($i = 1, 2, 3$) represents the order of MPF in each rosette shown in Figure 12.5

Therefore, the undetermined parameter vector in the impact localization approach is defined as $\theta = [x_p, y_p]^T$; θ_k is used to express each unknown parameter in θ ($k = 1, 2$).

As a result of realistic measurement errors and structural uncertainties, the calculated normalized magnitude $NA_{ij}{}^c$ and the measured normalized magnitude $NA_{ij}{}^m$ can be used to establish the likelihood relationship:

$$NA_{ij}^m = NA_{ij}^c + \xi_1 + \xi_2 \qquad (12.34)$$

where ξ_1, ξ_2 represent the uncertainties of structures and measurements, respectively; both of them are assumed as independent variables sampled from Gaussian distribution with a mean of 0. The variance of ξ_1 is $\sigma_{\xi 1}^2$ and the variance of ξ_2 is $\sigma_{\xi 2}^2$. Therefore,

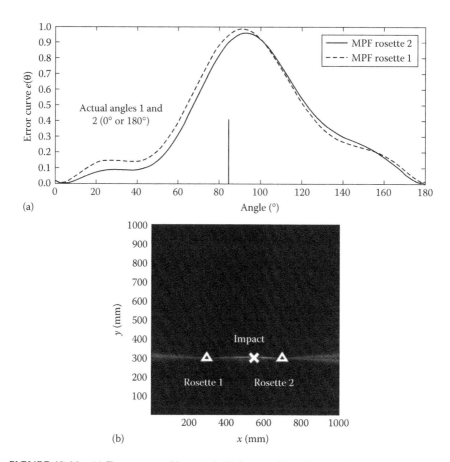

(a)

(b)

FIGURE 12.16 (a) Error curve of impact 4. (b) Impact 4 imaging result.

TABLE 12.3

Summary of the Experimental Results (mm)

Number	Actual Position	Estimated Position	Error
1	(490, 660)	(486, 667)	8.06
2	(320, 550)	(320, 536)	14.00
3	(180, 550)	(178, 546)	8.25
4	(550, 300)	Error	Error

the likelihood function, utilizing the normalized magnitude of the ith MPF in the jth rosette NA_{ij}^m and NA_{ij}^c, can be defined as

$$p_L(NA_{ij}^m \mid \theta, \sigma_\xi^2) = \frac{1}{\sqrt{2\pi\sigma_\xi^2}} \exp\left\{-\frac{1}{2\sigma_\xi^2}\left[NA_{ij}^m - NA_{ij}^c(\theta)\right]^2\right\} \qquad (12.35)$$

where $\sigma_\xi^2 = \sigma_{\xi 1}^2 + \sigma_{\xi 2}^2$ under the assumption that ξ_1 and ξ_2 are mutually independent.

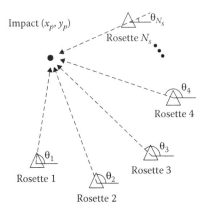

FIGURE 12.17   Illustration of the impact localization model and sensor network.

For all the MPFs in N_s rosettes, the likelihood function can be written as

$$p_L(\mathbf{M}|\boldsymbol{\theta},\sigma_\xi^2) = \prod_{i=1}^{3}\prod_{j=1}^{N_s} p_L(NA_{ij}^m|\boldsymbol{\theta},\sigma_\xi^2)$$

$$= \frac{1}{\left(\sqrt{2\pi\sigma_\xi^2}\right)^{N_s}} \exp\left\{-\frac{1}{2\sigma_\xi^2}\sum_{i=1}^{3}\sum_{j=1}^{N_s}\left[NA_{ij}^m - NA_{ij}^c(\boldsymbol{\theta})\right]^2\right\}$$

(12.36)

The likelihood function clarifies the probabilistic relationship between the measured normalized magnitude vector $\mathbf{M}$ and the unknown parameters $\boldsymbol{\theta}, \sigma_\xi^2$. For simplicity and convenience, the sum of squares in Equation 12.36 can be expressed as

$$F(\mathbf{M},\boldsymbol{\theta}) = \sum_{i=1}^{3}\sum_{j=1}^{N_s}\left[NA_{ij}^m - NA_{ij}^c(\boldsymbol{\theta})\right]^2$$

(12.37)

Therefore, the posterior probability density function (PDF) of the undetermined parameters $\boldsymbol{\theta}, \sigma_\xi^2$ can be given as

$$p(\boldsymbol{\theta},\sigma_\xi^2|\mathbf{M}) = \frac{p_L(\mathbf{M}|\boldsymbol{\theta},\sigma_\xi^2)p_\pi(\boldsymbol{\theta},\sigma_\xi^2)}{p(\mathbf{M})}$$

(12.38)

where:
$p_L(\mathbf{M}|\boldsymbol{\theta},\sigma_\xi^2)$ is the likelihood function
$p_\pi(\boldsymbol{\theta},\sigma_\xi^2)$ is the joint prior PDF of parameters $\boldsymbol{\theta}, \sigma_\xi^2$

The marginal posterior distribution for each undetermined parameter θ_k in $\boldsymbol{\theta}$ can be written as

$$p(\theta_k|\mathbf{M}) = \int \frac{p_L(\mathbf{M}|\boldsymbol{\theta},\sigma_\xi^2)p_\pi(\sigma_\xi^2)p_\pi(\boldsymbol{\theta})}{p(\mathbf{M})} d\boldsymbol{\theta}_{-k}d\sigma_\xi^2$$

(12.39)

where the subscript $-k$ denotes the multidimensional integration over all parameters of $\boldsymbol{\theta}$ except θ_k. For the joint prior PDF $p_\pi(\boldsymbol{\theta}, \sigma_\xi^2)$, it can be written as $p_\pi(\boldsymbol{\theta}, \sigma_\xi^2) = p_\pi(\boldsymbol{\theta}) p_\pi(\sigma_\xi^2)$ based on the assumption of the independence between $\boldsymbol{\theta}$ and σ_ξ^2.

Generally, it is not easy to obtain an analytical solution for the complicate multidimensional integration. However, under some special circumstances, the integration can be simplified via the selection of prior PDF. Consider the rewriting form of the likelihood function (Equation 12.36),

$$p(\mathbf{M}|\boldsymbol{\theta}, \tau) \propto \tau^{N_s/2} \exp\left(\frac{-\tau}{2} \cdot F(\mathbf{M}, \boldsymbol{\theta})\right)$$

where $\tau = 1/\sigma_\xi^2$ (τ is usually called "the precision" in Bayesian inference). Obviously, τ is in the Gamma family, which can be described as $p(n) \propto n^{\alpha-1}\exp(-\beta n)$ for $n > 0$, that is, $n \sim \Gamma(\alpha, \beta)$. Therefore, if the prior is still in the Gamma family, the posterior will have the same probability distribution form as prior; the posterior and prior are then called conjugate distributions [23]. The posterior PDF is simplified to a certain extent compared to those that are not conjugate with priors. In the following study, the precision parameter τ in the prior PDF is sampled from Gamma distribution. In addition, the prior PDF $p_\pi(\boldsymbol{\theta})$ is treated as "uninformative prior" in which $\boldsymbol{\theta}$ is sampled from a uniform distribution with the upper and lower limits of $\boldsymbol{\theta}_{max}$ and $\boldsymbol{\theta}_{min}$. It should be mentioned that an uninformative prior is used at the stage of the laboratory research due to the lack of location database. However, in engineering applications of SHM, the possible impact location databases for different components of a structure can be improved with the increase of monitoring cycles. The prior, which is then treated as "informative prior," can be easily incorporated into judgments as experienced information or expert knowledge to reduce initial uncertainties.

Although some simplifications of Equation 12.39 are delivered, it is still difficult to obtain an analytical solution for the complicate multidimensional integration; thus, a numerical method called Markov chain Monte Carlo (MCMC) is proposed in the "MCMC Method" section to obtain the marginal posterior distribution of each undetermined parameter.

MCMC Method

Monte Carlo method is known as a general term for the simulation thought, which realizes the probabilistic and statistical analysis using a large number of random independent variables [24]. However, it is difficult to draw independent samples in complicated joint posterior distributions. Therefore, another simulation method known as MCMC is put forward to solve this problem with a sequence of dependent samples generated from a first-order homogeneous Markov chain. This kind of Markov chain has two unique properties: First, the transition probability only relies on the current status of samples, and second, the Markov chain can converge to a stationary distribution equated with the target distribution after a certain number of iterations. These initial iteration samples will be ignored, described as a "burn-in" period [25].

Metropolis–Hastings Algorithm

The fundamental algorithm for MCMC simulation is the Metropolis–Hastings (MH) algorithm [26]. Suppose that each state of the Markov chain is sampled from a particular distribution $g(\theta)$; thus, after initializing the first state $\theta^{(1)}$, a candidate point θ^* is generated using a proposal distribution $h(\theta^*|\theta^{(t-1)})$ at state t. The next step is to determine the acceptance or rejection of the candidate point θ^*. To this aim, the acceptance ratio between θ^* and $\theta^{(n-1)}$ is defined first:

$$r = \min\left(1, \frac{g\left(\theta^*\right)h\left(\theta^{(t-1)}\middle|\theta^*\right)}{g\left(\theta^{(t-1)}\right)h\left(\theta^*\middle|\theta^{(t-1)}\right)}\right) \tag{12.40}$$

Then, a random variable u which is sampled from the uniform distribution, that is, $u \sim U(0,1)$, is generated to decide whether to actually accept or reject the candidate point θ^*. If $u \leq r$, $\theta^{(t)} = \theta^*$; otherwise, $\theta^{(t)} = \theta^{(t-1)}$. The presence of u ensures that the candidate θ^* may also be accepted, even though the probability of $\theta^{(t-1)}$ is larger than that of θ^*. This strategy can avoid local convergence effectively.

In practical application of MCMC simulation using the MH algorithm, the proposal distribution is often selected as a symmetric distribution, so that $h(\theta^*|\theta^{(t-1)}) = h(\theta^{(t-1)}|\theta^*)$. The symmetric distribution form simplify the calculation of the acceptance ratio r; therefore, a uniform distribution centered on $\theta^{(t-1)}$ with an interval of length $2L$ is usually selected as the proposal distribution:

$$h\left(\theta^*\middle|\theta^{(t-1)}\right) = \frac{1}{2L}, \left|\theta^* - \theta^{(t-1)}\right| < L \tag{12.41}$$

Obviously, if the value of the tuning parameter L is large, the candidate θ^* may jump far from the center of interval, so that the acceptance rate may decrease and the Markov chain tend to be highly correlated. If the value is small, it will take the Markov chain lots of time to converge. Here, a simple tuning strategy is given in order to keep the acceptance rate about 40%. This range can generate a Markov chain with relatively low autocorrelation and high convergence speed. If the candidate θ^* is rejected, divide the tuning parameter L by a constant 1.007; otherwise, multiply the tuning parameter L by a constant 1.01 during the burn-in period.

Gibbs Sampling

A drawback of the MH algorithm is that it may be difficult to tune the proposal distribution, especially in the multidimensional case. Therefore, Gibbs sampling algorithm is proposed, which utilizes full conditional distribution of each component conditioned on other components in the parameter vector [27]. The full conditional distribution of $\theta_k^{(t)}$ is $p(\theta_k^{(t)}|\mathbf{M},\boldsymbol{\theta}_{-k} = \boldsymbol{\theta}_{-k}^{(t-1)})$, which can be thought of the posterior distribution of θ_k under the assumption that the components of $\boldsymbol{\theta}_{-k}$ are known. Like the posterior distribution $p(\boldsymbol{\theta},\sigma_\xi^2|\mathbf{M})$ in Equation 12.38, the full conditional distribution is proportional to the product of prior distribution and likelihood function:

$$p\left(\theta_k^{(t)}\middle|\mathbf{M},\boldsymbol{\theta}_{-k} = \boldsymbol{\theta}_{-k}^{(t-1)}\right) \propto p_L\left(\mathbf{M}\middle|\boldsymbol{\theta},\sigma_\xi^2\right) p_\pi\left(\boldsymbol{\theta},\sigma_\xi^2\right)$$

To implement Gibbs sampling, the key lies in the specific form of the full conditional distribution, which depends on the form of the prior distribution and the likelihood function. As aforementioned, the full conditional distribution and the prior distribution belong to the same parameter family utilizing conditional conjugate prior which simplifies the calculation in the Gibbs sampling process.

The MCMC Algorithm Using MH Idea with Gibbs Sampling

MH algorithm and Gibbs sampling principles are used to generate two Markov chains of parameters θ_1 and θ_2 and the probability distributions of θ_1 and θ_2 are obtained via cumulation of iterations.

Task

Generate the posterior distribution $p(\theta_k)$ for $\theta = [x_p, y_p]^T$ using the likelihood function $p_L(\mathbf{M}|\theta,\sigma_\xi^2)$ and the prior distribution $p_\pi(\theta,\sigma_\xi^2)$. The likelihood function is given by the impact localization model shown in Equation 12.16 and the measured normalized magnitude $\mathbf{M}$ [28].

Initialization

Set the number of the total iterations N_T and the iterations in the burn-in period N_B, and initialize the Markov chain with the state $t = 0$.

Initialize the parameter values randomly selected in the prior distribution $p_\pi(\theta)$, $\theta^{(0)} = [\theta_1^{(0)}, \theta_2^{(0)}]^T$.

Sample the initial variance $\sigma_\xi^2(0)$ from the inverse Gamma distribution $IG(N_s/2 + 1, F(\mathbf{M},\theta^{(0)})/2)$.

Set the tuning parameters L_k ($k = 1, 2$).

Algorithm of iterations

Let $t = t + 1$, renovate the parameters one by one, for each parameter θ_k.

1. Generate a candidate $\theta_k^* = \theta_k^{(t-1)} + 2L_k \times U(-1,1)$.
2. Compute the acceptance rate r:

$$r = \frac{p_\pi(\theta^*)}{p_\pi(\theta^{(t-1)})} \exp\left\{-\frac{1}{2\sigma_\xi^2(t-1)}\left[F(\mathbf{M},\theta^*) - F(\mathbf{M},\theta^{(t-1)})\right]\right\}$$

where:
$\theta^* = [\theta_1^*, \theta_2^{(t-1)}]^T$ when updating θ_1
$\theta^* = [\theta_1^{(t)}, \theta_2^*]^T$ when updating θ_2

3. Randomly generate u from a uniform distribution, that is, $u \sim U(0,1)$.
   If $u \leq r$, set $\theta_k^{(t)} = \theta_k^*$, and adjust the tuning parameter $L_k = L_k \times 1.01$.
   If $u > r$, set $\theta_k^{(t)} = \theta_k^{(t-1)}$, and adjust the tuning parameter $L_k = L_k/1.007$.
4. Sample the variance $\sigma_\xi^2(t)$ from inverse Gamma distribution $IG(N_s/2 + 1, F(\mathbf{M},\theta^{(t)})/2)$ at state t.
5. Repeat steps 3 and 4 until $t = N_T$.

In addition, the adjustment of the tuning parameters will be ceased after the burn-in period, and subsequent values $\theta_k^{(t)}$ ($t = N_B + 1,\ldots, N_T$) will be recorded as samples of posterior distribution $p(\theta_k)$.

Experimental Studies for Bayesian Inference Strategy

Experiments are conducted to validate the direction-based impact localization strategy on an isotropic plate. The dimension of the plate is $1000 \times 1000 \times 2$ mm. A delta-type MPF rosette network is mounted on the surface of tested specimen. The test setup comprises a handheld instrumented hammer, a NI PXIe-1082 chassis with a PXIe-8133 controller and a PXI-5105 digitizer, a voltage amplifier with a voltage gain of 80, and a passband range from 500 Hz to 200 kHz. The signals from MPF rosettes are recorded at a sampling frequency of 1 MHz, which will be triggered when one of the signals reaches a threshold level of 500 mV, and the previous history of each signal is also acquired by 25% pre-trigger recording. An illustration of the isotropic plate and rosette placements is shown in Figure 12.18. The origin of the coordinates is set at the lower left corner of the plate.

Three typical impact cases are tested to validate the proposed strategy. The MCMC algorithm is performed to identify the impact locations. The initial values of x and y coordinates are sampled from uniform distribution $U(0, 1000)$. The total number of MCMC iterations is 100,000 and the "burn-in" period is 40,000. Figure 12.19 shows the Markov chain iterative processes of x and y coordinates in case 1. Figure 12.20 presents the probability distribution histograms formed by the rest 60,000 samples

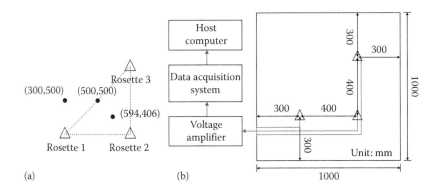

FIGURE 12.18 (a) Illustration of rosette placements. (b) Illustration of an experimental setup.

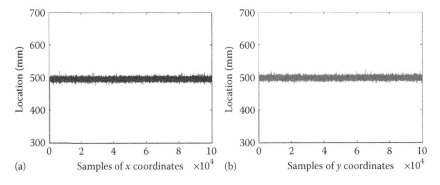

FIGURE 12.19 Markov chain of (a) x and (b) y coordinates in case 1 (500, 500).

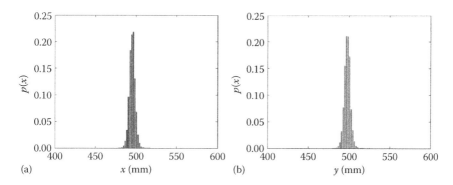

(a)

(b)

FIGURE 12.20 Probability distribution histograms of (a) x and (b) y coordinates in case 1 (500, 500).

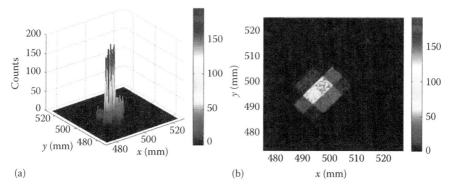

(a)

(b)

FIGURE 12.21 (a) 3D histogram and (b) 2D view of case 1 (500, 500) after one MCMC iteration.

of x and y coordinates in case 1. Figure 12.21a shows the 3D histogram for x and y coordinates sampled from the stationary distribution. Figure 12.21b illustrates the identified impact location. The identified impact location with maximum probability is (491.4, 495.0) mm, the expectation of the identified impact locations is (495.4, 497.9) mm, and the standard deviation of the identified impacts is (23.4, 23.8) mm.

As shown in Figure 12.21b, after a complete MCMC iterative process, the probabilistic distribution of the identified impact locations is decentral, because every state of Markov chain is sampled randomly from uniform distribution. Although the concrete form of the identified probabilistic distribution after each MCMC iterative process is different, the digital characteristics of undetermined parameters are the same. Therefore, 15 times of MCMC processes are carried out to acquire the average probabilistic distribution, shown in Figure 12.22. After 15 averages, the identified impact location with maximum probability is (501.0, 500.2) mm, the expectation of the identified impact locations is (500.5, 501.2) mm, and the standard deviation of the identified impacts is (19.7, 20.6) mm.

Similar identification results are obtained for other two impact cases with locations of (594, 406) mm (Figure 12.23) and (200, 500) mm (Figure 12.24), respectively. Experimental results of three impact cases are listed in Table 12.4.

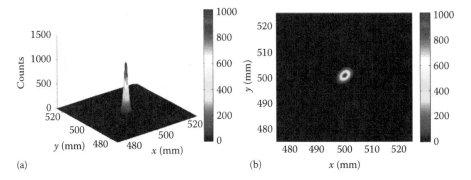

(a)   (b)

FIGURE 12.22   (a) 3D histogram and (b) 2D view of case 1 (500, 500) after 15 MCMC iterations.

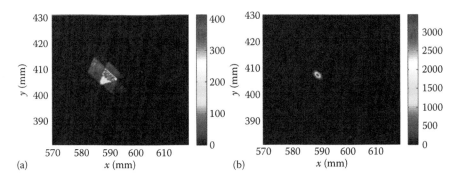

(a)   (b)

FIGURE 12.23   2D view of case 2 (594, 406) after (a) 1 and (b) 15 MCMC iterations.

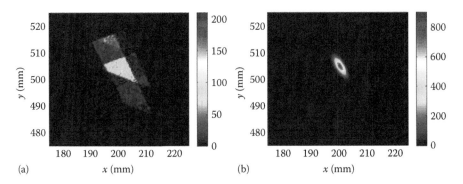

(a)   (b)

FIGURE 12.24   2D view of case 2 (200, 500) after (a) 1 and (b) 15 MCMC iterations.

TABLE 12.4
Summary of the Experimental Results

	MCMC Processes	Maximum Probability Location (mm)	Expectation of Locations (mm)	Standard Deviation of Locations (mm)
Case 1 (500, 500)	1	(491.4, 495.0)	(495.4, 497.9)	(23.4, 23.8)
	15	(501.0, 500.2)	(500.5, 501.2)	(19.7, 20.6)
Case 2 (594, 406)	1	(595.4, 403.8)	(589.3, 408.3)	(23.8, 23.5)
	15	(596.2, 401.4)	(589.3, 408.2)	(20.4, 20.9)
Case 3 (200, 500)	1	(206.2, 500.2)	(205.4, 501.0)	(26.0, 39.7)
	15	(205.4, 501.0)	(201.1, 505.1)	(18.9, 28.4)

CONCLUSION

This chapter proposes an imaging method for locating the impact using MPF rosettes. The basic of the localization procedure is measuring the angle of the principal strain, which corresponds to the wave propagation direction without the use of wave speed or time-of-flight information. The MPF rosette is made up of three MPFs, which have strong directivity in sensing the Lamb waves. In order to simplify the process of placing three MPFs onto the structure, a package procedure is proposed by combining the flexible printed circuit technique with the composite manufacturing process. After theoretical derivation and experimental validation, the directivity characteristic of the MPF rosette can be represented by an appropriate expression. Then, an imaging method using the error curve functions for locating the impact location is proposed, which can avoid solving the trigonometric equations and provide an informative imaging to show the estimation position. By the impact testing, the impact position can be clearly displayed with a high accuracy. However, when the impact is on or near the line between two rosettes, the method cannot be effective. In realistic impact localization, applications would require more MPF rosettes to solve the blind zone problem.

To tackle this problem, an impact localization method based on the Bayesian inference theory and the MCMC algorithm is proposed using a delta-type MPF rosette network to reduce the localization uncertainties caused by multiple rosettes. Unlike the traditional deterministic approaches, in which the identified impact location is indicated by deterministic coordinates, the presented localization approach can focus on the impact location gradually without the knowledge of wave velocity, enhance the robustness to uncertainties of measurements and structures, and implement a whole-structure covered impact detection. Moreover, the identified results are quantified by probability distributions of impact locations.

Experimental study for an isotropic plate is employed to validate the effectiveness of the proposed Bayesian impact localization strategy. The identified results show a high accuracy without any information about the wave velocity or structural properties.

In addition, the proposed method is designed on the wavefield in isotropic plates. However, the propagation direction is different with the principal strain direction in composite materials [29]. When the principal strain direction is measured by the proposed method, an angle compensation process should be added to evaluate the propagation direction. Future work will extend this study for impact localization in composite structure.

REFERENCES

1. Tribikram Kundu, Samik Das, Steven A Martin, Kumar V Jata. Locating point of impact in anisotropic fiber reinforced composite plates. *Ultrasonics*, 2008, 48(3):193–201.
2. Victor Giurgiutiu. Embedded NDT with piezoelectric wafer active sensors. In *Nondestructive Testing of Materials and Structures*. Springer: Dordrecht, the Netherlands, 2013, pp. 987–992.
3. Chao Zhang, Jinhao Qiu, Hongli Ji, Shengbo Shan. An imaging method for impact localization using metal-core piezoelectric fiber rosettes. *Journal of Intelligent Material Systems and Structures*, 2014. doi:10.1177/1045389X14551432.
4. Zhongqing Su, Xiaoming Wang, Li Cheng, Long Yu, Zhiping Chen. On selection of data fusion schemes for structural damage evaluation. *Structural Health Monitoring*, 2009, 8(3):223–241.
5. Luca De Marchi, Alessandro Marzani, Nicolò Speciale, Erasmo Viola. A passive monitoring technique based on dispersion compensation to locate impacts in plate-like structures. *Smart Materials and Structures*, 2011, 20(3):035021.
6. Michael Barbezat, Andreas J Brunner, Christian Huber, Peter Flüeler. Integrated active fiber composite elements: Characterization for acoustic emission and acousto-ultrasonics. *Journal of Intelligent Material Systems and Structures*, 2007, 18(5):515–525.
7. James W High, W Keats Wilkie. Method of fabricating NASA-standard macro-fiber composite piezoelectric actuators, National Aeronautics and Space Administration Report, Langley Research Center. NASA/TM-2003-212427, ARL-TR-2833, June, 2003.
8. Jinhao Qiu, Junji Tani, Naoki Yamada, Hirofumi Takahashi. Fabrication of piezoelectric fibers with metal core. *Proceedings of the SPIE Conference on Smart Structures and Materials*, SPIE: San Diego, CA, vol. 5053, 2003, pp. 475–483.
9. Jun Luo, Jinhao Qiu, Kongjun Zhu, Jianzhou Du, Hongli Ji, Xuming Pang. Temperature stability and fabrication of Pb $(Zn_{1/3}Nb_{2/3})$ O_3–Pb (Zr, Ti) O_3 fibers with Pt core. *Journal of Intelligent Material Systems and Structures*, 2012, 23(15):1735–1740.
10. Jinhao Qiu, Hongli Ji, Kongjun Zhu, Mangon Park. Response of metal core piezoelectric fibers to unsteady airflows. *Modern Physics Letters B*, 2010, 24(13):1453–1456.
11. Bian Yixiang, Qiu Jinhao. Dynamic admittance matrix of metal core piezoelectric fiber. *International Journal of Applied Electromagnetics and Mechanics*, 2011, 35(3):189–200.
12. Horn-Sen Tzou. *Piezoelectric Shells: Distributed Sensing and Control of Continua*. Solid Mechanics and Its Applications. Dordrecht, the Netherlands: Kluwer Academic Publishers, 1993.
13. Gael Sebald, Jinhao Qiu, Daniel Guyomar, Daisuke Hoshi. Modeling and characterization of piezoelectric fibers with metal core. *Japanese Journal of Applied Physics*, 2005, 44(8R):6156.
14. Ajay Raghavan, Carlos ES Cesnik. Modeling of piezoelectric-based Lamb wave generation and sensing for structural health monitoring. *Proceedings of the SPIE Conference on Smart Structures and Materials*, San Diego, CA, vol. 5391, SPIE Optical Engineering Press: Bellingham, WA, 2004, pp. 419–430.
15. Joseph L Rose. *Ultrasonic Waves in Solid Media*. Cambridge University Press, New York, 2004.

16. Francesco Ciampa, Michele Meo, Ettore Barbieri. Impact localization in composite structures of arbitrary cross section. *Structural Health Monitoring*, 2012. doi:10.1177/1475921712451951.

17. Jian Cai, Lihua Shi, Shenfang Yuan, Zhixue Shao. High spatial resolution imaging for structural health monitoring based on virtual time reversal. *Smart Materials and Structures*, 2011, 20(5):055018.

18. Jennifer E Michaels. Detection, localization and characterization of damage in plates with an in situ array of spatially distributed ultrasonic sensors. *Smart Materials and Structures*, 2008, 17(3):035035.

19. Masaki Morii, Ning Hu, Hisao Fukunaga, Jinhua Li, Yaolu Liu, Satoshi Atobe Alamusi, JinHao Qiu. A new inverse algorithm for tomographic reconstruction of damage images using Lamb waves. *Computers Materials and Continua*, 2011, 26(1):37.

20. Jian Liu, Jinhao Qiu, Weijie Chang, Hongli Ji, Kongjun Zhu. Metal core piezoelectric ceramic fiber rosettes for acousto-ultrasonic source localization in plate structures. *International Journal of Applied Electromagnetics and Mechanics*, 2010, 33(3):865–873.

21. Howard Matt, Francesco Lanza di Scalea. Macro-fiber composite piezoelectric rosettes for acoustic source location in complex structures. *Smart Materials and Structures*, 2007, 16(4):1489–1499.

22. Piotr Kijanka, Arun Manohar, Francesco Lanza di Scalea, Wieslaw J Staszewski. Damage location by ultrasonic Lamb waves and piezoelectric rosettes. *Journal of Intelligent Material Systems and Structures*, 2014. doi:10.1177/1045389X14544140.

23. William A Link, Richard J Barker. *Bayesian Inference: With Ecological Applications*. Academic Press: Burlington, MA, 2009.

24. Ronald W Shonkwiler, Franklin Mendivil. *Explorations in Monte Carlo Methods*. Springer Science + Business Media: New York, 2009.

25. W Keith Hastings. Monte Carlo sampling methods using Markov chains and their applications. *Biometrika*, 1970, 57(1):97–109.

26. Siddhartha Chib, Edward Greenberg. Understanding the metropolis-hastings algorithm. *The American Statistician*, 1995, 49(4):327–335.

27. Pierre Bremaud. *Markov chains: Gibbs Fields, Monte Carlo Simulation, and Queues*. Springer Science + Business Media: New York, 1999.

28. Jonathan M Nichols, William A Link, Kevin D Murphy, Colin C Olson. A Bayesian approach to identifying structural nonlinearity using free-decay response: Application to damage detection in composites. *Journal of Sound and Vibration*, 2010, 329(15):2995–3007.

29. Jinling Zhao, Jinhao Qiu, Hongli Ji, Ning Hu. Four vectors of Lamb waves in composites: Semianalysis and numerical simulation. *Journal of Intelligent Material Systems and Structures*, 2013. doi:10.1177/1045389X13488250.

13 Shape Memory Polymers and Their Applications

*W.S. Al Azzawi, Jayantha Ananda Epaarachchi,
M. Mainul Islam, and Jinsong Leng*

CONTENTS

INTRODUCTION

GENERAL

Scientists studying synthetic polymers found some intelligent polymers that have the capability of recovering their original shapes upon exposure to an external stimuli such as electricity (Asaka and Oguro 2000), magnetic field (Xulu et al. 2000), heat (Lendlein and Kelch 2002), moisture (Yang et al. 2006), and light (Jiang et al. 2006, Lendlein et al. 2005) that are potentially very useful for a variety of applications, including aerospace, automotive, and some related to biotechnology and biomedical engineering. Smart polymers are becoming increasingly more prevalent as scientists learn about the chemistry and triggers that induce conformational changes in polymer structures and devise ways to take advantage of and control them. In this relatively new area of polymer technology, the potential aerospace, mechanical, chemical, and biomedical engineering uses for smart polymers appear to be limitless. There have been a substantial development of a variety of shape memory polymer (SMP) materials during recent years, and the shape recovery of process has been significantly enhanced and multi-SMPs have been produced (Xie 2010).

The SMP materials have the ability to return from a deformed shape and recover their original shape once they are exposed to the suitable external stimulus. Among different types of stimulation methods such as magnetic field (Schmidt 2006), electrical field (Sahoo et al. 2005), light (Lendlein et al. 2005), pH (Han et al. 2012), and moisture (Yang et al. 2005), thermally stimulated SMPs have attracted special attention and have been used in a wide range of applications. Compared with shape memory alloys (SMAs), SMPs have inimitable advantages of being lightweight, inexpensive, low density, good manufacturability, high shape deformability, large recoverability, good biodegradability, and an easily tailorable glass transition temperature (Lendlein and Kelch 2002, Leng et al. 2011, Ohki et al. 2004). Table 13.1 shows the comparison of general physical properties between SMPs and SMAs as presented by Liu et al. (2007). These interesting characteristics have provided the SMPs the

TABLE 13.1

General Physical Properties of SMPs Compared with SMAs

	SMPs	SMAs
Density (g cm^{-3})	0.9–1.1	6–8
Extent of deformation (%)	Up to 800	<8
Young's modulus at $T < T_{tran}$ (GPa)	0.01–3	83 (NiTi)
Young's modulus at $T > T_{tran}$ (GPa)	$(0.1-10) \times 10^{-3}$	28–41
Stress required for deformation (MPa)	1–3	50–200
Stress generated during recovery (MPa)	1–3	150–300
Critical temperatures (°C)	−10 to −100	−10 to −100
Transition breath (°C)	10–50	5–30
Recovery speeds	<1 s to several minutes	<1 s
Thermal conductivity (W m^{-1} K^{-1})	0.15–0.30	18 (NiTi)
Biocompatibility and biodegradability	Can be biocompatible and/or biodegradable	Some are biocompatible (i.e., NiTi), not biodegradable
Processing conditions	<200°C, low pressure	High temperature (>1000°C) and high pressure required
Corrosion performance	Excellent	Excellent
Cost	<$10 per lb	~$250 per lb

Source: Liu, C et al., *Journal of Materials Chemistry*, 17(16), 1543–58, 2007. With permission.

novelty for many potential applications. Space deployable structures, airplane morphing wings, and biomedical devices (Figures 13.1 through 13.3) are some examples of the potential applications in which thermally stimulated SMPs are mainly used. However, because SMPs are polymeric materials, they have low stiffness and low recovery stress. For instance, the recovery stress of SMPs is 1–5 MPa, whereas that of SMAs is 0.25–0.75 GPa (Madbouly and Lendlein 2010). These drawbacks have imposed some limitations on the implementation of SMPs in real engineering applications for many years. Recently, SMPs have been reinforced with hard materials such as fibers and carbon nanotubes (CNTs). Such reinforcement emerged an advanced material called SMP composites (SMPCs) that possess both the targeted

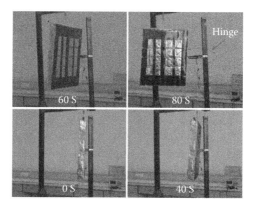

FIGURE 13.1 A deployment process of space solar arrays actuated by an SMPC hinge. (From Lan, X et al., *Smart Materials and Structures*, 18, 024002, 2009. With permission.)

FIGURE 13.2 Lockheed Martin's morphing unmanned combat aerial vehicle in its two different shapes. (From Love, M et al., Demonstration of morphing technology through ground and wind tunnel tests, in *Proceedings of the 48th AIAA/ASME/ASCE/AHS/ASC Structures, Structural Dynamics, and Materials Conference*, AIAA, pp. 1–12, 2007. With permission.)

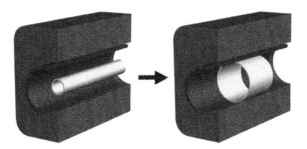

FIGURE 13.3   SMP stent undergoes a transition from a compressed shape when subjected to the body temperature. (From Wache, H et al., *Journal of Materials Science: Materials in Medicine*, 14, 109–12, 2003. With permission.)

shape memory effect (SME) and the high stiffness requirement in the engineering applications (Nishikawa et al. 2012).

THERMALLY INDUCED SMPs

The SME in the thermally stimulated SMPs is totally motivated and pivoted on the glass transition temperature (T_g). Furthermore, the mechanical properties are totally dependent on the material temperature. The SMPs are fully pliable with very low stiffness allowing the polymer to be strained up to 400% at above T_g; however, at below T_g, the SMPs have high stiffness and good rigidity. Figure 13.4 represents the relationship between the temperature and Young's modulus of the SMPs, in which a significant variation in Young's modulus can be noticed during the transition from glassy to the rubbery states and vice versa below and above the transition temperature.

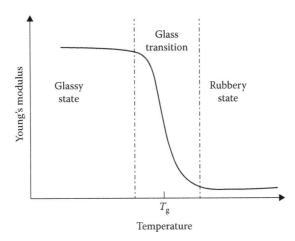

FIGURE 13.4   Young's modulus–temperature relation of the SMPs. (From Yarborough, CN et al., Shape recovery and mechanical properties of shape memory composites, in *Proceedings of the ASME International Mechanical Engineering Congress and Exposition*, American Society of Mechanical Engineers, pp. 111–19, 2008. With permission.)

One of the interesting features of the thermally stimulated SMPs is that if the material at the rubbery state is deformed to a temporary shape and is cooled below T_g while fixing the deformed shape, the induced strain is fixed into the polymer and the polymers will stay in this new state indefinitely or until heated above T_g again. Once the material is heated above T_g again, the stored strain relaxes and the material recovers its original shape. This process is repeated, for some polymers, up to 3000 cycles without degrading the properties (Yarborough et al. 2008).

SME Mechanism of Thermally Responsive SMPs

The internal chemical structure is responsible for the SME mechanisms of SMP materials. Basically, SMPs have two constitutive segments: the elastic segment and the transition segment, as shown in Figure 13.5a.

The two segments have two different roles during the SME cycle. Although the transition segment undergoes a significant change in its stiffness when exposed to the thermal stimulus, the elastic segment always maintains high elasticity during the cycle. Upon heating the material above the T_g, the transition segments become softer and the material becomes easy to deform (Figure 13.5b). Subsequently, in the next step of the cycle, when the material cools down with the existing external constraint, the transition segments recover their stiffness and prevent the elastic segments from returning back to their original shape even after the removal of the constraint resulting in a temporary shape (shape fixity) (Figure 13.5c). In the last step of the cycle, heating the material again above the softening temperature of transition segments reduces their stiffness again and enables the elastic segments to recover their original shape (shape recovery) (Figure 13.5d).

Above described mechanism can also be represented by the three-dimensional (3D) stress–strain–temperature diagram in Figure 13.6 (Ivens et al. 2011). (A) represents SMPs in permanent shape at high temperature $T_h > T_g$; (A–B) applying load at T_h; (B–C) cooling to $T_l < T_g$ under fixed applied load; (C–D) removing load at T_l;

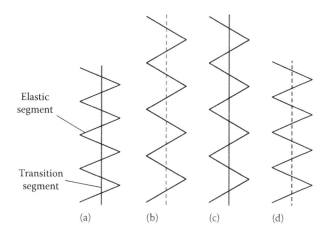

Elastic
segment

Transition
segment

(a)   (b)   (c)   (d)

FIGURE 13.5 (a–d) Schematic illustration of the mechanism in the SMP. (From Sun, L et al., *Materials and Design*, 33, 577–640, 2012. With permission.)

(D) some strain recovered due to the elastic spring back effect of the glassy state; (D–A) and (D–E) are two different recovery methods the material can recover with; (D–A) free stress strain recovery; and (D–E) fixed strain–stress recovery.

Westbrook et al. (2011a) presented another demonstration for the thermomechanical behavior of the SMP. Figure 13.7 shows the scheme of the shape memory cycle, including both shape fixity and shape recovery parts. In the first part of the cycle (shape fixity), the specimen is initially allowed to equilibrate at high temperature $T_{H1} > T_g$, and then it is deformed to a specific compressive strain using an external constrain. Thereafter, the specimen is allowed to relax before it cools down to $T_L < T_g$, where the shape fixity part of the thermomechanical cycle ends. In the second

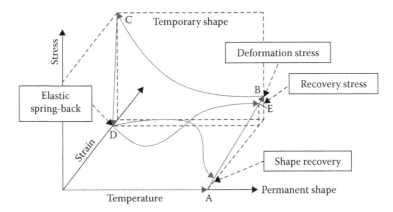

FIGURE 13.6 Stress–strain–temperature diagram illustrating the thermomechanical behavior of the SMP under different strain or stress recovery conditions. (A) Represents SMPs in permanent shape at high temperature $T_h > T_g$; (A–B) applying load at T_h; (B–C) cooling to $T_l < T_g$ under fixed applied load; (C–D) removing load at T_l; (D) some strain recovered due to the elastic spring back effect of the glassy state; (D–A) and (D–E) are two different recovery methods the material can recover with; (D–A) free stress strain recovery; and (D–E) fixed strain–stress recovery. (From Ivens, J et al., *Express Polymer Letters*, 5, 254–61, 2011. With permission.)

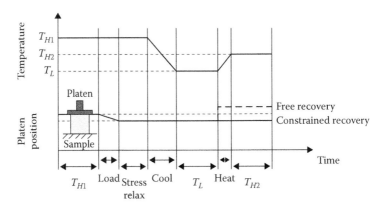

FIGURE 13.7 Thermomechanical history schematic of the SMPs. (From Westbrook, K et al. *Mechanics of Materials*, 43, 853–69, 2011a. With permission.)

part of the cycle (shape recovery), the specimen is heated to $T_{H2} \geq T_g$, where two types of recovery effects can be recognized depending on the presence of the constrain load (constrained recovery and free recover).

A systematic investigation of the thermomechanics of shape fixity and shape recovery was done by Liu et al. (2006) in which a small strain constitutive model for SMPs has developed to diagnose the uniaxial thermomechanical response of the SMPs under various constraint conditions. The developed model provides a good prediction for the strain and stress recovery responses of pre-deformed SMPs under a flexible external constraint with various compliance levels. Another thermomechanical model for SMPs was presented by Tobushi et al. (2001). In this model, the heating activation of large strain in the SMPs was investigated by modifying a linear model previously presented by Tobushi et al. (1997) to a nonlinear model, which is more efficient to present the large strain. The proposed model has efficiently expressed the thermomechanical response of the SMPs such as shape fixity, shape recovery, and stress recovery in a large strain up to 20% of maximum strain. In another study, Chen and Lagoudas (2008) developed a theory to describe the thermomechanical properties of SMPs under large deformation. The theory proposed that the coexisting active and frozen phases of the polymer and the transitions between them provide the underlying mechanisms for strain storage and recovery during a shape memory cycle. Also, this theory has been used an internal state variable which describes the volume fraction of the frozen phase.

However, research efforts in SMP modeling, to date, are still at the conceptual level to understand the basic behavior of materials under thermal condition. Moreover, factual implementation of the SMPs in real engineering application may be the most important aspect to realize. Based on the above discussions, it is not unexpected that the thermomechanical response of the SMPs grabs a significant deal of interest in engineering society.

ELECTRO-RESPONSIVE SMPs

The early-stage research on SMPs was focusing on thermal-responsive SMPs, which utilized thermal transitions of the polymer molecules as the switching mechanism. The SME in these materials is directly activated by Joule heating from an external source such as hot gas or hot water (Langer and Tirrell 2004). However, in most practical applications, hot gas or hot water is not convenient to do the required heating (Leng et al. 2011). Recently, an increasing interest has focused on electro-responsive SMPs by including special functional fillers in the SMPs. Incorporating conductive hard phases such as carbon nanoparticles (CNPs), carbon nanofibers, or CNTs in the SMP matrix emerges electro-active polymers. At first sight, it may seem that the electrical activation is a new SME triggering method; fundamentally, it is not a new one because the material is still triggered by Joule heating but using an indirect way. Xu et al. (2012) characterized the electrical properties and shape recovery efficiency of CNP-filled SMPC. The researchers concluded that the material conductivity varies significantly as a function of CNP concentration, and they pointed that there is an increase in material temperature with the increase material conductivity due to the Joule heating generated by the electrical current. In the same context,

Leng et al. (2009) investigated the electroactive thermoset styrene-based SMP nanocomposite filled with nanosized carbon powders. Considerable improvement in electrical conductivity was achieved by filling the SMP with 10%.vol of nanocarbon powders, and consequently a good electroactive shape recovery was reached by supplying the material with 30 V, which was enough to heat the material above the transition temperature.

Activating the SME through direct heating using external stimulus is practically inapplicable in some applications. Alternatively, internally induced Joule heating was found to be a more appropriate activation method. Different hard materials have been used to compose SMPCs, which are responsive to not only the electrical field but also the magnetic field.

LIGHT-INDUCED SMPs

A noncontact activation of the SME is an intrinsic requirement, especially in sheltered or noncontact environments, for example, inside the human body in biomedical applications. Electro-responsive SMPs incorporating special conductive particles are capable of showing indirect activation by generating a Joule heating inside the material when subjected to electrical voltage. However, this kind of SME triggering could be harmful for human health because of the electromagnetic field resulting from the electrical current (Leng et al. 2010).

Light-induced SMPs are another set of smart polymers that can be triggered indirectly when irradiated with light of appropriate wavelength. They may be categorized into two groups: the first group is naturally light-induced SMPs, which can be produced by incorporating reversible photosensitive functional fillers into the polymer (Lendlein et al. 2005; Monkman 2000). In this group, the temperature effect is not involved in any way in material stimulation process. In other words, the SMP's original shape can be achieved at ambient temperature, which is a highly desirable feature in biomedical applications to avoid the damage that may occur in the surrounding tissue within the human body if the implanted smart material is activated by heat (Jiang et al. 2006). In the second group of the light-induced SMPs, the shape triggering effect is totally based on the light–thermal transition mechanism to induce heat inside the material (Baer et al. 2007, Koerner et al. 2004, Leng et al. 2010; Maitland et al. 2002).

Noncontact and indirect activation methods represent a breakthrough in the field of smart SMP application, in which the triggering process can be remotely controlled. A few years ago, a light-active SMP has been developed and successfully used as an actuator biomedical field (Baer et al. 2007; Small IV et al. 2005). Further, some researchers have observed a two-way shape memory (TWSM) effect (Ahir and Terentjev 2005; Vaia 2005) on a multiwalled CNT (MWNT)–elastomer composite (Figure 13.8). The expansion and contraction, that is, shape recovery of the composite, depend on the limits of the composite that is strained. If the material is slightly changed, it will expand when it is exposed to infrared (IR) light, whereas it will shrink if it undergoes more than 10% strain under an identical exposure to IR light. These materials show an incredible potential again in developing breakthrough technological advancements in biomedical, aerospace, and space industries. In a separate

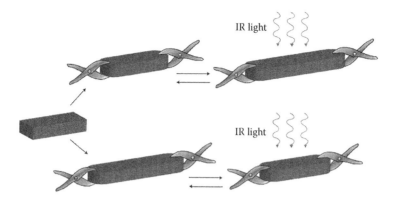

FIGURE 13.8 Bimodal and reversible actuation of an MWNT–elastomer nanocomposite induced by IR irradiation. Reversible expansion occurs at small pre-strains (top) and reversible contraction at large pre-strains (bottom). (From Vaia, R, *Nature Materials*, 4, 429–30, 2005. With permission.)

study, Leng et al. (2009) confirmed that light-active SMP filled with nanocarbon black enhanced the shape memory effect of the SMP.

pH-Sensitive SMP

A novel pH-responsive SMP prepared by Han et al. (2012) was found to be able to be deformed to a temporary shape at pH 11.5 and recover its original shape at pH 7. The mechanism of this effect is suggested as follows: at high pH value, the material is rigid because of the existing of both fixing and reversible cross-links (Liu and Urban 2010). However, in a neutral surrounding environment, the reversible cross-links dissociate, and the material becomes softer and ready to be deformed. Under the existing deforming force, changing the pH value to a higher value again freezes the deformation inside the material owing to the reforming of the reversible cross-links. Once the material is exposed to pH 7 again, the material relaxes and recovers its original shape as a consequence of dissociation of the reversible cross-links. Because the fluids in the human body are around pH 7, this fact gives the pH-responsive SMPs the required biocompatibility and opens the horizon for more potential applications in the biomedicine industry where exposing the material to pH 7 inside the human body triggers the original shape of the material.

SHAPE MEMORY EFFECTS

One-Way Shape Memory and TWSM Concepts

The SME in SMPs illustrated in the thermomechanical cycle of Figure 13.6 shows that once the material recovers its original shape in the shape recovery process, it will never be able to repeat the cycle again unless it is subjected to a new external programming step. Therefore, this is referred to as one-way shape memory (OWSM) effect. Accordingly, this sort of SMPs may not be applicable in some applications that seek repeatable actuation cycles. Hence, developing SMPs that can spontaneously

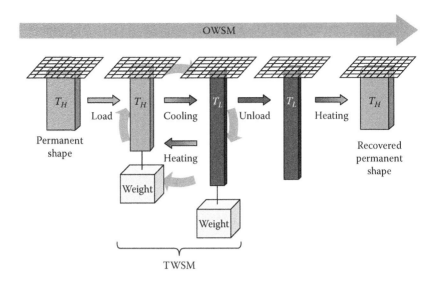

FIGURE 13.9   Schematic showing the difference between the OWSM and the TWSM. (From Westbrook et al. *Materials and Structures*, 20, 065010, 2011b. With permission.)

repeat its deformation recovery cycle has grabbed the researchers' attentions. In 2008, the development of a novel semicrystalline poly(cyclooctene) (PCO)-based SMP by Chung et al. (2007) allows for an SMP to exhibit a TWSM effect. According to Chung et al. (2007), cooling-induced crystallization of cross-linked PCO under a tensile load results in significant elongation and subsequent heating to melt the network reversing this elongation (contracting), yielding a net TWSM effect.

In Figure 13.9, the material is deformed under a constant weight at T_H (deformed shape I). Then, the material is cooled to T_L with the same load; the elongation during the cooling is due to the formation of crystalline domains which are stress free upon formation and result in an increase in material stretch to obtain an equilibrium force balance in the material. The material reaches a new deformed state under a constant weight at T_L (deformed shape II). Once it is heated back to T_H under the same weight, the material recovers to the deformed shape (I). The process is repeatable to switch the two deformed shapes (I and II) by heating to T_H or cooling to T_L, as long as the constant weight is on. Furthermore, the material can achieve the OWSM effect by removing the external load after cooling at T_L. This can result in some contraction in material due to the spring-back effect of the material in the glassy state. Subsequently, reheating the material to T_H allows full shape recovery with respect to the initially undeformed configuration.

Developing TWSM Effect

As explained previously, the OWSM effect in SMPs is an inherent feature. To enhance the performance of SMPs and make them possess the TWSM effect, different techniques have been adopted. The way proposed by Chung et al. (2007) requires a constant externally applied load in order to achieve the TWSM effect. However, in SMP applications, it is desirable to have the TWSM effect without the requirement of a constant externally applied load (free-standing) TWSM effect (Westbrook et al. 2011b). Therefore, the researchers have targeted this goal using different techniques. Imai (2014) proposed two operating methods: double SMP layer and single SMP layer. The double SMP layer method uses the SME of two kinds of SMPs with different glass transition temperatures (T_{g1} and T_{g2}, $T_{g1} < T_{g2}$). The two kinds of SMPs have reverse shapes of the memory with respect to each other. By heating these SMPs, the memory shape related to T_{g1} appears first, and then disappears around T_{g2} because the memory shape related to T_{g1} is canceled by the memory shape related to T_{g2}. In the single SMP layer method, the researcher used the SME of a single kind of SMP and the thermal contraction of the material. In this method, the TWSM effect is achieved as follows: After a memory shape appears above T_g, the SMP is cooled below the T_g. Then, the SMP shrinks thermally, especially near the T_g. This is used to return from the generated memory shape to the prememory shape (the shape prior to the memory shape). Thus, this method is effective only when the SMP undergoes tensile deformation in the memory shape.

TWO-LAYER TECHNIQUE OF TWSM POLYMER

Another type of TWSM polymer was developed by Chen et al. (2008), in which polymer laminated composites prepared by the layer technique with the OWSM polymers. First, a 1.0 mm-thick active layer from PHAG5000-based shape memory polyurethane prepared with 100% strain, and a 1.0 mm-thick substrate layer from PBAG600-based polyurethane without deformation were prepared in advance. Second, the two layers were combined to form a two-layer laminated film by adhesive solution. After the laminated film has been held for more than 48 h at ambient temperature, a novel polymer laminate exhibiting TWSM behavior, that is, bending on heating and reverse bending on cooling, has achieved.

TWO-WALLED CYLINDER TECHNIQUE FOR TWSM POLYMERS

A different technique was proposed by Wang et al. (2012) to obtain the TWSM effect by a special type of SMP composite structures. In this technique, a composite cylinder has made by confining a cylindrical SMP core with a different SMP film having different T_g (Figure 13.10). The glass transition temperature of the SMP film (T_{g1}) is higher than that of the SMP substrate (T_{g2}). The thermomechanical process includes the following steps:

1. Both the film and the substrate are in the rubbery state at the temperature above T_{g1}, in which the composite cylinder can have a large deformation under the applied load.
2. Keeping that deformation and decreasing the temperature to be lower than T_{g2}, both the SMP film and the core are fixed, and thus, the compressed deformation is kept even if it removes the applied load.

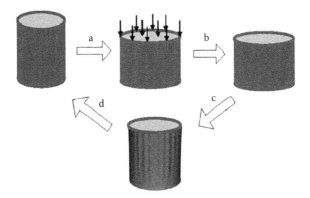

FIGURE 13.10 Schematic illustration of the proposed SMP cylinder composite: (a) applying load at $T>T_{g1}$; (b) decreasing temperature below T_{g2}; (c) heating above T_{g2}; and (d) heating above T_{g1}. (From Wang, Z et al., *Materials Letters*, 89, 216–18, 2012. With permission.)

3. Reheating and increasing the temperature to be higher than T_{g2} but lower than T_{g1}, the SMP core has a tendency to recover the original state (radial contraction), whereas the SMP film still keeps the frozen state. The mismatched deformation leads to the micro-buckling of the structure and forms a gear-like shape.
4. Further heating to above T_{g1}, the SMP film is also unfrozen, and the structure reversely recovers its original state.

The mismatching between the two materials' transition temperatures is the key idea behind the TWSM effect proposed in this research. Figure 13.11 shows experimental results of the shape recovery rate and storage modulus of the SMP composite cylinder as a function of temperature. It is clear that the TWSM effect can be achieved in the temperature range of 381–409 K.

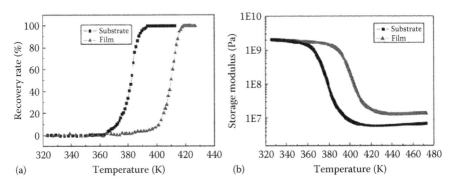

FIGURE 13.11 Experimental curves of the frozen strain recovery and storage modulus versus temperature: (a) recovery rate–temperature function and (b) storage modulus–temperature function. (From Wang, Z et al. *Materials Letters*, 89, 216–18, 2012. With permission.)

PROPERTIES AND APPLICATIONS OF SMPs AND THEIR COMPOSITES

In parallel with the development in the design and production technologies in recent decades, materials science revolutionized the idea of conventional materials to more innovative materials. Shape memory materials (SMMs) presented a fascinating discipline in materials science and provided access to nontraditional functions. SMPs have emerged recently as a major competitor to the other types of SMMs and rapidly grew in a varied range of applications, especially in the aerospace industry because of their lower density and high strain recoverability. However, implementing the SMP, in its net form, in a real structural application, may have some limitations because of its low stiffness as shown in Table 13.1. To enhance the mechanical properties without losing the inherent SME, different reinforcement materials have been incorporated with the SMPs to produce the more advanced SMPCs.

BIOMEDICAL AND SPACE INDUSTRY APPLICATIONS

As previously described, there is a broad range of environmental stimuli that can activate the SMPs. However, thermal stimulus could be the most effective one because different stimuli (e.g., electricity, magnetism, light, and moisture) are fundamentally belong to thermal response (Leng et al. 2011). As the SMPs are categorized to contain cross-linking polymers (chemical or physical), which are responsible for the SME in the material. The SME is attributed to the transition of these cross-linking polymers from a rubbery state ruled by the entropic energy to a glassy state ruled by the internal energy (Qi et al. 2008). The chemical stability, biocompatibility, and biodegradability that SMPs possess (Cabanlit et al. 2007) make them perfect selection for some biomedical devices. Figure 13.12 shows an SMP vascular stent, which is a small tubular scaffold that can be thermally activated locally inside the body to treat the arterial stenosis problems. Another interesting biomedical implementation is the surgical sutures (Figure 13.13). In this application, the temporary shape of the

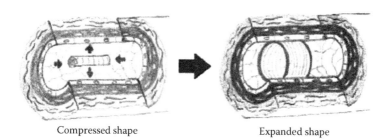

Compressed shape   Expanded shape

FIGURE 13.12 Intravascular SMP-based stent deployment prior to application (left) and after rest (right). (From Serrano, MC and Ameer, GA, *Macromolecular Bioscience*, 12, 1156–71, 2012. With permission.)

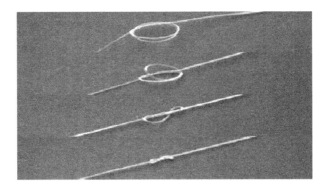

FIGURE 13.13 SMP fiber used as wound closure. (From Lendlein, A and Langer, R, *Science*, 296, 1673–6, 2002. With permission.)

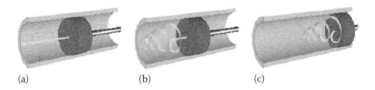

(a)   (b)   (c)

FIGURE 13.14 Laser-activated SMP device for clot removal: (a) temporary straight shape; (b) laser-activated corkscrew form; and (c) capture of the thrombus. (From Small, IV et al., *Optics Express*, 13, 8204–13, 2005. With permission.)

fiber is programmed by stretching about 200%; when the temperature is raised above T_g, the structure shrinks (recovers original shape) and tightens the knot. In the further application, SMPs can be effectively used to remove dangerous blood clots from the blood vessels. Figure 13.14 shows blood clot removal process steps using a laser-activated SMP device.

Additionally, because SMPs exhibit another impacting set of properties such as low density, low cost, easy manufacturability, high deformability, and high strain recoverability, these properties make them ideal from the structural point of view. Today, SMPs are playing a major role in space exploration engineering in which deployable structures like solar panels (Figure 13.1) activated by SMPC hinge (Figure 13.15) and lunar habitat (Figure 13.16) can be deformed to a compact form before they are loaded to the launching space vehicle, and then recover their original shape once they are in position in the outer space upon exposure to the suitable stimulus (Hinkle et al. 2011). Moreover, the future technology of morphing aircraft structures such as morphing wings (Figure 13.17) and variable camber wings (Figure 13.18) is another example of SMP applications in which the lightweight and SME are intrinsic features.

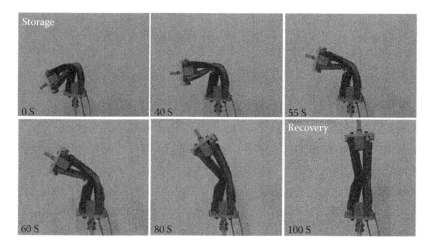

FIGURE 13.15 Shape recovery process of the SMPC hinge. (From Lan, X et al., *Smart Materials and Structures*, 18, 024002, 2009. With permission.)

FIGURE 13.16 Deployment process of expandable lunar habitat. (From Hinkle, J et al., 2011. With permission.)

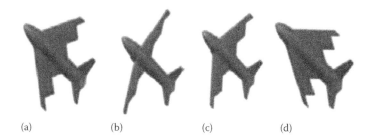

(a)   (b)   (c)   (d)

FIGURE 13.17 Morphing aerial vehicle in its four different shapes: (a) takeoff; (b) cruise; (c) loiter; and (d) dash. (From Jee, S-C, *Development of Morphing Aircraft Structure Using SMP*, DTIC Document, 2010. With permission.)

FIGURE 13.18   Configuration of the variable camber wing. (From Leng, J et al., *Progress in Materials Science*, 56, 1077–135, 2011. With permission.)

SMP COMPOSITES

The polymeric nature of the SMPs and the weakness in their stiffness and the recovery stress (e.g., the recovery stress of SMPs is 1–5 MPa, whereas that of SMAs is 0.25–0.75 GPa [Madbouly and Lendlein 2010]) have resulted in considerable difficulty to incorporate these materials in real structural applications. Reinforcement concept has solved the problem, added another positive feature for these materials, and opened the door for new horizon of applications. Recently, many researches have been conducted to enhance the technical properties and improve the functionality of SMPCs using different kinds of reinforcements. The reinforcement endeavors are divided into two main categories: (1) particles [e.g., CNTs (Cho et al. 2005), (2) carbon black particles (Ivens et al. 2011)] and (3) continuous fibers (Abrahamson et al. 2003; Lan et al. 2009). These researches have revealed that CNT reinforcement results in conductivity improvement and a 50%–100% increase in recovery stress. However, they reduce the shape recovery effect. Alternative carbon black particles are less effective (Ivens et al. 2011). In general, particles or short fibers develop some particular functions (e.g., electrical and thermal conductivity, magnetic responsiveness and stiffness on a microscale). The enhancement of mechanical properties is very limited. However, continuous-fiber reinforcement dramatically improves the mechanical properties of the SMPCs (Zhang and Ni 2007).

FIBER-REINFORCED SMPCs

Ohki et al. (2004) employed the glass fiber to develop SMPC and improved the mechanical weakness of the pure SMP. They concluded that there is a development in the tensile strength of the SMPC with the increase of the glass fiber weight fraction (Figure 13.19). Furthermore, they noted an important improvement in the material resistance to cycle loading and crack propagation. Similarly, fiber glass reinforcement was investigated by Ivens et al. (2011) who reported a 30-fold increase in the SMPC recovery stress with unidirectional glass fiber reinforcement compared to the neat SMP. The researchers investigated the recovery and relaxation stress of the net SMP and three different fiber volume fractions of SMPCs: 16.5% (UDG1200), 7.7% (WGF540), and 5% (TWC300). Figure 13.20 shows the improvement achieved in the recovery and relaxation stress with the increase of the fiber volume fraction.

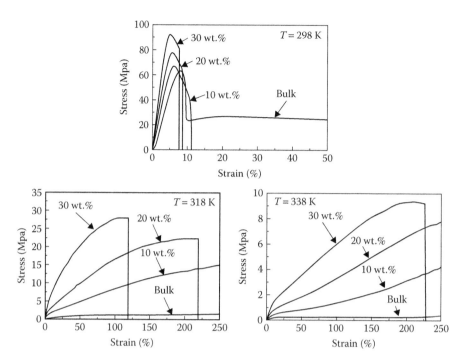

FIGURE 13.19 Stress–strain curves in static tensile test for four materials bulks, 10, 20, and 30 wt% at different testing temperature. (From Ohki, T et al. 2004. Mechanical and shape memory behavior of composites with shape memory polymer. *Composites Part A: Applied Science and Manufacturing*, 35, 1065–73. With permission.)

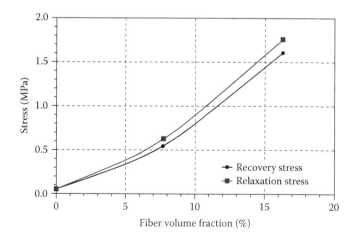

FIGURE 13.20 Recovery and relaxation stresses as a function of the fiber volume fraction. (From Ivens, J et al., *Express Polymer Letters*, 5, 254–61, 2011. With permission.)

However, cyclic loading test of the UDG1200 SMPC has revealed a significant drop in yield stress between the first and the second cycle of loading. Though, in the subsequent second, third, and fourth cycles, the drop in yield stress is much smaller (Figure 13.21).

Table 13.2 presents some experimental data reported by the researchers; it shows the relation between the composite fiber volume fractions and the composite mechanical properties. From the table, it is clear that with the increase of the deformation strain, there is a decrease in the recovery ratio, and this is certainly due to the increase in the elastic spring-back effect.

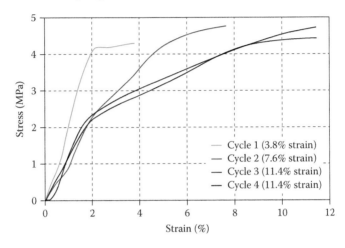

FIGURE 13.21   Stress–strain curves for UDG1200 SMPC under cyclic loading. Each curve presents the loading curve of a cycle. (From Ivens, J et al., *Express Polymer Letters*, 5, 254–61, 2011. With permission.)

TABLE 13.2
Numerical Data Presented by Ivens et al., Obtained Experimentally

	TMA Results		**Peak Stress (MPa)**	**Relaxation Stress (MPa)**	**Spring-Back Effect (%)**	**Recovery Stress (MPa)**
	Strain (%)	**Temperature (°C)**				
SMP	10	75	0.085 ± 0.005	0.05 ± 0.01	0.025 ± 0.005	0.05 ± 0.01
TWC300	10	75	0.5	0.3	0.07	0.3
WGF540	10	75	1.0 ± 0.1	0.62 ± 0.05	0.16 ± 0.03	0.55 ± 0.03
	10	65	3.65 ± 0.45	1.95 ± 0.15	0.45 ± 0.35	1.65 ± 0.3
		75	2.4 ± 0.2	1.75 ± 0.12	0.25 ± 0.10	1.60 ± 0.06
UDG1200		85	2.2 ± 0.2	1.75 ± 0.15	0.15 ± 0.05	1.65 ± 0.12
	20	75	5.9 ± 1.3	3.9 ± 0.5	2 ± 1	3 ± 1
	20	75 (2)	4.5 ± 0.7	3.4 ± 0.4	1.5 ± 1.5	2.5 ± 0.7

Source: Ivens, J et al., *Express Polymer Letters*, 5(3), 254–61, 2011. With permission.
Note: 75° (2) indicates the second deformation–reheating cycle.

Lu et al. (2010) investigated the outcome of adding short carbon fiber to the SMP. They reported a significant improvement in the thermomechanical and conductivity properties of the SMP, and also pointed that fibrous fillers could significantly develop the mechanical properties of the SMPCs, whereas particulate fillers degrade the essential recovery property of the composite. Nishikawa et al. (2012) conducted a periodic cell simulation of the thermomechanical cycle of thermally activated SMP-based composites to analyze the influence of fiber arrangement (e.g., fiber volume fraction, fiber aspect ratio, and fiber end position) on the shape fixity and shape recovery of the composite. From the 3D thermomechanical diagram of SMPs shown in Figure 13.6, the strain–temperature diagram of SMPCs under different fiber volume fractions and different fiber lengths was detected as shown in Figures 13.22 and 13.23, respectively.

Fejős and Karger-Kocsis (2013) quantified the shape memory properties of the carbon-fiber-reinforced SMP. They found out that the recovery stress of the SMPC can be strongly improved by the carbon fiber reinforcement, and the more carbon fiber incorporated, the higher stress recovery achieved. For some bending applications in which small bending radius is important, carbon or glass fibers may not be efficient reinforcement materials because of their brittleness. Aramid fibers is another important material that can effectively be used in SMP reinforcement (Wei et al. 1998). This type of fibers shows lower tensile strength and lower modulus of elasticity, and can be efficiently used in bending applications in which smaller radius is required (Yarborough et al. 2008).

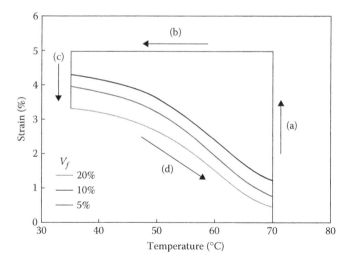

FIGURE 13.22 Strain–temperature relation when the fiber volume fraction was varied. (A) represents SMPs in permanent shape at high temperature $T_h > T_g$; (A–B) applying load at T_h; (B–C) cooling to $T_l < T_g$ under fixed applied load; (C–D) removing load at T_l; (D) some strain recovered due to the elastic spring back effect of the glassy state; (D–A) and (D–E) are two different recovery methods the material can recover with; (D–A) free stress strain recovery; and (D–E) fixed strain–stress recovery. (From Nishikawa, M et al., *Composites Part A: Applied Science and Manufacturing*, 43, 165–73, 2012. With permission.)

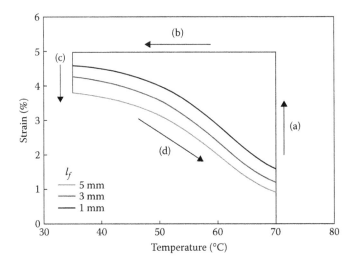

FIGURE 13.23 Strain–temperature relation when the fiber length was varied. (A) represents SMPs in permanent shape at high temperature $T_h > T_g$; (A–B) applying load at T_h; (B–C) cooling to $T_l < T_g$ under fixed applied load; (C–D) removing load at T_l; (D) some strain recovered due to the elastic spring back effect of the glassy state; (D–A) and (D–E) are two different recovery methods the material can recover with; (D–A) free stress strain recovery; and (D–E) fixed strain–stress recovery. (From Nishikawa, M et al., *Composites Part A: Applied Science and Manufacturing*, 43, 165–73, 2012. With permission.)

MODELING

MODELING OF SMP

The fascinating characteristics, and the evolving interest in this member of the smart materials family have motivated many researchers to develop constitutive models to describe their mechanical behavior in recent years. Researchers' efforts have categorized into three models. First, the macroscopic or phenomenological models have appeared as a significant tool for simulations of SMP applications; therefore, more details will be given in the following about this model, which was presented by Baghani et al. (2014). Second, the microscopic or physical models proposed to describe the small and finite deformation of the SMP. Third, the rheological models have been used in most of earlier modeling researches. Though it is a simple model, it gave good agreement with the experimental results. However, due to the complicated behavior of the SMPs, these types of studies are relatively insufficient, and this field of research is still in progress (Alexander et al. 2014; Baghani et al. 2012b; Leng et al. 2011; Liu et al. 2006).

Srivastava et al. (2010) investigated the numerical simulation capability to model the response of the thermally stimulated SMPs, and they tried to develop a thermomechanically coupled large deformation constitutive theory for them. The researchers used an SMP synthesized via photopolymerization (ultraviolet curing) of the monomer tertbutyl acrylate with the cross-link agent. In this research, a large strain compression experiment on SMP was conducted, a thermomechanically coupled large deformation constitutive theory was formulated, and a finite element user material subroutine for the ABAQUS/

Standard program was written to implement the theory. The experimental work has shown that at different strain rates and at different temperatures, the SMP shows two noticeably different responses at the temperatures below and above glass transition temperature. Also, the numerical results that the researchers have obtained showed that the formulated theory accurately matches the experimental results. Similarly, Diani et al. (2012) modeled the thermomechanical behavior of the SMP using a generalized Maxwell model to describe the viscoelastic behavior of the material and implemented the developed model in a commercial finite element code. The efficiency of the developed model to simulate and ultimately predict the shape storage and shape recovery of the SMP material was evaluated against the experimental SMP thermomechanical torsion data in a large deformation regimen, in which a good agreement is found.

Satisfying the second law of thermodynamics, Baghani et al. (2012a) developed the model presented by Liu et al. (2006) and proposed evolution laws for internal variables of SMPs not only in cooling but also in heating. The researchers used a representative volume element (RVE) and assumed that the material in the RVE is a mixture of two phases: frozen and active. Furthermore, for small material strains, they assumed that the strain can be additively decomposed into four components: $\varepsilon = \varphi_a \varepsilon^a + \varphi_f \varepsilon^f + \varepsilon^{is} + \varepsilon^T e$, where ε^a and ε^f denote the elastic strains in the active and frozen phases, respectively, whereas ε^{is} and ε^T stand for the inelastic stored strain and the thermal strain, respectively. To validate the proposed model, the researchers simulated the torsion of rectangular bars and circular tubes. The simulation was done by implementing the developed model within a user-defined material subroutine (UMAT) (UMAT) in ABAQUS/Standard. Finally, the predicted results are compared with the experimental results reported in the literature; a good agreement was found, which validated the proposed model.

Ghosh et al. (2013) developed a beam theory for a small strain continuum model of SMPs using the evolution model previously developed by Ghosh and Srinivasa (2011), which was based on the Euler–Bernoulli beam theory. Their study was based on two incorporated networks: (1) the permanent (backbone) network, which is responsible for the shape recovery, and (2) the temporary transient network, which is responsible for the shape-setting phenomenon. Based on the elastic-predictor inelastic-corrector procedure, the researchers also presented a finite element formulation for the SMP beam theory model and a boundary value problem for three-point bending, which has solved four cases: (1) three different beam material cases (elastic, plastic, and thermoplastic); (2) time step convergence and mesh density convergence; (3) strain recovery thermomechanical cycle; and (4) stress recovery thermomechanical cycle. The behavior of the finite element model for these four cases complied with the validation for the ideal elastic case, as well as showed the different kinds of evolution of inelastic strain for the plastic and thermoplastic cases.

In a recent work, Baghani et al. (2014) employed the Euler–Bernoulli beam theory to present the exact analytical model for deflection of SMP beam in a thermomechanical cycle. During the modeling, it has been assumed that the active and frozen phases of the SMP can be transformed to each other through external stimuli of heat; also they decomposed the strain into four parts as follows:

$$\varepsilon = \left(\varphi_a \varepsilon_a + \varphi_f \varepsilon_f + \varepsilon_{is}\right) + \varepsilon_T = \varepsilon_m + \varphi_T \tag{13.1}$$

where:

   φ_a and φ_f are the volume fractions of the active and frozen phases, respectively, and $\varphi_a + \varphi_f = 1$

   ε_m and ε_T represent the mechanical and thermal strains, respectively

   ε_a and ε_f represent the elastic strains in the active and frozen phases, respectively

   ε_{is} is the inelastic stored strain evolving during thermally activated changes by the following laws (Baghani et al. 2012a):

$$\varepsilon'_{is} = \varphi'_f \left(k_1 \varepsilon_a + k_2 \frac{\varepsilon_{is}}{\varphi_f} \right); \begin{cases} k_1 = 1, & k_2 = 0; \quad \dot{T} < 0 \\ k_1 = 0, & k_2 = 1; \quad \dot{T} > 0 \\ k_1 = 0, & k_2 = 0; \quad \dot{T} = 0 \end{cases} \tag{13.2}$$

where primes and dots stand for derivatives with respect to the temperature and time, respectively. k_1 and k_2 define the process (heating or cooling). With the general assumption of the Euler–Bernoulli beam theory, beam bending with small strain and rotations is derived as

$$\varepsilon = \varepsilon_m + \varepsilon_T; \quad \varepsilon_m = -ky \tag{13.3}$$

where k stands for the beam neutral axis curvature. The equilibrium equations give

$$\sum F_x = 0, \quad \sum M_z = 0 \tag{13.4}$$

Loading the beam at T_h, path A–B of the SMP cycle (Figure 13.6) results in a bending of SMP beam. The material is fully in the active mode; thus, the structure responds as a fully elastic material:

$$\sigma_l = -\frac{M_o y}{I} \tag{13.5}$$

where:

   M_o represents the moment distribution at T_h

   I denotes the cross-sectional moment of inertia of the beam about the neutral axis

During beam cooling from T_h to T_l (path B–C in Figure 13.6), with a constant external applied moment, the following expression for cooling stress is arrived (Baghani et al. 2014):

$$\sigma_c = \Psi_c^{-1} \left[\int_{T_h}^{T} E \varepsilon'_m \Psi_c dT - \frac{M_o y}{I} \right] \tag{13.6}$$

where:

$$\Psi_c(\varphi) = \left[\varphi(E_a / E_f - 1) + 1 \right]^{E_a / E_a - E_f}$$

Now, considering Equations 13.3 and 13.4 gives

$$M = -\int_A \Psi_c^{-1} \left[\int_{Th}^{T} -Ek'y\Psi_c dT - \frac{M_o y}{I} \right] y dA \tag{13.7}$$

where A is the cross-sectional area. Fixing the moment in the cooling process ($M = M_o$), the following expression is calculated for the beam curvature k_c during cooling:

$$k_c = \frac{M}{E_a I} + \frac{M}{I} \int_{T_h}^{T} \frac{\Psi'_c}{E\Psi_c} dT = \frac{M}{E_a I} \left(1 + \frac{E_a}{E_f} \varphi \right) \tag{13.8}$$

During the cooling process ($k_1 = 1$, $k_2 = 0$), the stored strain from Equation 13.2 can be written as (Baghani et al. 2014)

$$\varepsilon_{is}^c = \frac{My}{E_a I} \varphi \tag{13.9}$$

Unloading the SMP beam at T_l (path C–D in Figure 13.6), the unloading stress can be expressed as

$$\sigma_u = E_f \left(\varepsilon_m - \varepsilon_{is}^{cf} \right) = E_f \left(-k + \frac{M}{E_a I} \right) y \tag{13.10}$$

During the heating from T_l to T_h (path D–A or D–E in Figure 13.6), from Equation 13.2, $k_1 = 0$, $k_2 = 1$, the following expression for the stored strain is obtained:

$$\varepsilon_{is}^h = \varepsilon_{is}^{cf} e^{\int (\varphi'/\varphi) dT} = \varepsilon_{is}^{cf} \varphi \tag{13.11}$$

And the stress in the heating step (σ_h) is given as

$$\sigma_h = \Psi_h^{-1} \left[\int_{T_l}^{T} E \left(\varepsilon'_m - \varepsilon_{is}^{cf} \varphi' \right) \Psi_h dT + C \right] \tag{13.12}$$

where $\Psi_h(\varphi) = \varphi(1/E_f - 1/E_a) + 1/E_a$. Using Equations 13.3, 13.4, and 13.12 yields

$$M_h = \Psi_h^{-1} \int_A \left[\int_{T_l}^{T} E \left(k'_h y + \varepsilon_{is}^{cf} \varphi' \right) \Psi_h dT \right] y dA \tag{13.13}$$

Recasting the above equation, the following curvature formula during the cooling process is arrived:

$$k_h = \frac{M_h}{I} \left(\frac{E_f - E_a}{E_f E_a} \right) (1 - \varphi) + \frac{M}{E_a I} \varphi \tag{13.14}$$

This model presents an effectual and precise analytical expression for the SMP beam in the bending condition. The Euler–Bernoulli beam theory in the thermomechanical cycle is implemented to provide mathematical expressions for the internal variables, stress, and beam curvature distribution during each step of the SMP thermomechanical cycle. Comparing the analytical results obtained from this model with the results obtained numerically using a special subroutine embedded in ABAQUS/Standard software, a good agreement is achieved, which makes this model an efficient tool to examine the effect of changing any of the material or geometrical parameters on smart structures containing SMP beams.

MODELING OF SMPCs

Incorporating hard fibers into the SMPCs dramatically improves the mechanical properties and makes the material more applicable in structural applications. Therefore, modeling of SMPCs in different thermomechanical circumstances is an essential job. Recently, many researches have been done for modeling the SMPCs in which different approaches have been used. Based on the continuum thermodynamic consideration and the composite bridging model, Tan et al. (2014) developed a 3D constitutive model to simulate the stress–strain–temperature relationship of unidirectional elastic carbon-fiber-reinforced SMPCs. In the modeling, they assumed that the material is a mixture of elastic reinforcement and SMP matrix, which in turn is further divided into a continuum mixture of a glassy and a rubbery phase, and the fraction of each phase is complementary variation depending on the temperature.

$$\phi_g = \frac{V_{\text{gla}}}{V_{\text{matrix}}} = \frac{V_{\text{gla}}}{V_{\text{gla}} + V_{\text{rub}}}, \quad \phi_r = 1 - \phi_g$$

$$\psi_g = \frac{V_{\text{gla}}}{V}, \quad \psi_r = \frac{V_{\text{rub}}}{V}, \quad \psi_f = \frac{V_{\text{fiber}}}{V}, \quad \psi_m = \psi_g + \psi_r \qquad (13.15)$$

$$V = V_{\text{matrix}} + V_{\text{fiber}} = V_{\text{gla}} + V_{\text{rub}} + V_{\text{fiber}}, \quad \psi_g + \psi_r + \psi_f = 1$$

where:
  V, V_{gla}, V_{rub}, V_{fiber} stand for SMPC volume, glassy phase volume, rubbery phase volume, and fiber reinforcement volume, respectively
  ψ_g, ψ_r denote the volume fractions of the glassy and rubbery phases, respectively
  ψ_f denotes the fiber reinforcement volume fraction

Tan et al. (2014) divided the total strain of SMPCs into five parts: glassy phase strain, rubbery phase strain, fiber strain, stored strain, and thermal strain, which is as follows:

$$\varepsilon_i = \varepsilon_i^f \psi_f + \varepsilon_i^g \psi_f + \varepsilon_i^r \psi_f + \varepsilon_T + \varepsilon_{gs}^n \qquad (13.16)$$

where:

The subscripts g, r, f stand for the glassy phase, rubbery phase, and fiber, respectively

ε_T represents the thermal strain

ε_{gs}^n denotes the strain stored and released during the cooling and heating process

Correspondingly, they divided the stress into five parts as follows:

$$\sigma_i = \sigma_i^T + \sigma_i^{rec} + \sigma_i^c$$

$$\sigma_i^c = \psi_f \sigma_i^f + \psi_g \sigma_i^g + \psi_r \sigma_i^r \qquad (13.17)$$

$$\sigma_i^m = \phi_g \sigma_i^g + \phi_r \sigma_i^r$$

where:

c and m denote the composite and matrix, respectively

σ_i^{rec} represents the recovery stress

To integrate the effect of elastic fiber reinforcement in this constitutive model, the researchers adopted the theory of composite bridging model. Accordingly, a unidirectional fiber-reinforced composite micromechanics model must follow the relationship below:

$$\sigma_i^c = \psi_f \sigma_i^f + \psi_m \sigma_i^m$$

$$\varepsilon_i^c = \psi_f \varepsilon_i^f + \psi_m \varepsilon_i^m \qquad (13.18)$$

where:

ψ_m is the volume fraction of the matrix

σ_j denotes the external stress exerted on the SMPCs

ε_i^m, ε_i^f, σ_i^m, σ_i^f denote the strains and stresses of SMP matrix and fiber reinforcement, respectively

Based on the assumption that the fiber and the matrix are perfectly bonded together, the two internal stresses σ_i^m and σ_i^f can be related to a bridging matrix $\left[A_{ij} \right]$ as follows:

$$\sigma_i^m = \left[A_{ij} \right] \sigma_j^f \qquad (13.19)$$

Considering the fiber constitutive elasticity relation and Equation 13.19, the internal strain in fiber can be derived as follows:

$$\varepsilon_i^f = \left[S_{ij}^f \right] \sigma_j^f = \left[S_{ij}^f \right] \left[B_{ij} \right] \sigma_j^c \qquad (13.20)$$

where:

$\left[S_{ij}^f \right]$ is the flexibility matrix of fiber

$\left[B_{ij} \right]$ is the inverse matrix of $\left(\psi_f \left[I \right] + \psi_m \left[A_{ij} \right] \right)$

To calculate the strain in the matrix at the glassy state, the matrix has very high hardness below T_g; thus, Hooke's law can be utilized as follows:

$$\varepsilon_i^g = S_g : \sigma_i^g \tag{13.21}$$

Above T_g, SMP is in the rubbery phase in which a nonlinear hyperelastic constitution equation is required. However, according to Tan et al. (2014), the nonlinearity can be ignored for small strains and the strain in the rubbery phase can also be calculated using Hooke's law as follows:

$$\varepsilon_i^r = S_r : \sigma_i^r \tag{13.22}$$

When the temperature drops below T_g, a part of the rubbery phase (V_{froz}) with its content of deformation translates into the glassy phase gradually. It is assumed that the elastic fiber reinforcement is evenly distributed in the SMP matrix. During the process of strain storage, the fiber, which is evenly distributed in the rubbery phase, its deformation will be fixed too as this part of the rubbery phase translates into the glassy phase. During the shape-storing process of the SMPCs (heating above T_g, applying the external load, cooling below T_g again), the stored strain can be expressed as

$$\varepsilon_{gs}^n = \frac{1}{1 - \psi_f} \left[\psi_{g1} \varepsilon_{gs} + \left(\psi_{g2} - \psi_{g1} \right) \varepsilon_r \right] \tag{13.23}$$

where:

ψ_{g1} and ε_{gs} denote the volume fraction and strain of the frozen phase at the initial temperature

ψ_{g2} and ε_{gs}^n denote the volume fraction and strain of the frozen phase at the current temperature

It can be seen that the carbon fiber content has no impacts on the stored strain. The remaining part of strains in the strain decomposition equation (13.23) is the thermal strain (ε_T). The total thermal strain of SMPCs is constitutively defined as

$$\varepsilon_T = \psi_r \varepsilon_T^r + \psi_g \varepsilon_T^g + \psi_f \varepsilon_T^f \tag{13.24}$$

Assuming that the rubbery phase, glassy phase, and carbon fiber are homogeneous materials, and their thermal strains are only related to the thermal expansion coefficient separately, the thermal strain can be simply defined as follows:

$$\varepsilon_T = \int_{T_o}^{T} \left[\psi_r \left(T \right) \alpha_r + \psi_g \left(T \right) \alpha_g + \psi_f \left(T \right) \alpha_f \right] dT \tag{13.25}$$

where α_r, α_g, and α_f denote the thermal expansion coefficients of the rubbery phase, glassy phase, and fiber, respectively.

To show the validity of the developed model under different thermomechanical loading conditions, a comparison between the experimental and theoretical results is shown in Figure 13.24, in which a good agreement is found between the two set of results. Accordingly, the model has been used to predict the stress recovery and strain releasing as a function of temperature as shown in Figures 13.25 and 13.26.

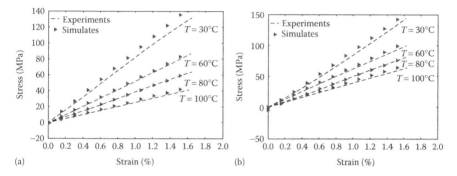

FIGURE 13.24 Comparison between experiments and simulations: (a) fiber content 7.5%; (b) fiber content 8.2%. (From Tan, Q et al., *Composites Part A: Applied Science and Manufacturing*, 64, 132–8, 2014. With permission.)

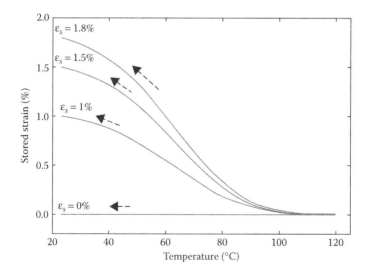

FIGURE 13.25 Numerical simulations of the stored strain release with increasing temperature. (From Tan, Q et al., *Composites Part A: Applied Science and Manufacturing*, 64, 132–8, 2014. With permission.)

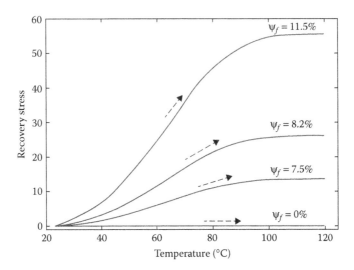

FIGURE 13.26 Numerical simulations of the stress–temperature behavior from free shape recovery process with different fiber contents (0%, 7.5%, 8.2%, 11.5%). (From Tan, Q et al., *Composites Part A: Applied Science and Manufacturing*, 64, 132–8, 2014. With permission.)

CONCLUSIONS

Since its development in the last decade, SMPs have been gaining widespread attention for new product innovation. They are lightweight, have high strain shape recovery ability, and are easy to process, and their required properties can be tailored for a variety of different applications. Such properties have enabled a variety of applications such as deployable space structures, life-saving biomedical devices, and adaptive optical devices (Leng et al. 2009, 2011; Xie 2010). Recently, a number of medical applications have been considered and investigated for polyurethane-based SMPs. Two major unique properties are of particular interest to the medical world. One is that these materials are found to be biocompatible, nontoxic, and nonmutagenic in the human body. Another attractive aspect is the glass transition temperature T_g of these materials, which can be tailored for shape restoration/self-deployment of different clinical devices when inserted in the human body. Most interestingly, light-activated SMP actuators can be activated by remotely sending IR lights through fiber optic cables. At the moment the utilization of light-activated SMPs is limited to microactuators, and a wealth of opportunities for using these materials for breakthrough technologies is still to be discovered.

ACKNOWLEDGMENTS

The authors would like to thank, Centre of Excellence in Engineered Fibre Composites (CEEFC) of University of Southern Queensland, Australia for unlimited support available for on going SMP research, The Iraqi Ministry of Higher

Education and Scientific Research, and the College of Engineering, University of Diyala, Iraq for providing a scholarship and the financial support for a post-graduate student, Harbin Institute of Technology (HIT), China for providing SMP materials for on-going SMP research at CEEFC.

REFERENCES

Abrahamson, ER, Lake, MS, Munshi, NA and Gall, K. 2003. Shape memory mechanics of an elastic memory composite resin. *Journal of Intelligent Material Systems and Structures*, 14(10): 623–32.

Ahir, SV and Terentjev, EM. 2005. Photomechanical actuation in polymer–nanotube composites. *Nature Materials*, 4(6): 491–5.

Alexander, S, Xiao, R and Nguyen, TD. 2014. Modeling the thermoviscoelastic properties and recovery behavior of shape memory polymer composites. *Journal of Applied Mechanics*, 81(4): 041003.

Asaka, K and Oguro, K. 2000. Bending of polyelectrolyte membrane platinum composites by electric stimuli: Part II. Response kinetics. *Journal of Electroanalytical Chemistry*, 480(1): 186–98.

Baer, GM, Small, W, Wilson, TS, Benett, WJ, Matthews, DL, Hartman, J and Maitland, DJ. 2007. Fabrication and in vitro deployment of a laser-activated shape memory polymer vascular stent. *Biomedical Engineering Online*, 6(8): 43–51.

Baghani, M, Mohammadi, H and Naghdabadi, R. 2014. An analytical solution for shape-memory-polymer Euler–Bernoulli beams under bending. *International Journal of Mechanical Sciences*, 84: 84–90.

Baghani, M, Naghdabadi, R, Arghavani, J and Sohrabpour, S. 2012a. A constitutive model for shape memory polymers with application to torsion of prismatic bars. *Journal of Intelligent Material Systems and Structures*, 23(2): 107–16.

Baghani, M, Naghdabadi, R, Arghavani, J and Sohrabpour, S. 2012b. A thermodynamically-consistent constitutive model for shape memory polymers. *International Journal of Plasticity*, 35: 13–30.

Cabanlit, M, Maitland, D, Wilson, T, Simon, S, Wun, T, Gershwin, ME and Van de Water, J. 2007. Polyurethane shape-memory polymers demonstrate functional biocompatibility in vitro. *Macromolecular Bioscience*, 7(1): 48–55.

Chen, S, Hu, J, Zhuo, H and Zhu, Y. 2008. Two-way shape memory effect in polymer laminates. *Materials Letters*, 62(25): 4088–90.

Chen, Y-C and Lagoudas, DC. 2008. A constitutive theory for shape memory polymers. Part I: Large deformations. *Journal of the Mechanics and Physics of Solids*, 56(5): 1752–65.

Cho, JW, Kim, JW, Jung, YC and Goo, NS. 2005. Electroactive shape-memory polyurethane composites incorporating carbon nanotubes. *Macromolecular Rapid Communications*, 26(5): 412–16.

Chung, T, Romo-Uribe, A and Mather, PT. 2007. Two-way reversible shape memory in a semicrystalline network. *Macromolecules*, 41(1): 184–92.

Diani, J, Gilormini, P, Frédy, C and Rousseau, I. 2012. Predicting thermal shape memory of crosslinked polymer networks from linear viscoelasticity. *International Journal of Solids and Structures*, 49(5): 793–9.

Fejős, M and Karger-Kocsis, J. 2013. Shape memory performance of asymmetrically reinforced epoxy/carbon fibre fabric composites in flexure. *Express Polymer Letters*, 7: 528–34.

Ghosh, P, Reddy, JN and Srinivasa, AR. 2013. Development and implementation of a beam theory model for shape memory polymers. *International Journal of Solids and Structures*, 50(3/4): 595–608.

Ghosh, P and Srinivasa, A. 2011. Modeling and parameter optimization of the shape memory polymer response. *Mechanics of Materials.*

Han, X-J, Dong, Z-Q, Fan, M-M, Liu, Y, Li, J-H, Wang, Y-F, Yuan, Q-J, Li, B-J and Zhang, S. 2012. pH-induced shape-memory polymers. *Macromolecular Rapid Communications*, 33(12): 1055–60.

Hinkle, J, Lin, JH and Kling, D. 2011. Design and materials study of secondary structures in deployable planetary and space habitats. In *Proceedings of the 52nd AIAA/ASME/ ASCE/AHS/ASC Structures, Structural Dynamics, and Materials Conference*, AIAA: Denver, CO, April 4–7,

Imai, S. 2014. Operating methods for two-way behavior shape memory polymer actuators without using external stress. *IEEJ Transactions on Electrical and Electronic Engineering*, 9(1): 90–6.

Ivens, J, Urbanus, M and De Smet, C. 2011. Shape recovery in a thermoset shape memory polymer and its fabric-reinforced composites. *Express Polymer Letters*, 5(3): 254–61.

Jee, S-C. 2010. *Development of Morphing Aircraft Structure Using SMP*, DTIC Document. MSc Thesis, Air University, Montgomery, AL.

Jiang, H, Kelch, S and Lendlein, A. 2006. Polymers move in response to light. *Advanced Materials*, 18(11): 1471–5.

Koerner, H, Price, G, Pearce, NA, Alexander, M and Vaia, RA. 2004. Remotely actuated polymer nanocomposites—Stress-recovery of carbon-nanotube-filled thermoplastic elastomers. *Nature Materials*, 3(2): 115–20.

Lan, X, Liu, Y, Lv, H, Wang, X, Leng, J and Du, S. 2009. Fiber reinforced shape-memory polymer composite and its application in a deployable hinge. *Smart Materials and Structures*, 18(2): 024002.

Langer, R and Tirrell, DA. 2004. Designing materials for biology and medicine. *Nature*, 428(6982): 487–92.

Lendlein, A, Jiang, H, Junger, O and Langer, R. 2005. Light-induced shape-memory polymers. *Nature*, 434(7035): 879–82.

Lendlein, A and Kelch, S. 2002. Shape-memory polymers. *Angewandte Chemie International Edition*, 41(12): 2034–57.

Lendlein, A and Langer, R. 2002. Biodegradable, elastic shape-memory polymers for potential biomedical applications. *Science*, 296(5573): 1673–6.

Leng, J, Lan, X, Liu, Y and Du, S. 2009. Electroactive thermoset shape memory polymer nanocomposite filled with nanocarbon powders. *Smart Materials and Structures*, 18(7): 074003.

Leng, J, Lan, X, Liu, Y and Du, S. 2011. Shape-memory polymers and their composites: Stimulus methods and applications. *Progress in Materials Science*, 56(7): 1077–135.

Leng, J, Zhang, D, Liu, Y, Yu, K and Lan, X. 2010. Study on the activation of styrene-based shape memory polymer by medium-infrared laser light. *Applied Physics Letters*, 96(11): 111905.

Liu, C, Qin, H and Mather, P. 2007. Review of progress in shape-memory polymers. *Journal of Materials Chemistry*, 17(16): 1543–58.

Liu, F and Urban, MW. 2010. Recent advances and challenges in designing stimuli-responsive polymers. *Progress in Polymer Science*, 35(1): 3–23.

Liu, Y, Gall, K, Dunn, ML, Greenberg, AR and Diani, J. 2006. Thermomechanics of shape memory polymers: Uniaxial experiments and constitutive modeling. *International Journal of Plasticity*, 22(2): 279–313.

Love, M, Zink, P, Stroud, R, Bye, D, Rizk, S and White, D. 2007. Demonstration of morphing technology through ground and wind tunnel tests. In *Proceedings of the 48th AIAA/ ASME/ASCE/AHS/ASC Structures, Structural Dynamics, and Materials Conference*, AIAA, pp. 1–12.

Lu, H, Yu, K, Sun, S, Liu, Y and Leng, J. 2010. Mechanical and shape-memory behavior of shape-memory polymer composites with hybrid fillers. *Polymer International*, 59(6): 766–71.

Madbouly, SA and Lendlein, A. 2010. Shape-memory polymer composites. In A. Lendlein, ed. *Shape-Memory Polymers*. Springer-Verlag, Berlin, Germany, pp. 41–95.

Maitland, DJ, Metzger, MF, Schumann, D, Lee, A and Wilson, TS. 2002. Photothermal properties of shape memory polymer micro-actuators for treating stroke. *Lasers in Surgery and Medicine*, 30(1): 1–11.

Monkman, G. 2000. Advances in shape memory polymer actuation. *Mechatronics*, 10(4): 489–98.

Nishikawa, M, Wakatsuki, K, Yoshimura, A and Takeda, N. 2012. Effect of fiber arrangement on shape fixity and shape recovery in thermally activated shape memory polymer-based composites. *Composites Part A: Applied Science and Manufacturing*, 43(1): 165–73.

Ohki, T, Ni, Q-Q, Ohsako, N and Iwamoto, M. 2004. Mechanical and shape memory behavior of composites with shape memory polymer. *Composites Part A: Applied Science and Manufacturing*, 35(9): 1065–73.

Qi, HJ, Nguyen, TD, Castro, F, Yakacki, CM and Shandas, R. 2008. Finite deformation thermo-mechanical behavior of thermally induced shape memory polymers. *Journal of the Mechanics and Physics of Solids*, 56(5): 1730–51.

Sahoo, NG, Jung, YC, Goo, NS and Cho, JW. 2005. Conducting shape memory polyurethane-polypyrrole composites for an electroactive actuator. *Macromolecular Materials and Engineering*, 290(11): 1049–55.

Schmidt, AM. 2006. Electromagnetic activation of shape memory polymer networks containing magnetic nanoparticles. *Macromolecular Rapid Communications*, 27(14): 1168–72.

Serrano, MC and Ameer, GA. 2012. Recent insights into the biomedical applications of shape-memory polymers. *Macromolecular Bioscience*, 12(9): 1156–71.

Small IV, W, Wilson, T, Benett, W, Loge, J and Maitland, D. 2005. Laser-activated shape memory polymer intravascular thrombectomy device. *Optics Express*, 13(20): 8204–13.

Srivastava, V, Chester, SA and Anand, L. 2010. Thermally actuated shape-memory polymers: Experiments, theory, and numerical simulations. *Journal of the Mechanics and Physics of Solids*, 58(8): 1100–24.

Sun, L, Huang, WM, Ding, Z, Zhao, Y, Wang, CC, Purnawali, H and Tang, C. 2012. Stimulus-responsive shape memory materials: A review. *Materials and Design*, 33: 577–640.

Tan, Q, Liu, L, Liu, Y and Leng, J. 2014. Thermal mechanical constitutive model of fiber reinforced shape memory polymer composite: Based on bridging model. *Composites Part A: Applied Science and Manufacturing*, 64: 132–8.

Tobushi, H, Hashimoto, T, Hayashi, S and Yamada, E. 1997. Thermomechanical constitutive modeling in shape memory polymer of polyurethane series. *Journal of Intelligent Material Systems and Structures*, 8(8): 711–18.

Tobushi, H, Okumura, K, Hayashi, S and Ito, N. 2001. Thermomechanical constitutive model of shape memory polymer. *Mechanics of Materials*, 33(10): 545–54.

Vaia, R. 2005. Nanocomposites: Remote-controlled actuators. *Nature Materials*, 4(6): 429–30.

Wache, H, Tartakowska, D, Hentrich, A and Wagner, M. 2003. Development of a polymer stent with shape memory effect as a drug delivery system. *Journal of Materials Science: Materials in Medicine*, 14(2): 109–12.

Wang, Z, Song, W, Ke, L and Wang, Y. 2012. Shape memory polymer composite structures with two-way shape memory effects. *Materials Letters*, 89: 216–18.

Wei, Z, Sandstroröm, R and Miyazaki, S. 1998. Shape-memory materials and hybrid composites for smart systems. Part I: Shape-memory materials. *Journal of Materials Science*, 33(15): 3743–62.

Westbrook, K, Castro, F, Ding, Y and Jerry Qi, H. 2011a. A 3D finite deformation constitutive model for amorphous shape memory polymers: A multi-branch modeling approach for nonequilibrium relaxation processes. *Mechanics of Materials*, 43(12): 853–69.

Westbrook, K, Mather, PT, Parakh, V, Dunn, ML, Ge, Q, Lee, BM and Qi, HJ. 2011b. Two-way reversible shape memory effects in a free-standing polymer composite. *Smart Materials and Structures*, 20(6): 065010.

Xie, T. 2010. Tunable polymer multi-shape memory effect. *Nature*, 464(7286): 267–70.

Xu, B, Zhang, L, Pei, YT, Luo, J, Tao, S, De Hosson, JTM and Fu, YQ. 2012. Electro-responsive polystyrene shape memory polymer nanocomposites. *Nanoscience and Nanotechnology Letters*, 4(8): 814–20.

Xulu, PM, Filipcsei G and Zrinyi M. 2000. Preparation and responsive properties of magnetically soft poly (N-isopropylacrylamide) gels. *Macromolecules*, 33: 1716–19.

Yang, B, Huang, W, Li, C and Li, L. 2006. Effects of moisture on the thermomechanical properties of a polyurethane shape memory polymer. *Polymer*, 47(4): 1348–56.

Yang, B, Min Huang, W, Li, C and Hoe Chor, J. 2005. Effects of moisture on the glass transition temperature of polyurethane shape memory polymer filled with nano-carbon powder. *European Polymer Journal*, 41(5): 1123–8.

Yarborough, CN, Childress, EM and Kunz, RK. 2008. Shape recovery and mechanical properties of shape memory composites. In *Proceedings of the ASME International Mechanical Engineering Congress and Exposition*, American Society of Mechanical Engineers, pp. 111–19.

Zhang, C-S and Ni, Q-Q. 2007. Bending behavior of shape memory polymer based laminates. *Composite Structures*, 78(2): 153–61.

Index

Note: Page numbers followed by f and t refer to figures and tables, respectively.

479

Printed and bound by CPI Group (UK) Ltd, Croydon, CR0 4YY

01/11/2024

01782617-0017